Classic and High-Enthalpy Hypersonic Flows

Classic and High-Enthalpy Hypersonic Flows presents a complete look at high-enthalpy hypersonic flow from a review of classic theories to a discussion of future advances centering around the Born-Oppenheim approximation, potential energy surface, and critical point for transition. The state-of-the-art hypersonic flows are defined by a seamless integration of the classic gas dynamic kinetics with nonequilibrium chemical kinetics, quantum transitions, and radiative heat transfer. The book is intended for graduate students studying advanced aerodynamics and taking courses in hypersonic flow. It can also serve as a professional reference for practicing aerospace and mechanical engineers of high-speed aerospace vehicles and propulsion system research, design, and evaluation.

Features

- Presents a comprehensive review of classic hypersonic flow from the Newtonian theory to blast wave analogue.
- Introduces nonequilibrium chemical kinetics to gas dynamics for hypersonic flows in the high-enthalpy state.
- Integrates quantum mechanics to high-enthalpy hypersonic flows including dissociation and ionization.
- Covers the complete heat transfer process with radiative energy transfer for thermal protection of earth reentry vehicle.
- Develops and verifies the interdisciplinary governing equations for understanding and analyzing realistic hypersonic flows.

Classic and High-Enthalpy Hypersonic Flows

Joseph J.S. Shang

CRC Press
Taylor & Francis Group
Boca Raton London New York

CRC Press is an imprint of the
Taylor & Francis Group, an **informa** business

Designed cover image: Pressure waves of air flowing off an airplane. © Shutterstock | Fouad A. Saad.

First edition published 2023
by CRC Press
6000 Broken Sound Parkway NW, Suite 300, Boca Raton, FL 33487-2742

and by CRC Press
4 Park Square, Milton Park, Abingdon, Oxon, OX14 4RN

CRC Press is an imprint of Taylor & Francis Group, LLC

© 2023 Karen L. Shang

Library of Congress Cataloging-in-Publication Data
Names: Shang, Joseph J. S., author.
Title: Classic and high-enthalpy hypersonic flows / Joseph J.S. Shang.
Description: First edition. | Boca Raton : CRC Press, [2023] | Includes
bibliographical references and index. |
Identifiers: LCCN 2022055040 (print) | LCCN 2022055041 (ebook) |
ISBN 9781032079813 (hbk) | ISBN 9781032079875 (pbk) | ISBN 9781003212362 (ebk)
Subjects: LCSH: Aerodynamics, Hypersonic. | Enthalpy.
Classification: LCC TL571.5 .S53 2023 (print) | LCC TL571.5 (ebook) |
DDC 629.132/306—dc23/eng/20230131
LC record available at https://lccn.loc.gov/2022055040
LC ebook record available at https://lccn.loc.gov/2022055041

ISBN: 978-1-032-07981-3 (hbk)
ISBN: 978-1-032-07987-5 (pbk)
ISBN: 978-1-003-21236-2 (ebk)

DOI: 10.1201/9781003212362

Typeset in Times
by codeMantra

To my beloved wife of 60 years, Karen

Contents

SECTION 1   Classic Hypersonic Flow Theories

SECTION 2   High-Enthalpy Hypersonic Flow

Preface

Gas dynamic kinetics is the foundation for the physical description of all classic aerodynamic phenomena from subsonic, transonic, supersonic, to hypersonic flow domains. The gist of this theory is that the interactions between elementary gas particles involve binary elastic collisions among gas molecules. Building on this premise, a series of illuminating and outstanding theories have been developed spanning the complete spectrum of gas motions and have withheld the test of time. Meanwhile, defining the transport properties of gas medium clearly demarcates the distinctive inviscid and viscous flow behaviors.

For supersonic flows, the zone of dependence generates a complex shock and expansion waves system by convergent and divergent limiting characteristic surfaces. The piecewise continuous shock jump condition is established by the Rankine-Hugonoit relations. As the Mach number increases further into the hypersonic domain, the flow field structure approaches a nearly invariant pattern and also displays significant discrepancies from the classic theories. The source of disparities is attributable to the high-enthalpy gas behaviors that are generated by the extremely strong shock compression within a very thin shock layer envelope over the solid surface. The gas thermodynamic state must transform from a rarefied ambient condition to reach a state that is comparable to the surface of sun. The chemical reacting gas medium simply cannot attain the equilibrium state, especially when dissociation and ionization take place. The hypersonic flow now becomes an interdisciplinary endeavor that includes consequences of nonequilibrium chemical kinetics, quantum mechanics, and radiative energy transfer.

This book delineates the integration of required multiple disciplines for analyzing high-enthalpy hypersonic flows. First, the nonequilibrium chemical kinetics is fused progressively from the classic theories to the law of mass action, the master equation of probability formulation, and finally to the first-principle approach. The observed climbing-the-ladder and big-bang ionizing process is linked through the quantum states of the internal degrees of freedom of gas atoms and molecules at discrete energy levels. Under this circumstance, the elementary gas particle interactions involve inelastic collisions. To develop particle dynamics on the microscale scale, the knowledge of quantum mechanics is applied to obtain the physics-based modeling using the principle of detailed balance and experimental database. Similarly, the instantaneous radiative energy transfer by bond-bond, bond-free, and free-free quantum transitions is molded using the radiation rate equation via the emission, absorption, and scattering mechanisms of electron dynamics. Impressive achievements have been made by sustained research and development efforts over the past five decades and reflected by successful space explorations. However, the basic knowledge and a better understanding of the last frontier of aerodynamics must be gained through the *ab initio* approach aided by high-performance computational technology. The challenge for future advancement shall be built by exploring further the Born-Oppenheimer approximation, potential energy surface, and critical point for transition.

Joseph J.S. Shang

Acknowledgement

In a distinguished scientific career tied to the vast field of computational physics, it is in hypersonic flows that the author found both the highest personal affinity for study as well as the ambition to master. Dr. Shang completed the manuscript in the 12 chapters the reader sees here, just days before his unexpected passing.

He would wish to express most profound and humble gratitude to his teachers, colleagues, and students for over five decades of research and education.

In the realization of this text on hypersonic flows, he would (and did) express deep appreciation to his very kindly publisher Ms. Kyra Lindholm of CRC Press. He also thanks highly skillful editorial staff including Ms. Sunayina Dadhwal and Ms. Kathrin Immanuel and to Mr. Jay Clouse grateful acknowledgement for careful work in developing some highly complex graphics. Also gratitude to Ms. Carly R. Limtiaco who provided support that proved to be critical in the final phases of the editorial process.

CCS and AKS, his sons,
Livermore, California, April 2023

Author

Dr. Joseph J.S. Shang (1936–2021) was a pioneer in computational fluid dynamics, electromagnetics, and aero-electromagnetic dynamics. He was born in Nanking, China in 1936 to a well-known family of Hunan province. In 1958, he graduated from the National Cheng-Kung University with a B.Sc. in Mechanical Engineering with honors. In May 1960, he became a graduate student at the Aerospace Engineering Department, Polytechnic University of New York under the tutelage of Professor Antonio Ferri and earned his M.Sc. in 1962, followed by his Ph.D. in Aerospace Engineering from The Ohio State University under the direction of Professor Odus Burggraf.

In 1967, he formally joined U. S. Air Force Research Laboratory as a Visiting Scientist and later an Aerospace Research Scientist. He made pioneering and lasting contributions to computational fluid dynamics, specifically in the development of computing procedures for solving three-dimensional Navier–Stokes equations, strong inviscid–viscous interaction including flow separation, non-equilibrium reacting hypersonic flows, turbulent modeling, self-sustained oscillatory flows, high-performance vector, parallel data processing, as well as large-scale numerical simulation of aerodynamic performance of aircrafts. Dr. Shang also made original and enabling contributions to computational electromagnetics for radar signature prediction and micro-patch antenna technology.

He was an Emeritus Research Professor at Wright State University, had 34 years research experience in Air Force Research Laboratory (AFRL), and retired as the leader of the Center of Excellence in Aerodynamic Research.

He served as technical advisor to National Aeronautics and Space Administration, Department of Energy, Von Karman Institute for Fluid Dynamics, University of Cincinnati, University of Illinois, and The Ohio State University. He was the recipient of the Air Force Primus Award, Basic Research Award, General Benjamin Foulois Awards, Star Team Awards, Scientific Achievement Awards, Distinguished Alumnus Award of The Ohio State University, and Secretary of the Air Force Exceptional Civilian Service Award.

Dr. Shang published nearly 400 articles and conference papers, 14 book chapters, and has three books published by John Wiley and Sons, Inc., Cambridge University Press, and Taylor and Francis Group of CRC Press.

Section I

Classic Hypersonic Flow Theories

1 Unique Features of Hypersonic Flow Fields

1.1 GENERAL REMARKS

Traditionally aerodynamic flow fields are classified by the velocity of air motion relative to the local speed of sound as the subsonic, transonic, supersonic, and hypersonic regimes. In the subsonic flow regime, a perturbation in the flow field can be transmitted freely throughout the entire flow field, and the deflection of flow with respect to the free stream is controlled by the vorticity of the flow. Once the flow velocity exceeds the speed of sound, the zone of dependence plays a dominant role and the propagation of disturbance is confined by the reach of the local speed of sound. The outer boundary of the zone of dependence is either extended to the unbound and unperturbed freestream or confined by the limiting characteristic surface. Depending on the flow inclination with respect to the oncoming freestream; either the converging limiting characteristic surfaces coalesce into a shock wave or diverge into a system of expanding waves. Therefore beyond the supersonic flow regime, the flow field structure is exclusively dependent upon the orientation of disturbance propagation. Hypersonic flows reside on the upper limit of the velocity spectrum of the aerodynamics regime and possess identical aerodynamic features as that of supersonic flow. Any flow deflection induced by the submerged body to the undisturbed free stream, independent from the particular shape, will always produce a shock wave. However, hypersonic flows always encounter stronger shock wave compression that tightly confines within the shock wave envelope known as the shock layer which leads to drastically different thermodynamic properties by the dissociation and ionization of air. Even in a weakly ionized condition, the ionized air generates an electromagnetic field to be distinguished from all other flow regimes.

The flow deflection from the free stream is very significant in hypersonic flows, and the entire flow field formation is determined by the leading edge configuration to be classified as the slender or blunt body. Nevertheless, the flow fields of both configurations always associate with a strong shock wave upstream to the body either with an attach or detach shock, as it displays in Figure 1.1 by the schlieren photographs around distinctive configurations at a Mach number of 5.0. The compressed flow domain is bounded by the shock envelope and the interface of the submerged configuration, referred to as the shock layer. At the very high Mach number ranges, the pressure and temperature jump across the shock by strong shock compression are proportional to the product of the square value of the freestream Mach number and the sine of the shock angle, $M_\infty \sin \beta$. However the density jump across the shock wave must satisfy the mass conservation law attaining to a constant asymptote which

DOI: 10.1201/9781003212362-2

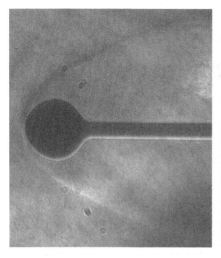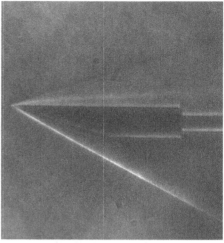

FIGURE 1.1   Hypersonic flow structures of blunt and slender bodies.

is a sole function of the gas thermodynamic property immediately downstream to the shock wave.

From these physical observations, the hypersonic flow fields frequently abridge the rarefied and continuum regimes emerging as one of the many unique features. A huge amount of pieces of knowledge have been accumulated to understand and define the demarcation of the intermediate gas transition from the free molecular, rarefied, to continuum domains.

1.2 FREE MOLECULE, RARIFIED, AND CONTINUUM GAS DOMAINS

All gas dynamic phenomena have its root in the behavior of multiple particle interaction through collisions, the most important parameters are the mean free path of kinetic theory and the mean time elapsing between particle collisions. The interactions between particles are random actions, in which the interaction processes do not depend on the initial condition of collisions. The elapsed time between collisions is often referred to as the relaxation time, and clearly, it's inversely proportional to the number of collisions. The motion of particles between successive impacts is free from any mutual particle influences, for this reason, the distance of the traversing particle during the relaxation time is called the free path. The average values actually determine the distinctive domains of rarefied or denser gas [1–3].

In the outer space of the atmosphere and beyond the thermosphere layer where the mean free path is approaching infinity, intermolecular collision lose its importance to the global gas behavior. Under this condition, the gas motion is considered to be the free molecule flow. In this environment, the gas density is very low and each collision acts independently from all other collisions. Especially, the individual particle reflected from a solid surface does not collide with others until they are outside the region of the medium interface. The boundary layer disappears and the particle adjacent to the solid surface remains in a dynamic equilibrium state or the gas is considered to be the Maxwellian [2]. The tangential velocity component of the colliding particles may be small but finite; the slip flow condition will occur. A discontinuity of temperature also develops at the medium interface and will manifest in the global gas behavior. The main effect of extremely rarefied hypersonic flow is that the shock upstream to the solid surface merges with flow near the configuration [3].

The rarefied gas phenomenon begins to dominate as the number density of the gas is increased further from the free molecule regime. The slip velocity relative to the wall has been considered to be proportional to the velocity gradient adjacent to the interface by the coefficient of slip. The coefficient has the same order as the mean free path of the gas; some experimental observations have been made to find that under standard atmospheric pressure, there is a yield of a range of values around 10^{-6} cm [1].

In analogy with the slip velocity, the heat transfer in the rarefied gas has a discontinuity of temperature which has been assumed to be proportional to the outer normal of its gradient on the interface. Again the proportional constant is called the temperature jump distance. Another theory has also been developed to represent the interchange energy for a molecule that strikes the medium interface by the accommodation coefficient for the temperature jump. The basic idea takes into consideration of the energy exchange by the energy level difference of the incident stream and the wall condition. The actual energy transfer at the medium interface in the rarefied gas is proportional to this energy level difference. The coefficient has been found from experiments to depend on its initial value or its past history. The magnitudes of the accommodation coefficients have values from 0.62 to 0.81 for the oxygen and nitrogen molecules [1].

The continuum gas regime is reached usually as the mean free path is on the same order of magnitude as the typical gas molecule dimension and is sufficiently small in comparison with the characteristic length of the flow field. In this regime, the number density of the gas particle is sufficiently high that the intermolecular collisions dominate over collisions with the medium interface. More importantly, the number density of colliding particles is sufficiently high, and the statistical results between the microscopic and macroscopic gas dynamics become meaningful. Under the collision dynamic equilibrium state, all the thermodynamic variables of gas such as the pressure, temperature, and density are unambiguously and completely definable by the statistical mechanics. A vast amount and illuminating accomplishments of aerodynamic science by basic research have been carried out and are the focus of the following discussion.

1.3 MEAN FREE PATH

In order to understand the detailed physics of gas dynamics at the microscopic scales, the following experimental observations are presented to put these parameters in an appropriate perspective [2]. The number of gas molecules in a cubic centimeter at the standard atmospheric condition is $\sim 2.687 \times 10^{19}$. The radius of a hydrogen molecule assumed to be a rigid sphere is about 1.365×10^{-8} cm, the value is similar to most gas molecules. The number of collisions between molecules within a cubic centimeter is found to be 2.05×10^{29} per second, and the collision frequency is 1.50×10^{10} per second. Therefore, the distance traveled between collisions or the mean free path of the molecule is 1.16×10^{-5} cm, and the mean time between successive collisions is 6.6×10^{-11} seconds [2].

In general, the mean free path is not dependent on either the mass or the temperature of the molecule. Although the radius of gas molecules is different but is of the same order of magnitude, thus the mean free path of any gas in the standard atmospheric condition should be on the same order of 10^{-5} cm. By comparing the mean free path and the assumed rigid spherical molecules of all gas; the ratio approaches a value around a thousand times greater. For this reason, the collision process of gas molecules is chaotic and there may not be an appreciable correlation between the initial value or velocities among colliding molecules. In a rarefied gas condition, the assumption of molecular chaos is invalid.

The path of the center of mass of finite-size gas particle follows an irregular trajectory consisting of straight free path between collisions. The lengths of these free paths vary widely, the average values do have a finite value to be referred to as the mean free path. The frequency of a great many mutual particle impacts over a sufficiently long period is called the collision rate. The rate of moving gas particles can also be put in terms of probability as the simplest and most useful viewpoint. The mean free path and the collision rate establish a simple and important relationship with each other.

The mean free path or the mean collision free path λ, is the average distance traveled between gas particles collision. From the gas kinetic theory, it can be determined by the total number of collisions in unit volume in time n_c and the arithmetical mean molecule speed u_m in an equilibrium condition.

$$n_c = 4n^2\sigma^2\sqrt{\pi kT/m} \tag{1.1}$$

$$u_m = \sqrt{8kT/\pi m} \tag{1.2}$$

where the symbol m denotes the unit mass of a specific gas molecule, k is the Boltzmann constant ($k = 1.3807 \times 10^{-23}$ J/K), and σ is the collision section. For the moment, the collision cross section is introduced here as the target area for impact between particles, or the range of influence for particle interactions. It has a value of $\sigma = 3.75 \times 10^{-8}$ cm^2 at the standard atmospheric condition.

The average distance traversed by a molecule before colliding with another in unit volume and time is simply the quotient of the two variables.

$$\lambda = 1/\sqrt{2}\, n\pi\sigma^2 \tag{1.3}$$

The distance between particle impacts is the most rigorous way of defining the mean free path or mean collision free path of the gas molecule collision process. This distance is a very significant for gas kinetic theory, because all the pertinent transport coefficients for a gas can be estimated by the free path concept with some simplified assumptions when the gas is not in dynamic equilibrium state.

For air at the standard condition, the mean free path yields a value of $\lambda \approx 4\alpha \times 10^{-5}$ cm [3]. The proportional constant α is different from collisions between molecules and the impacts with the medium interface, as well as, for different modes of the internal degree of excitation. An analysis of the collision process indicates that the value of α is of order unity.

In most practical aerodynamic applications and from the result of the kinetic theory of gas, the mean free path has also been obtained by an order of magnitude analysis. It approximated by the ratio of kinematic viscosity and the speed of sound to become [4]

$$\lambda = 1.26\sqrt{\gamma}\, v/c \tag{1.4a}$$

or

$$\lambda \sim v/c = \mu/\rho c \tag{1.4b}$$

In Equation (1.4a), the specific heat ratio $\gamma = c_p/c_v$, is given by the specific heat of constant pressure $c_p = (dq/dT)_p$ and constant volume $c_v = (dq/dT)_v$ of the reversible heat transfer processes.

1.4 KNUDSEN NUMBER

From all experimental observations, the determination of different gas domains depends strongly on the different collisions between particles and their impact on the medium interface. Especially gas flowing over an obstacle is mostly derived from the encounters between the gas particles and the interface. For this reason, the ratio of the mean free path to some characteristic dimension of the investigated gas dynamic problem has an important effect on determining the properties of a flow field. The separation of gas dynamics into various domains is most accurate according to the degree of rarefaction by the Knudsen number. The basic parameter is the ratio of the mean free path λ and the characteristic length of the flow field in consideration as shown in the equation:

$$K_n = \lambda/l \tag{1.5a}$$

The characteristic length can be either the merged configuration dimension or the boundary layer thickness attached to the configuration. From Equation (1.4b), the Knudsen number in the free stream can be expressed by the unperturbed Mach number $M_\infty = u_\infty/c_\infty$ and Reynolds number $\mathrm{Re}_\infty = \rho_\infty u_\infty l/\mu_\infty$;

$$K_n \approx M_\infty/\mathrm{Re}_\infty \tag{1.5b}$$

From Equation (1.5b), it may conclude that it is not possible to consider the incompressible flow to be a rarefied gas because even if the Mach number vanishes, the rarefied gas flow always exists in the low Reynolds number condition. In the continuum realm, where the density is sufficiently high and the intermolecular collisions are dominated over the collision with the medium interface; the Knudsen number will be much lower than unity $K_n \ll 1.0$. In the free molecule flow, the gas is highly rarefied and the collision processes just reverse, and the Knudsen number becomes very large; $K_n \gg 1.0$. The Transition regime between the two extremes is known as the rarefied gas regime, usually defined by the Knudsen number in the range <10.0 and greater than unity; $1.0 < K_n < 10.0$.

In the continuum regime, the particle number density is sufficiently high, and the laws of probability and the methods of statistics are applicable to obtain a meaningful microscopic description of gasdynamics from the random initial condition and untraceable multibody body interactions. Furthermore, the gas kinetic theory is capable to describe transport processes furnishing general relationships between diffusion and gas species concentration, shear stress, and rate of strain, as well as heat flux and temperature gradient. The Navier-Stokes equations together with canonical relationships of thermodynamics are always applicable to relate the mean quantities of fluid dynamics. It is the main domain from which all gasdynamics theories are derived and all the illuminating and lasting landmark accomplishments are documented.

In principle, the continuum flow theory becomes inapplicable when very large fluctuations prevent a proper mean values resolution. Under this circumstance, the information derived from the statistical information becomes meaningless. The uncertainty not only depends on the Knudsen number and the length of the statistical sampling period but also on the ergodic property [3]. For the rarefied gas regime, direct numerical simulation by solving the Boltzmann equation has been made possible with the aid of high-performance computing technology [5]. The basic and successive investigation is built on the Monte Carlo algorithm, which mimics perfectly the random particle collision process [6]. In fact, all Monte Carlo methods exploit the statistical nature of multiple collisions at the microscopic scale.

In the hypersonic shock layer, the Knudsen number is modified as:

$$K_{ns} = \left(\rho_\infty/\rho_s\right)\left(c_\infty/c_s\right)\left(\mu_s/\mu_\infty\right)K_n \qquad (1.5c)$$

Most hypersonic flows operate in the upper atmosphere from the troposphere and extend into the stratosphere and mesosphere where the gas is highly rarefied, as depicted in Figure 1.2. The Reynolds number is always low and the Mach number is higher than supersonic flows, while the Knudsen number under the ambient condition is generally greater than unity. However, the local densities, speed of sound, and molecular viscosity encounter substantial jumps in value across the shock by strong shock compression. From the increased gas temperature, the density, speed of sound, and molecular viscosity coefficient will be increased. The net result always leads to a lower Knudsen number, thus the gas downstream to the shock has a significantly reduced degree of gas rarefaction within the shock layer. The gas in most sustained flights in the hypersonic shock layer becomes a continuum medium through extremely strong shock compression.

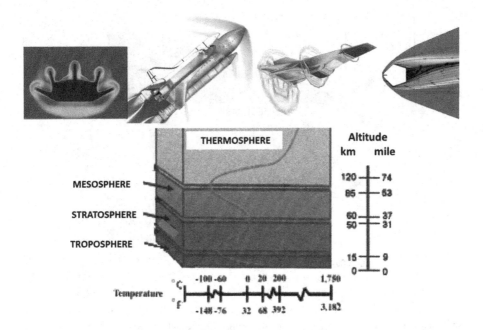

FIGURE 1.2 Hypersonic flight regime.

1.5 NONEQUILIBRIUM CHEMICAL REACTIONS

As the temperature increases with an increasing Mach number, the air composition also rapidly changes accordingly [7]. At the standard atmospheric condition with the pressure of one atm and the temperature of 298 K, the air constitutes 78.1% of nitrogen molecule N_2, and 21% of oxygen molecule O_2, the rest 0.9% by weight is a mixture of rare gas molecules of argon, helium, and water vapor. The chemical composition of air starts to change by chemical reaction at a temperature of over 2,500 K; the oxygen molecules are first dissociated together with a slight amount of nitric oxide NO formation. When the temperature arises over 6,000 K, the nitrogen will dissociate and all other molecular components begin to ionize, The air composition consist of 41.0% N_2, 28.0% of nitrogen atom N, 29.0% of oxygen atom O, the reminder chemical species are composed of a tracing amount of oxygen ion, nitric oxide ion, and electrons, $(e^-, NO^+, O^\pm, N^+)$ [4]. Most importantly the ionized gas transforms the high-temperature air into an electrically conducting medium. During reentry to earth with temperature of around 10,000 K, only 0.19% N_2 remains, while 74% O, and 20% N are dissociated. The ionized species are similar to that at the temperature of 6,000 K with only additional positively charged nitrogen molecules, N_2^+ [4]. The chemical reactions, together with the dissociation and ionization processes in the hypersonic flow field do not occur for air at moderate temperatures.

Figure 1.3 presents the distinguishable differences by a computational simulation between the shock-layer structures between the perfect and high-enthalpy

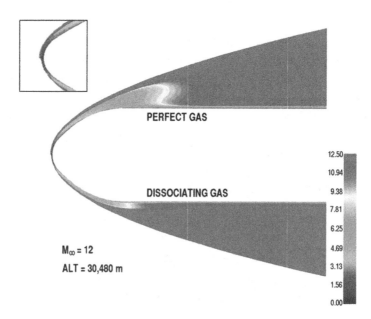

FIGURE 1.3   Perfect and chemical reacting hypersonic shock layer.

dissociating gas flowing over a hypersonic blunt body. The air composition is altered by the chemical reactions in the high-enthalpy gas model but not by the perfect gas. The kinetic energy of the oncoming stream is also converted and redistributed into the internal structure of the air medium to substantially reduce the static temperature. These visible differences reduce the shock wave standoff distance and the shock layer thickness, but the resulting aerodynamic heating and performance are no less the observable results.

The presence of different species concentration and steep temperature gradient in the thin hypersonic flow shock layer indicates all chemical reactions must taking place in an extremely short duration traveling from the shock front to the stagnation point. Under this environment, the convective and diffusion rate of the flow field exceeds the chemical reaction rate. As the consequence, the equilibrium chemical reaction process is not attainable which is easily illustrated by the heat release rate and the Damkohler number distribution along the stagnation streamline of a blunt body [8]. This parameter measures the relative rates of chemical reaction over molecular diffusion, in other words, it displays the dominance of the diffusion phenomenon in hypersonic flows. The detailed chemical reactions and diffusion phenomenon is depicted in Figure 1.3 from a computational simulation of a hypersonic flow over a blunt body at a Mach number of 27.09 and a Reynolds number of 1.38×10^4. The nonequilibrium chemical reactions including dissociation and ionization of air are based on chemical kinetic modeling. The numerical result duplicates all key physics of hypersonic flows and is verified by available experimental data. It exhibits the well-known fact that in order to achieve the ionized gas state the chemical reaction is endothermic immediately downstream to the bow shock. Then the chemical radicals recombine at the stagnation point; the local chemical reaction becomes exothermic.

Meanwhile, the Damkohl number distribution indicates an overwhelming faster diffusion speed over the chemical reaction rate within the shock layer. The exception only appears in the stagnation region. In fact, the Damkohl number exhibits multiple extremes and one of which even exceeds the value of 2014 to show a much faster chemical reaction rate than diffusion at the medium interface. In essence, the nonequilibrium chemical reactions and the diffusion phenomenon dominate the hypersonic flow (Figure 1.4).

The presence of dissociated and ionized gas components reveals that the internal degrees of freedom of gas must be considered for hypersonic flows. In other words, the detailed internal atomic and molecular structure of gas particles needs to be taken into consideration. The international degrees of freedom from rotation and vibration, to electronic excitation of molecules or atoms, occur at the instantaneous quantum transitions. Thus, the realistic gas particle interaction on a microscopic scale is no longer accurately described by the classic gas dynamic kinetics, which is built on an elastic collision model. In principle, an improved kinetic theory can be derived by adopting the inelastic collision model, but a more challenging circumstance is invoked by the nonequilibrium thermodynamic condition under which the link between microscopic and macroscopic states becomes invalid. Then the practical thermodynamic properties of a gas mixture such as the temperature, pressure, and density become indefinable. The gas in hypersonic flows therefore can only be accurately labeled as in the high-enthalpy state.

The observed dissociation and ionization process implies the quantum transition occurs in the high-enthalpy gas. The knowledge of quantum mechanics becomes a required knowledge for understanding and analyzing hypersonic flows. During the quantum jumps across discrete energy levels, only the excess energy during ionization can be emitted and absorbed at continuous spectra through an electromagnetic wave or radiation to its surrounding environment. The radiative energy transfer rate has contributed up to 30% of the total heat transfer at the peak heat loading condition of an earth reentry space vehicle [6]. In essence, the nonequilibrium thermodynamic

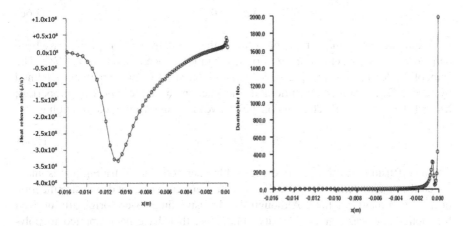

FIGURE 1.4 Heat transfer rate and Damkohler number distribution along the stagnation streamline of blunt body.

state of hypersonic flows includes nonequilibrium chemically reacting, quantum transition, and radiation has propelled the mainly aerodynamic endeavor into a multidisciplinary arena.

1.6 TRANSPORT PROPERTIES

In gas mixture the diffusion phenomenon arises from the collision nonequilibrium state. Based on the gas kinetic theory of diluted gases, the diffusion coefficient is not affected by the internal degree of freedom in molecules [9]. In a chemical reacting system, diffusion controls the mass, momentum, and energy transfer processes. The net rate of a species generation or depletion appears as a source term in the mass conservation law for the heterogeneous reacting system. The chemical change occurs only in a confined region of hypersonic flows, such as the shock layer or over an ablating surface. The diffusion phenomenon is one of the major added complications in the study of hypersonic flows. In a heterogeneous gas mixture, the velocities of individual species can be significantly different from each other. The species diffusion velocity v_i is defined as the relative velocity with respect to a local mass-average velocity of the mixture, u [9,10].

$$u = \Sigma \rho_i v_i / \Sigma \rho_i \qquad (1.6a)$$

In Equation (1.6a), the density of the i species is designated as ρ_i. The relative diffusion velocity of a chemical species u_i is defined by the difference between the mass-averaged value and the species diffusion velocity with respect to a stationary frame of reference;

$$u_i = v_i - u \qquad (1.6b)$$

By definition, the sum of mass diffusion velocities of all species is identical to zero, namely;

$$\Sigma \rho_i u_i \equiv 0 \qquad (1.6c)$$

For multi-species chemical reacting systems, the concentrations of species have been defined as mass concentration, $\alpha = \rho_i / \rho$, or molar concentration, $x_i = \rho_i / M_i$. The symbol M_i denotes the specific molecular weight, and the mole fraction is defined by $\beta_i = x_i / \Sigma x_i$. Again by definition, all the fractions have the following properties; $\Sigma \alpha_i = 1.0$ and $\Sigma \beta_i = 1.0$. The relationship between the mass and mole fraction is

$$\beta_i = \alpha_i M_i / \Sigma \alpha_i M_i \qquad (1.6d)$$

All the diffusion mass fluxes are expressed by thermodynamic driving forces under the collision nonequilibrium state. Force diffusion is also present due to stratification or an electromagnetic field. Although the diffusive fluxes were originally derived for monatomic gases at low-density conditions, they have been applied to polyatomic gases with small errors. The most common diffusive mass flux is generated by the gradients of the species concentration, pressure, and temperature of the flow field [10].

The ordinary diffusion mass flux is produced by the gradient in mass concentration;

$$\rho_i u_i = \left(c^2/\rho RT\right)\Sigma M_i M_j D_{ij}\left[\beta_i \Sigma \left(\partial G_i/\partial \beta_k\right)\nabla \beta_k\right] \tag{1.6e}$$

The pressure diffusion mass flux is

$$\rho_i u_i = \left(c^2/\rho RT\right)\Sigma M_i M_j D_{ij}\left\{\beta_i M_j\left[\Sigma\left(v_j/M_j\right) - (1/\rho)\right]\nabla p\right\} \tag{1.6f}$$

The thermal diffusion mass flux

$$\rho_i u_i = -D_i^T \nabla \ln(T) \tag{1.6g}$$

In Equations (1.6e)–(1.6g), c, R, p, and T denote the speed of sound, the universal gas constant, pressure, and temperature. The symbol G_i is Gibbs free energy; $G_i = E_i + pV - TS$.

In hypersonic flow, the forced diffusion is exerted by the electromagnetic field generated by the ionized gas known as Ambipolar diffusion [11]. In plasma, the electron is accelerated by a unit electric charge $e = 1.6022 \times 10^{-19}$ C, in the negative direction by the electric field E with a magnitude of eE/m_e leading to the electronic drifting velocity. A drifting velocity also attains by the positively charged ion but in the positive electric field direction, eE/m_+. Due to the great difference in mass between an electron and an ion, the ion drift velocity is much slower than that of electron.

The drift velocities of the charged particles are connected to the electrical field by a proportional constant and called the mobility; $\mu_e = e/m_e v$ and $\mu_+ = e/m_+ v$. The averaged collision frequency for momentum transfer is designated as v. The forced diffusion induced by the electromagnetic field on the ionized air is proportional to the concentration gradient of the charged particle number density ∇n. The diffusion coefficient is then determined by the random velocity u and the mean free path λ between collisions. The rate of the charged particle mass flow per unit area or the mass flux generated by the force diffusion becomes,

$$\Gamma = -D\nabla n; \; D \sim \lambda u \tag{1.7a}$$

In general, the electron mobility is much greater than the ion mobility, $\mu_e \gg \mu_+$ and the Ambipolar diffusion coefficient is approximated as;

$$D_a = \left(D_e \mu_i + D_i \mu_e\right)/\left(\mu_e + v_i\right) \tag{1.7b}$$

The forced diffusion by electromagnetic force can be given

$$\rho_i u_i = -\left(c^2/\rho RT\right)\Sigma M_i M_j D_a\left\{\beta_i M_j\left[E - \Sigma\left(\rho_k/\rho\right)E_k\right]\right\} \tag{1.7c}$$

The energy flux vectors are driven by the identical mechanisms of mass and momentum fluxes in a multi-species gas mixture. In the case the average velocity of any species differs from the gas mass-averaged velocity, the individual species will carry specific enthalpy, equal to $\rho_i u_i h_i$ across the control surface. This additional energy flux is contributed by the inter-diffusion process; Onsager's reciprocal relation of the irreversible thermodynamic process implies that if temperature gives rise to the Soret effect in diffusion, the concentration gradient must produce a heat flux. The coupling effect, known as the Dufour energy flux produces an additional contribution to the energy transfer. However, in most current hypersonic research, both the Soret effect of mass diffusion and the Dufour effect of energy flux are neglected. Thus the heat flux for a multi-species system including the radiative transfer is:

$$q = -k\nabla T + \rho\Sigma\alpha_i u_i h_i + q_r \tag{1.8}$$

The elevated transport properties in hypersonic flows impact more significantly mass, momentum, and energy transfer and the redistribution processes in direct contrast to all other flow speed regimes. Nevertheless, all these phenomena are still in the regime of the traditional aerodynamic discipline.

1.7 INTERNAL DEGREE OF FREEDOM

The classic gas kinetic theory treats the gas particles as elastic spheres, but detailed atomic and molecular structures of air in hypersonic flow are no longer ignorable. For a particle, either a polyatomic or a diatomic molecule the modes of excitation consist of translation, rotational, vibration, and electron excitations. The electronic excitation can be further separated into dissociation and ionization. Individual properties of gas particles in a highly excited state must be evaluated by quantum mechanics. The distinction between the classic gas kinetic theory and the elastic collisional mechanism resides on the internal degree of freedom of the colliding partner. The internal structure must possess the internal mode of translational motion, the rotational model, vibrational model, together with the electron mode. By solving all internal model with the Schrödinger wave equation, the solution of the quantum equation, the energy is given by quantum state results, and the transition is also instantaneously and in discrete energy levels.

Planck in 1901 discarded the ancient maxim that governs the energy distribution by considering the energy of an atom is not a fixed value but depended on the frequency of its oscillatory motion [12]. From the spectroscopic measurement of the hydrogen atom, a relationship between the frequency and the atomic line of hydrogen in the visible region of the spectrum was discovered [13]. This finding leads to a discrete energy level between quantum states from the first Bohr orbit or the ground state is given by $n = 1$. The quantum transition of energy between $n > 1$ and ground state is;

$$e_n - e_1 = \frac{2\pi^2 Z^2 em}{h^2}\left(1 - \frac{1}{n^2}\right) \tag{1.9a}$$

where Z is the number of elementary electric charges of an atom, and the symbol h is the Planck constant $h = 6.62 \times 10^{-34}$ Js.

The governing wave mechanics of quantum mechanics is the Hamiltonian of a conservative system for kinetic and potential energy V. The second-order partial differential equation of quantum mechanics of a single particle with mass m, and potential function V is the Hamiltonian operator which forms the,

$$\frac{h^2}{2m}\nabla^2\varphi - v(r)\varphi + ih\frac{\partial\varphi}{\partial t} = 0; \ D\varphi = \lambda\varphi \tag{1.9b}$$

From pure mathematics' view point, the quantum numbers are the eigenvalues of the partial differential equation for each degree of freedom of molecule or atom. The eigenfunction φ of the three-dimensional wave equation is sometimes called the probability amplitude function which is required to be a single-valued, finite, and continuous for all physically possible values in space [12,14]. The solutions of the wave equation represent a set of discrete stationary energy states for an atom, characterized by the quantum numbers n, l, and m in three-dimensional space. The description for the electronic configuration of a molecule using the quantum number is similar to that for an atom. Usually, the quantum numbers of a molecule consist of three methods. The first group is defined by the Born-Oppenheimer approximation for which the nuclei are held stationary and only the electronic motions are considered. Second, only a single quantum number is used to define the vibration state of the nuclei. Third, the finer details of the electronic motion may not be separable and be considered together as the $^1\Sigma$ states [14].

The energy of a gas molecule is measured above its zero-energy or ground state and is the sum of the energy of translational, rotational, vibrational, and electronic degrees of freedom.

$$e_i = e_t + e_r + e_v + e_e \tag{1.10a}$$

For atoms, the total energy consists of only the translational and electron energy:

$$e_i = e_t + e_e \tag{1.10b}$$

The inception of the quantum concept of gas dynamics by Planck has a deep root in electromagnetics. The energy states are quantum-restricted, in other words, the energy is transferred in discrete units and the transfer process is instantaneous because there is no continuity between the energy states.

Figure 1.5 depicts the discrete quanta of the translation, rotation, vibration, and electronic excitations. Only the translational degree of freedom is a nearly continuous spectrum of energy and extends to infinity, the vibrational quantum states have finite numbers until molecular dissociation occurred. However, the quantum transition can take place by jumping to the next immediate next level and known as the climbing step process. A quantum transition may also jump from a ground state to a fully excited state, and in this process, it's often called as the big bang.

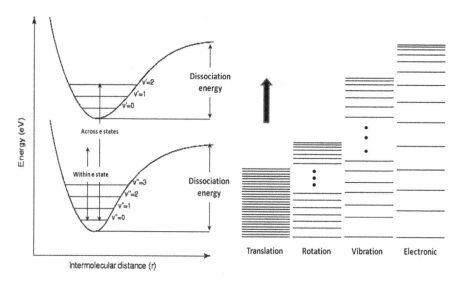

FIGURE 1.5   Quantum states of the internal degree of freedom of gas.

According to de Broglie's hypothesis, the wavelength of a particle of mass m and the particle's velocity u are related by $mu = h/\lambda$. In free space with a bounding dimension l, there is an integral number of half wavelengths, $l = n\lambda/2, n = 1,2,3,....$ The allowed energy level for the translational degree of freedom is then,

$$e_t = \frac{h^2}{8m}\left[\left(\frac{n_x}{l_x}\right)^2 + \left(\frac{n_y}{l_y}\right)^2 + \left(\frac{n_z}{l_z}\right)^2\right] \tag{1.10c}$$

All the energy levels of translational motion are so closely packed relative to the datum that the energy levels are practically continuously distributed.

The quantum solution of the rotational *motion* of a diatomic molecule is modeled by a rigid dumbbell. If the molecule is rotating freely, the acceptable energy levels are

$$e_r = j(j+1)h^2/8\pi^2 , j = 0,1,2,... \tag{1.10d}$$

where I is the moment of inertia about two orthogonal axes of a rotating diatomic molecule. It has been found any level of j can be constructed by $2j + 1$, thus the appropriate degeneracy of the rotational motion of a diatomic molecule shall be set as such [14]. For molecular nitrogen, oxygen, and nitric monoxide, the characteristic temperatures of rotation excitation θ_r are 2.86, 2.07, and 2.42 K. Therefore, the rotational degree of freedom of air is fully exited at the standard atmospheric condition. For this reason, the rotational excitation is always considered to be in equilibrium with the translational degree of freedom.

The simplest model for a diatomic molecule in vibration excitation is a spring connected two masses to reflect that the restoring force is proportional to the square of the separation distance. The potential energy of the vibrational model is $V = (r - r_e)^2 f/2$ and f is the spring constant. The quantum mechanics solution for the discrete vibrational energy levels for a simple harmonic vibrator is:

$$e_v = (v + 1/2)hv, v = 0,1,2,3,\ldots \quad v = \sqrt{f/2\pi^2 m} \tag{1.10e}$$

The simple harmonic vibrator is actually inconsistent with the behavior of a diatomic molecule because it can dissociate into separate atoms. The correction comes from the anharmonic vibrator which permits energy associated with sufficient vibration energy to cause atoms separation and more detail will be deferred. The difference between the characteristic temperatures of dissociation and vibration of oxygen is substantial, $\theta_{de} = 59,000$ K and $\theta_v = 1,230$ K [14].

The electronic or nuclei energy is from the ionic or the homopolar bonding (diatomic molecule) that is intimately connected to the electronic structure of the atom. A third kind of nuclear attraction is the van der Waals binding, but the quantum energy levels have degenerated. Most electronic ground states of diatomic molecules are multiplet due to electron spin. For the present purpose, the electronic contribution to the internal energy can only be given as a function of its degeneracy and characteristic electronic temperature;

$$e_e = e_e\left(g_i, \theta_e, \theta_e/T\right), i = 0,1,2,\ldots \tag{1.10f}$$

In Equation (1.10f), g_i is the degeneracy and θ_e denotes the characteristic temperatures for electronic excitation. In general, the next electronic energy level lies far above the ground state and the ionization potential is in the order of 10 eV (eV $= 1.6022 \times 10^{-12}$ enrgy) [15].

The quantum theory recognizes that the energy is not continuous but through discrete quanta, and the value of quantum transition depends on the natural frequency of wave motions $e = hv$. The physics is clearly demonstrated by the emission spectrum of gas because electromagnetic wave emission always occurs concurrently with quantum jumps. Therefore, the radiative energy transfer and the quantum mechanics represent two unique new features in hypersonic flows

1.8 EQUATION OF STATE

Classic gas dynamics is built on the foundation of the gas kinetic theory from collisions of a large group of gas particles without further knowledge of the particle's internal structure. The interaction between colliding partners is governed by classic mechanics. The link between microscopic dynamics to macroscopic gas motion is established by statistical means. When the gas reached a state of dynamic equilibrium, the thermodynamic property will be independent of time and determines the global behavior of the gas flow. The two aspects of the gas kinetic theory and thermodynamics are closely intertwined for gas dynamics.

The shock waves are an essential feature of hypersonic flows, the property of the gas medium needs to be unambitiously determinable by thermodynamics. For this purpose, the equation of the gas state must be known which encompasses all thermo-dynamic variables. The fundamental equation of state is given by the relationships between enthalpy with entropy and pressure of the gas;

$$h = h(S, p) \tag{1.11a}$$

The thermodynamic variables such as the temperature, pressure, and density of gas are immediately obtainable by definition from thermodynamic formulas or canonical laws of thermodynamics;

$$T = (\partial h/\partial S)_p; \quad \rho = 1/(\partial h/\partial p)_s \tag{1.11b}$$

The subscripts of Equation (1.11b) designate the differentiation with respect to a particular invariant of a reversible process. The internal energy of a system with a volume V consists of molecules n_1, n_2, ... n_i of a gas mixture; then the internal energy of the gas mixture is a function of pressure, temperature, and gas species; $e = e(p.T, n_1, n_2, n_3, \ldots, n_i)$.

The following group of thermodynamics variables occurs frequently in aerodynamics, including the relationship between thermodynamic variables and the Gibbs potential or free energy;

$$h = e + pV$$
$$g = h - TS \tag{1.11c}$$

Any infinitesimal change in the internal energy can be written mathematically as

$$de = (de/dV)_{S,n} \, dV + (de/dS)_{V,N} \, dS + (de/dn_i)_{S,V} \, dn_i \tag{1.12a}$$

The first two terms associated with the change in the internal energy in volume and entropy for a constant mass mixture of invariant chemical composition lead to the canonical relationships;

$$p = -(\partial e/\partial V)_{S,n_i}; \quad T = (\partial e/\partial S)_{V,n_i} \tag{1.12b}$$

The First law of thermodynamics for the system of volume V in a reversible process is then

$$de = TdS - pdV \tag{1.12c}$$

The concept of a perfect gas makes the gas medium to be the simplest work medium. The concept is basically derived from experimental observations that the gas obeys the Boyle law and the Joule-Thomsen effect vanishes [4]. The basic assumptions restrict the colliding particles or molecules to only possess translational motion and without heat loss through expansion. For perfect gas, the characteristic tempera-tures will be the same for all gas components and are frequently called the absolute temperature. This simplification is widely adopted for the classic aerodynamics investigation for which the equation of the state is

$$p = \rho RT \tag{1.13}$$

Where R is known as the universal gas constant, for air at standard condition, it has the value of $R = 287 \ m^2/s^2 \ K$. The further simplification of the perfect gas assumption is considered the internal energy is a function of temperature only, or calorically perfect gas, the specific heats for S reversible become constants:

$$c_v = (dq/dT)_v \approx de/dT; \quad c_p = (dq/dT)_p \approx dh/dT \tag{1.14}$$

The ratio of the specific heat $\gamma = c_p/c_v$ in the air has a constant value of $\gamma = 1.4$ at the standard condition. Realistically the specific heat ratio is a variable with respect to chemical reactions and the composition of a gas mixture of excited internal degrees of freedom. Its value will approach unity when the internal degrees of molecules are excited at a higher temperature environment.

REFERENCES

1. Kennard, E.H., *Kinetic theory of gases, with an introduction to statistical mechanics*, McGraw-Hill, New York, 1938.
2. Chapman, S. and Cowling, T.G., *The mathematical theory of nonuniform gases*, Cambridge University Press, London, 1964.
3. Liepmann, H.W. and Roshko, A., *Elements of gasdynamics*, John Wiley & Sons, London, 1957.
4. Chernyi, G.G., *Introduction to hypersonic flow*, Academic Press, New York, 1961.
5. Shang, J.S., Surzhikov, S.T., and Yan, H., Hypersonic nonequilibrium flow simulation based on kinetic models, *Frontier in Aerospace Engineering*, Vol. 1, No. 1, 2012, pp. 1–12.
6. Park, C., Review of chemical kinetics problems of future NASA missions, I Earth entries, *Journal of Thermophysics and Heat Transfer*, Vol. 7, No. 3, 1993, pp. 385–388.
7. Hayes, W.D. and Probstein, R.F., *Hypersonic flow theory*, Academic Press, New York, 1959.
8. Shang, J.S., Landmarks and new frontiers of computational fluid dynamics, *Journal of Advances in Aerodynamics*, Vol. 1, No. 5, 2019, pp. 1–36.
9. Hirschfelder, J.O., Curtiss, C.F., and Bird, R.B., *Molecular theory of gases and liquids*, 2nd Ed., John Wiley & Sons, New York, 1954.
10. Bird, R.B., Stewart, W.E., and Lightfoot, E.N., *Transport phenomena*, John Wiley & Sons, New York, 1960.
11. Shang, J.S., *Computational electromagnetic-aerodynamics*. John Wiley & Sons, Hoboken NJ, 2016.
12. Leighton, R.B., *Principles of modern physics*, McGraw-Hill Inc., New York, 1959.
13. Zel'dovich, Y.B. and Raizer, Y.P., *Physics of shock waves and high-temperature hydrodynamic phenomena*, Dover Publication, Mineola, New York, 2002.
14. Clarke, J.F., and McChesney, M., *The dynamics of real gases*, Butterworths, Washington, DC, 1964.
15. Raizer, Y.P., *Gas discharge physics*, Springer-Verlag, Berlin, 1991.

2 Aerodynamic Governing Equations

2.1 BOLTZMANN EQUATION

The gas kinetic theory solves directly the particle dynamics of a given system. On the Newtonian frame, the random motions of a huge amount of interacting particles are nearly an unsolvable multi-body problem that requires specific knowledge of initial values. In fact, this required information is generally unknown and the interaction process is completely random. Thus, the kinetic theory must apply the laws of probability and the methods of statistics to achieve its objective. The solutions obtained are always probabilistic in nature, the macroscopic quantities of gas dynamics in the continuum domain must be generated by the averaged or the means values from the gas kinetic theory [1].

In general, the encounters between gas particles determine the distribution of particle velocities. The collision interactions among all particles without internal structure are assumed to be governed by Newtonian mechanics, and by means of the binary collision process to become a two-body problem. Most importantly, the average distance traveled between collisions is considered to be very large in comparison with the particle dimension that breaks the connection between the initial conditions of these colliding particles. Therefore, the chaotic motion has no correlation existing between the particle velocity and position. Finally, all particles are assumed as solid elastic spheres and have spherical symmetry.

Boltzmann treats the general motion of gas particles by the spatial distribution functions of each particle, which generally is not in dynamic equilibrium and varies from one location to another. The distribution function considers the gas particle or molecule just like a mass point and ignores their internal structure, and the interaction among these mass points arises by occasional collisions. Thus, the multibody problem is resolvable only by utilizing the methods of statistical mechanics. The ultimate statistical features of the dynamic system in statistical equilibrium are independent of its initial condition or satisfy the ergodic surmise [1].

The Boltzmann-Maxwell equation is developed on the basis of statistic mechanics in which the Liouville theorem is the cornerstone. The general dynamical theory is described by a number of coordinates, usually denoted as $q_1, q_2, q_3, ..., q_n$. The number is often referred to as the degrees of freedom of the particle. Associating with these coordinates are the generalized momenta as $p_1, p_2, p_3, ..., p_n$. The q_i's and p_i's are the six independent variables in phase space together with time for describing the dynamic system. For a three-dimensional space, there are six degrees of freedom of the generalized coordinate q_i, and the generalized momenta p_i during a particular motion, these variables can vary only as functions of time. In classical

DOI: 10.1201/9781003212362-3

mechanics for a conservative system, the *Hamiltonian function H* is the sum of the kinetic and potential energies. This theorem has been proved by the equations of motion by the Hamiltonian form H [1,2]. According to the Heisenberg uncertainty principle of quantum mechanics, $h \approx \Delta q \cdot \Delta p$, the Planck's constant, h has the value of $h = 6.62 \times 10^{-27}$ erg·seconds, it is the demarcation between Newtonian and quantum mechanics. However, in the absence of all external forces, a completely free moving particle can have unquantized energy [3]. The very concept of a classical trajectory directly contradicts explicitly to quantum mechanics, but still within the accuracy limit imposed by the Heisenberg principle.

For particles without internal structure, the Hamiltonian is the energy function representing the kinetic energy and the potential energy, which is designated by the symbol V_o. It acquires the following form in three-dimensional space as follows:

$$H = \sum_i \left(p_{1i}^2 + p_{2i}^2 + p_{3i}^2 \right) \bigg/ 2m + V_o\left(q_1, q_2, q_3\right) \tag{2.1}$$

The Liouville theorem for the equation of motion in the Hamiltonian form can be given as

$$dq_i/dt = \partial H/\partial p_i; \quad dp_i/dt = \partial H/\partial q_i \tag{2.2}$$

Equation (2.2) also are known as the canonical equations of Hamilton which constitute a group of second-order equations of motion. There are modifications required in deriving the Boltzmann equation when quantum theory is substituted by classic mechanics. Conceptually, a thermal equilibrium state is considered a finite series of possible quantum states. From the statistical point of view, the individual state is just one of the complete fundamental series of stationary states of the particles. These states that have the same energy are called multiplicity or the statistical weight [2].

Boltzmann has shown that it is more convenient to describe an equivalent distribution function in the phase space of different systems at the same moment. More importantly, the formulation is a general description of the laws for the conservative property of a dynamic system. The distribution function is defined for the number of particles to have velocities in the range of the specular velocity c_i, and positions in the range of dx_i around x_i as; $f(c_i, x_i, t)$. The velocity c_i is the peculiar velocity of the particle. The distribution function or the probability function has the fundamental significance to a particle with velocities in the range dc_i, which satisfies the following conditions:

$$\iiint f(c_i, x_i, t) dc_i dc_j dc_k = 1$$

$$\iiint Nf(c_i, x_i, t) dc_i dc_j dc_k = N \iiint f(c_i, x_i, t) dc_i dc_j dc_k = N \tag{2.3}$$

It is important to note that the velocities and coordinates are independent of each other in the phase space and both are varying with respect to time. In the original formulation of the distribution function, the cardinal assumption is strictly for the translational motion in the classic energy mode and the magnitudes of velocity are permitted for any values from minus to plus infinity. The energy exchange is limited to linear momentum; it is the so-called Elastic collision. However, when each particle allows an internal structure change, the energy exchange by this collision model must be specified by the appropriated quantum numbers. Under this condition, the process is termed an Inelastic collision.

In this way, the formulation is a general description of the laws for conservative property of a dynamic system. The statistical features of a single system can be achieved by sampling a huge assembles of distribution functions at any given moment in different phases. The distribution function is defined for a number of particles to have an infinitesimal amount of change of the specular velocity dc_1 about c_i, similarly the positions in the range of dx_i around x_i as $f(x_i,c_i,t)dx_1 \ldots dx_n \, dc_1 \ldots dc_n$. The specular velocity c_i is the actual velocity relative to some standard frame of reference or the peculiar velocity of the particle.

At a time level of t, the number of particles in the phase space $dx_i dc_i$ with the degrees of freedom x_i and c_i is $f(x_i,c_i,t)dx_i dc_i$. After a time elapse of dt, the particle will occupy the new position $x_i + c_i dt$, if an external force F_i is acted on the unit mass of the particle, such as the gravitational force, the particle will have increased velocity to $c_i + F_i dt$. Then the number of particles in the region will be $f(x_i + c_i dt, c_i + F_i dt, t + dt)dx_i dc_i$. The equation of change corresponds to the first-order approximation to the full system of equations by the chain rule of differentiation. The collision process will lead to the depletion or replenishment of the distribution function as

$$f(x_i + c_i dt, c_i + F_i dt, t + dt) - f(x_i,c_i,t) = \left[\partial f(x_i,c_i,t)/\partial t \right]_c \qquad (2.4a)$$

Expand the left-hand-side of Equation (2.3) by Taylor's series in time and set the dt to the limiting value of zero to get the Boltzmann equation for the distribution function,

$$\partial f/\partial t + c_i \, \partial f/\partial x_i + F_i \, \partial f/\partial c_i = \left[\partial f(x_i,c_i,t)/\partial t \right]_c \qquad (2.4b).$$

As it has mentioned before, the specular velocity exists in a range from minus to plus infinity. In the original formulation of the distribution function, each particle may possess an internal structure associated with appropriate quantized states. Then the collision process involves the interactions of internal energy and chemical reaction, and the energy exchange leads to a quantum transition where the emission and absorption of the electromagnetic wave take place. Under this circumstance, the interaction between particles is referred to as an inelastic collision. The added complexity in formulation for the inelastic collisions requires an enormous amount of basic research efforts and incisive insights. The challenge is formidable and almost beyond our present knowledge reach, but it must be pursued for our basic knowledge gained for the future.

For the classic gas kinetic theory, the colliding particles are assumed to be rigid spheres, so the interaction between particles only involves the translation velocity of colliding partners and is a simple elastic collision. The infinitesimal change generated in time to the distribution function by the collision process, the right-hand side of Equation (2.4b), has been evaluated by elastic-sphere particles leading to

$$\left[\partial f(x_i,c_i,t)/\partial t\right]_c = \iint \left[f(c_i')f(x_i') - f(c_i)f(x_i)\right]g\sigma d^3x_i d^3c_i \qquad (2.5)$$

where g is designated for the relative specular velocity $g = c_i' - c_i$, and σ is the collision cross section or the probability of collisions may occur. The collision integral of Equation (2.5) represents the effect on the distribution function by the completing collision process that leads to depleting and replenishing of the conservative property of a dynamic system. In 1972, the Boltzmann equation of gas kinetic theory acquires the following form:

$$\partial f/\partial t + c_i \, \partial f/\partial x_i + F_i \, \partial f/\partial c_i = \iint \left[f(c_i')f(x_i') - f(c_i)f(x_i)\right]g\sigma d^3x_i d^3c_i \quad (2.6)$$

The Boltzmann formulation is an integro-differential equation; it gives the distribution of gas molecules with all available energy states. The equation is very difficult to solve because the particle dynamics on microscopic scale is nearly impossible to specify completely and accurately.

This challenge is met by the direct simulation Monte Carlo (DSMC) method, which was initialed in the middle 1960s [4]. The Numerical algorithm actually exploits the statistical nature of multiple molecule elastic collisions on the microscopic scale. The basic idea is to track a large number of statistically represented particles but not the astronautically huge number of individual gas molecules. The approximation reduces tremendous computational resources but also introduces statistical error that leads to the well-known issue of statistical stationary, which means how many sampling points are sufficient to reach a statistically meaningful result. The simulation is appropriate in the intermediate rarefied-continuum gas domain and when the value Knudson number is up to the value of ten.

The foregoing derivation of the Boltzmann equation is purely based on the physical argument, but the formulation has a rigorous foundation based on the Liouville theorem based on the condition that the magnitude of intermolecular force is small in comparison with the electronic binding in molecules or atoms. This approximation is valid for an electrically neutral gas without the presence of a remote force field and when the density is low. Nevertheless, the Boltzmann equation is the foundation for all conservation laws applied in fluid dynamics [1–3,5].

2.2 BINARY ELASTIC COLLISION

In general, the encounters between gas particles determine the distribution of particle velocities. The collision interactions among all particles without the internal structure are assumed to be governed by Newtonian mechanics, and by means of the binary collision process to become a two-body problem. Most importantly, the average distance traveled between collisions is considered to be very large in comparison with the particle dimension that breaks the connection between the initial conditions for these particles. Therefore, the chaotic motion has no correlation existing between the particle velocity and position [1–3]. Finally, all particles are assumed as solid spheres having spherical symmetry. Thus any gas particle whether a diatomic or polyatomic molecule, its dynamics include only translational motion with very densely packed discrete energy states that can be considered to have a continuous spectrum. On the other hand, the inner structure of a molecule consists of rotational, vibrational, dissociation, and ionization excitations and is referred to as the internal degrees of freedom. The energy contents of these internal excitations can be very significantly greater in comparison with the translational mode and are redistributed by discrete energy levels or quantum states. The energy transfer from excitation from the lower quantum state or cascading downward is conducted by quantum jumps. It is clear; the solutions by the kinetic theory for gas dynamics at high enthalpy conditions with excited internal degrees of freedom will be inaccurate.

From the kinetic theory of gas, the interaction of particles is limited to binary encounters, for which only collisions between pairs of molecules are significant. This assumption also is based on investigations of a dilute gas and by comparing with experiments conducted in laboratories that have amply justified the assertion. A detailed description of the binary dynamic exchange process by elastic collision is warranted to gain a better understanding of the collision mechanism, especially for establishing the concept of dynamic equilibrium conditions. For the binary collision model, the interaction between particles is restricted to momentum exchange and the internal energy of particles in mutual interaction by elastic collision is an invariant.

In the collision process, the measure of probability for momentum transfer occurring by a localized phenomenon is the collision cross section. In classical mechanics, the mutual cross section of two particles interact is the area transverse by their relative motion within which colliding and scattering from each other. The cross section is defined as

$$\sigma = 1/n\lambda \tag{2.7}$$

where the λ is the mean free path and n is the number density of the gas, the momentum transfer cross section thus has a physical unity of m^2. For elastic spheres by direct contract, the cross section is simply $\sigma = \pi d^2$. However, when random collisions among particles involve multiple incident angles, the probability of momentum exchange is now dependent on the differential cross section. The differential cross section is defined as the differential limit of all the collision probabilities which can be assessed. The integrated total cross section is the result of the integrant of the differential cross section over all scattering angles.

From the dynamic of binary collision, the number of particles of a given state within the phase space after collision either depleting or replenishing by the interaction are functions of the incoming velocity and the angles of φ. This angle specifies the direction of the line connecting centers for colliding elastic spheres, and the impact parameters and its differential element are $b = d \sin\varphi$ and $db = d \cos\varphi d\varphi$.

The magnitude of the differential cross section is the function of the incident angle, on the spherical polar coordinate it's the zenith angle, and the impact parameter b is the perpendicular offset of the trajectory of the incident particle. The incident angle and the impact parameter have a unique functional dependence, and the elementary area of the cross section in the plane of the impact parameter b is;

$$d\sigma = bdbd\varphi \qquad (2.8)$$

The symbol φ is the azimuthal angle of the spherical polar coordinate system. The differential angular range of the scattered particle at angle θ has the elementary solid angle, $d\Omega = \sin\theta d\theta d\varphi$. The solid angle of the incident particle stream may have infinitely many directions and traditionally is described as the vector in terms of spherical coordinates as shown in Figure 2.1. The total surface area of a unit radius is enclosed by the solid angle above the control surface and the polar angle θ, which is measured from the surface outward normal, and the azimuthal angle φ from the axis of the vector projects onto the surface. The domains of the two angulars measured for a hemisphere is $0 \le \theta \le \pi/2$ and $0 \le \varphi \le 2\pi$, thus the solid angle $\bar{\Omega}$ is the projection area of the vector.

The differential cross section is $d\sigma/d\Omega$. The total cross section is the integrant of the differential cross section over the angles;

$$\sigma = \int_0^{2\pi}\int_0^{\pi} (d\sigma/d\Omega)\sin\theta d\theta d\varphi \qquad (2.9)$$

For the binary collision by elastic sphere, the trajectory on the center of mass frame of reference is depicted in Figure 2.2. The coordinate origin of the spherical polar coordinate system is located at the center of one of two identical spheres; the two polar angles zenith and azimuthal are denoted as θ and φ respectively. The relation between the dimension of colliding particles and collision probabilities is best explained by the concept of a collision cross section. This condition is valid in

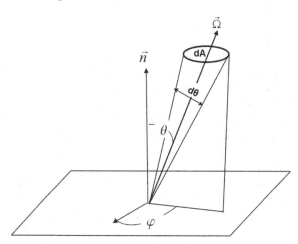

FIGURE 2.1   Solid angle of incident particle stream.

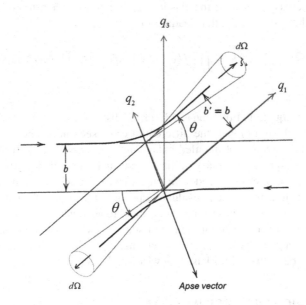

FIGURE 2.2   Trajectory of mass center in a binary collision.

low-density conditions and the differential cross section for collision is $bdbd\varphi$, which is given by Equation (2.8). The deflection angle after the collision is often given as χ and the impact parameter b is just the diameter of the elastic-sphere particles. The collision cross section of elastic sphere is independent of both the relative velocity $g = c_i - c_j$, and the incremental deflection angle of elastic collisions on the center-of-mass frame of reference to have a value of $d\Omega = \sin\theta d\theta d\varphi$. The solid angle of binary collision is then is represented as

$$\Omega = \int_0^{2\pi} \int_0^{\pi} \sin\theta \, d\theta \, d\varphi \tag{2.10}$$

The rate of change in time for the distribution function is the difference between the total number of particles by depletion and addition from the colliding particles;

$$\left[\partial f(x_i, c_i, t)/\partial t\right]_c = \iint f(c_i')(c_j') g(d\sigma/d\Omega)' \, d\Omega d^3 c - \iint f(c_i)(c_j) g(d\sigma/d\Omega) \, d\Omega d^3 c \tag{2.11}$$

By assuming that the differential cross section for the colliding and scattering mechanics are identical, the collision integral of the Boltzmann distribution equation can be written as

$$\left[\partial f(x_i, c_i, t)/\partial t\right]_c = \iint \left[f(c_i')(c_j') - f(c_i)f(c_j)\right] g(d\sigma/d\Omega) d^3 c_i \tag{2.12}$$

The collision cross section for the elastic sphere is independent to both the relative velocity and deflection angle, thus leading to

$$\partial f/\partial t + c_i\,\partial f/\partial x_i + F_i\,\partial f/\partial c_i = \iint \left[f(c_i')f(x_i') - f(c_i)f(x_i) \right] g(d\sigma/d\Omega)\sin\theta\,d\theta\,d\varphi d^3 c$$

$$(2.13)$$

The Boltzmann equation, Equation (2.13) for the distribution function is based on the binary elastic collision model; the differential cross section has been derived from the ideas of the principle of detailed balancing in microscopic collision dynamics [3]. In order to expand the binary collision into inelastic collision, there are many urgent basic research efforts needed to implement the interactions between atoms, molecules, and atom-molecule to evaluate especially for these differential cross sections for energy transfer. These needs are critical to account for the realistic internal energy transfer between different quantum states, as well as, when chemical reactions. The challenge in basic research for chemical physics is paramount and formidable but must be answered for basic knowledge advancement.

2.3 DYNAMIC EQUILIBRIUM STATE

Two important facts are critical for applying Boltzmann distribution function. The first one is that the ergodic surmise must be held for a fixed force field for the ultimate statistic features for the behavior of a dynamical system in statistical stationary. The requirement ensures the statistical result in general, which is independent of its initial condition [2]. Second, it must follow naturally to reach the condition of a dynamic system in collision equilibrium.

The dynamic equilibrium state is a special case but also the most likely state of gas particle motion on microscopic scale in which the collisions between molecules no longer affect the distribution function. This special state of microscopic dynamics can be viewed as a perturbation generated by one type of interaction that will be canceled by an opposite and equal encounter. The inverse action against the direct process is a special case of the principle of detailed balance [2,3]. The dynamic equilibrium state of a gas dynamics system is that the contribution by the collision integral must vanish.

$$\left[\partial f(x_i,c_i,t)/\partial t \right]_c = \iint \left[f(x_i,c_i',t)(x_j,c_j',t) - f(x_i,c_i,t)f(x_j,c_j,t) \right] g(d\sigma/d\Omega)d^3 c_i = 0$$

$$(2.14)$$

Therefore $f(x_i,c_i',t)f(x_j,c_j',t) = f(x_i,c_i,t)f(x_j,c_j,t)$ is the sufficient condition for dynamic equilibrium. However, it does not follow immediately that it's the necessary condition or dynamic equilibrium. In gist, the integrand must have a finite integrant to be zero over the entire range of the integral.

A class of molecule in a unit volume is associated with a quantity $Q = \log nf$, the variation of Q is a measure of their contribution by the relevant class of molecule collisions to the change of state. The mean value is defined as the Boltzmann H function,

$$H = \int_{-\infty}^{+\infty} f(x_i,c_i,t)\log f(x_i,c_i,t)d^3c = \int_{-\infty}^{+\infty}\left\{[\log f(x_i,c_i,t)+1]\left[\partial f(x_i,c_i,t)/dt\right]\right\}d^3c \quad (2.15)$$

According to Boltzmann's H-theorem, and by defining a number H to be independent of the position but only as a function of velocity c_i and time, $f = f(c_i,t)$. In fact, Equation (2.15) is directly related to the entropy from the kinetic theory of gas or thermodynamics of the system [1]. Substitute the collision integral from Equation (2.14) to get the rate of change of H as

$$\partial H/\partial t = 1/4 \iint \log\left(f_i f_j / f_i' f_j'\right)\left(f_i' f_j' - f_i f_j\right)d^3c_i d^3c_j \quad (2.16).$$

It is observed that Equation (2.16) is a definite integral of distinctly asymmetric variables of integration, and the subscript i for the distribution function is just a shorthand notation for different velocities of colliding particles. The integral is definitely either negative or zero because the value of $\log\left(f_i f_j / f_i' f_j'\right)$ is positive or negative according to $f_i f_j > f_i' f_j'$ or $f_i f_j < f_i' f_j'$. Thus the value of H can never increase, to be known as Boltzmann's H-theorem. In a uniform and steady state, the rate of change for the distribution function must vanish thus the conclusion is also applied to the rate of the H function,

$$\iint \left(f_i' f_j' - f_i f_j\right)dx_i\, dx_j = \iint \left[f(x_i',c_i',t)f\left(x_i',c_j',t\right)-f(x_i,c_i,t)f(x_j,c_j,t)\right]d^3c_i d^3c_j \quad (2.17)$$

Therefore $f\left(x_i',c_j',t\right)f\left(x_i',c_j',t\right)= f(x_i,c_i,t)f(x_j,c_j,t)$ is the necessary and sufficient condition for equilibrium, and the subscripts i and j are designated the different velocities for colliding particles.

The equilibrium condition of the Boltzmann equation is essentially required that the logarithmic of the distribution function $\log f(x_i,c_i,t)$ is a "summational invariant" of any colliding encounter.

From the macroscopic point of view, a chaotic system eventually will always pass into a state of equilibrium. In other words, the equilibrium state will be reached automatically. From the wave mechanical viewpoint, the chaotic relative phase velocity constitutes the essential characteristic of the equilibrium state. For this reason, the equilibrium state is the most probable distribution of energies in equilibrium. However, in chemical reactions, the chemical equilibrium condition is that of a dynamic balance between the forward and the reverse reactions.

2.4 MAXWELL DISTRIBUTION FUNCTION

The linkage between microscopic particles behavior of a gas system and the measurable properties of aerodynamic interest is a global averaging procedure. The basic approaches are either addressing specifically the identification of the thermodynamic variables individually when a distribution function of the gas particle system is known or by transforming the governing equations from the probabilistic into determination formulations by a globally summing process to reach a statistic assembly.

The Maxwell distribution for molecule dynamics at the microscopic scales was established in 1859 and the proof was carried out later by Boltzmann in 1872. It is the most general and the only analytic solution to the Boltzmann equation under the dynamic equilibrium condition. The Maxwell distribution function is valid for gas particles only possessing the translational degree of freedom. When a gas mixture is in collision dynamic equilibrium, its distribution function becomes the asymptote; all other distributions will always adjust by molecular collisions toward the Maxwell distribution. It also has been shown by kinetic theory that the Maxwellian distribution corresponds to the isentropic change of the state of thermodynamics [1,2]. It must recall, the velocity distribution function is valid for any diatomic or polyatomic molecule gas mixture with only the translation motion in the equilibrium state and is devoid of any other internal states or chemical reactions.

$$f(x_i, c_i) = (m/2\pi kT)^{3/2} \exp(-mu^2/2kT) \tag{2.18}$$

In Equation (2.18), The Boltzmann constant has a value of $k = 1.381 \times 10^{-16}$ erg/°C. An important result of the mean molecular speed in the equilibrium state is determined by the original form of the Maxwell distribution. It may be interesting to note that the distribution function is symmetrical with respect to all velocity components, thus it has no preferred direction in the gas. The total number of molecules in a unit volume on the spherical coordinate is obtained by integrating over the entire specular velocity range from the negative to the positive values at the infinities.

$$n_i du_i = n_i (m_i/2\pi kT)^{3/2} \iint e^{-m_i u_i^2/2kT} u_i^2 \cos\varphi \, d\varphi \, d\theta \, du_i$$

$$= 4\pi n_i (m_i/2\pi kT)^{(3/2)u_i^2} e^{-m_i u_i^2/2kT} \, du_i \tag{2.19}$$

The arithmetic mean molecular speed now can be found by performing the integral over all the allowable values of velocity components and dividing by the total number of molecules;

$$n_i \bar{u}_i = \int_0^\infty n_i u_i du_i \tag{2.20}$$

The arithmetic mean molecular speed is an average of the molecular velocity in six possible directions on the Cartesian coordinate. The result is related to the pressure of a system by the bombardment of gas particles on each other and the solid surface.

$$\bar{u}_i = \left(8kT/\pi m_i\right)^{1/2} \tag{2.21}$$

The mean value of the molecular velocity in an equilibrium state reveals that it's proportional to the square root value of the gas temperature and inversely proportional to the particle mass.

The Maxwell distribution function for energy can also be found as [1]

$$f(x_i,c_i) = 2\left[e/\pi(kT)^3\right]^{1/2} \exp(-e/kT) \tag{2.22}$$

Under thermodynamic equilibrium conditions, the energy of other possible internal degrees of freedom for rotational, vibrational, and electronic excitation can be described similarly. The basic formulation provides a base for the partition functions from statistic thermodynamics [3]. For each quantized state of distinctive internal degrees of freedom, the energy distribution functions in dynamic equilibrium conditions can be given as

$$f(x_i,c_i) = \sqrt{1/\pi ekT} \, \exp(-e/kT) \tag{2.23}$$

where the symbol e represents the characteristic energy of each individual mode of internal quantized excitation. The Maxwell distribution is the crucial linkage between the microscopic molecule dynamics and the macroscopic variables in gas dynamics under equilibrium conditions.

2.5 MAXWELL TRANSFER AND EULER EQUATIONS

The Macroscopic description of gasdynamics in thermodynamic variables, and the conservation laws for mass, momentum, and energy can be recovered by the Maxwell transfer equation from the Boltzmann equation [5]. The integral transformation from particle dynamics to macroscopic gas dynamics has also been referred to as the method of moments. The dependent variables of the conservation laws such as the density, gas velocity of organized motion, pressure, and temperature can also be linked to the microscopic motion of gas particles if the distribution function is known. Any global gas dynamic quantities associated with a group of particles or molecules can be obtained by summing over the entire phase space at a given time and an elementary volume

$$n\bar{Q}dv = \int nQf(x_i,c_i,t)\,dc_i dv \tag{2.24}$$

By applying the identical integration over the Boltzmann equation, Equation (2.13) without the presence of an external force field, the global description of gas dynamics through the Maxwell general transfer equation acquires the following form by the vector notation:

$$\partial(nQ)/\partial t + \nabla(nu\bar{Q}) = \iiint n^2(Q-Q')ff'\sigma^2\Omega\cos\varphi d\varphi dc^2 \tag{2.25}$$

Where the elementary volume is $\sigma^2\Omega\cos\varphi d\varphi dt$. According to the assumption of molecular chaos, the increments $d\varphi$ and dt are infinitesimals. Thus the integrant of

Equation (2.25) includes all the number of collisions between molecules by binary collision, within the elementary volume over a unit of time. In the formulation, the line of particle impact is orientated within the elementary solid angle $d\varphi$.

Following the forgoing average procedure, the integration is able to establish the bridge from the microscopic particle dynamics to the global gas dynamic in the well-known thermodynamic variables. Under the dynamic equilibrium condition, the Euler equations can be obtained straightforwardly by introducing the Maxwell distribution function to the Boltzmann equation. The essential step is by taking the moments of the microscopic quantity with a distribution function and performing the integration over the entire velocity space. As a consequence, the process is also known as the transformation of various variables Q by the method of moments of order n; $\int Q^n f[x_i, c_i, t] d^3c$.

The integration is carried over the entire phase space with the vanished far field integral limits at infinity, such that

$$\int Q(\partial f/\partial t) d^3c = \partial\left(\int Q f d^3c\right)\Big/\partial t - \int f(\partial Q/\partial t) d^3c$$

$$= \partial(nQ)/\partial t - n(\partial Q/\partial t)$$

$$\int Qu(\partial f/\partial x_i) d^3c = \partial\left(\int Q u f d^3c\right)\Big/\partial x_i - \int fu(\partial Q/\partial x_i) d^3c$$

$$= \partial(nQu)/\partial x_i - nu(\partial Q/\partial x_i) \tag{2.26}$$

and

$$\int Q(\partial f/\partial c_i) d^3c = \iint [Qf]_{-\infty}^{\infty} dv dw - \int f(\partial Q/\partial u) d^3c$$

$$= -n \, \partial Q/\partial c_i$$

The general integral transformation of the Boltzmann equation without the force field and under the dynamic equilibrium condition can be given as;

$$\int Q\big(\partial f/\partial t + c \cdot \partial f/\partial x_i\big) d^3c = \partial(nQ)/\partial t + \Sigma \partial(nQu)/\partial x_i - n\Big[\partial Q/\partial t + \Sigma u\big(\partial Q/\partial x_i\big)\Big]$$

$$\tag{2.27}$$

The equation of change, Equation (2.27) is a generalization by Enskog expansion to the Boltzmann equation to acquire the gas dynamics formulation in macroscopic scales. The distribution function is dependent only upon the peculiar velocity of particles; thus the terms in the square bracket on the right-hand side of Equation (2.27) do not appear. The general transfer equation or the equation of change now reduces to a simpler form to appear as [5]:

$$n \partial \bar{Q}/\partial t + u \cdot \nabla \bar{Q} = -[\nabla(n\bar{u}\bar{Q}) + \Delta Q \tag{2.28}$$

The ΔQ term is contributed by the collision integral of the Boltzmann equation, under the dynamic equilibrium state, it vanishes. The dependent variables for the resultant equations are then yielded the macroscopic property for the mass, momentum, and energy of the gas. The zero-order moment of the transfer equation is simply by integrating the distribution function with a scalar variable, such as the number of particle densities n, over the entire velocity space. The conservation laws for the mass, momentum equation, and energy just by assigning $Q = mn$, $Q = mu$, and $Q = m\left(u_i^2 + u_j^2 + u_k^2\right)/2$.

The integral result by Equation (2.28) for $Q = mn$ yields the species continuity equation:

$$\partial\rho/\partial t + \nabla \cdot (\rho u) = 0 \qquad (2.29a).$$

The well-known gas density ρ is the essentially the average particle number density multiplied by the mass of the particular gas species.

Similarly, the conservation momentum equation is acquired by letting the quantity as $Q = mu$.

$$\rho\,\partial u/\partial t + \nabla \cdot (\rho uu) = -\nabla p \qquad (2.29b)$$

The flux vector $\bar{Q}$ from the integral transformation actually has nine elements constituting a tensor or the pressure tensor. The normal stress or diagonal components of the tensor are coincided with the coordinate and are perpendicular to the surface of the two other coordinates. The off-diagonal elements are the shear stresses. The static pressure at a location of the gas is defined to be the mean values of the normal pressure to the three planes of the coordinate system, $p = \left(p_{ii} + p_{jj} + p_{kk}\right)/3 = \rho C^2/3$ [5]. According to Newton's second law, the force per unit area is consistent with the theory of elastic collision for which the pressure is opposite to the gas paticle motion.

The pressure in gas is thus interpreted as the mean rate of the molecular momentum per unit area moving with the mass flow. Under the dynamic equilibrium condition, only the pressure remains as an isentropic process. The concept of pressure is derived from the bombarding of gas molecules on a medium interface. Based on this line of reasoning, the internal degrees of freedom such as the rotation, vibration, and dissociation modes, will not contribute to pressure. Only the free electrons after ionization will generate a partial pressure

The conservation of energy equation in a dynamic equilibrium state is by assigning $Q = m\left(u_i^2 + u_j^2 + u_k^2\right)/2$, and the transformed variable $\bar{Q}$ becomes $\bar{Q} = m\left(u_i^2 + u_j^2 + u_k^2 + C^2\right)/2$ two. The integrand C^2 is raised from the random molecular agitations. The definition of temperature by the kinetic theory of gas is assumed that molecules only have the translation mode. In addition, the temperature is measured on an absolute scale, which is based on the second law of thermodynamics. Therefore, the temperature of a gas mixture is independent of the properties of the gas mixture. The energy conservation law becomes;

$$\rho\,\partial e/\partial t + \nabla \cdot (\rho eu) = 0 \qquad (2.29c)$$

In the kinetic theory, the temperature is defined as translatory energy as $RT = C^2/3$. Recall the definition of temperature is given as $p = (p_{ii} + p_{jj} + p_{kk})/3 = \rho C^2/3$. It yields immediately,

$$p = \rho C^2/3 = \rho RT \tag{2.30}$$

where R is known as the universal gas constant. As a consequence of the Avogadros law, the constant R has a universal value for all gases; $R = 82.06\,\text{atm}/\text{k}° = 83.15 \times 10^6\,\text{cm dyne}/\text{k}°$ [2]. By Dalton's law, the total static pressure of a gas mixture will be the sum of the partial pressures of all species in a gas mixture $p = \Sigma p_i$. The results of number particle density and the averaged molecular oscillatory motion from the transfer equation, Equation (2.28) are directly relatable to the thermodynamic properties of gas in density, pressure, and temperature.

The Euler equation, Equations (2.29a)–(2.29c) was derived by Leonard Euler in 1757. It is a system of hyperbolic partial differential equations which governs only isentropic flows or inviscid flows in which the thermodynamic state is subjected only to isentropic changes. From the mathematical point of view, the piecewise continuous solutions are admissible for the hyperbolic partial differential equations system. This unique feature is completely compatible with physics; in supersonic/hypersonic flow regimes, any propagating perturbation in the flow field is limited by the reach of the local speed of sound. In other words, it has the zone-of-dependence characteristics to permit shock waves as discontinuities.

2.6 DYNAMIC NONEQUILIBRIUM STATE

When the colliding gas particles system departs from the dynamic equilibrium state, gradients of velocity, pressure, and temperature will unavoidably occur in a flowfield. The equation of change by integrating the Boltzmann equation over the phase space with respect to the peculiar velocity must include the ΔQ term in Equation (2.28). All the deterministic global gas-dynamic phenomena are affected by the non-vanishing collision integral of the Boltzmann equation. Under the dynamic nonequilibrium condition, the investigation of the gas dynamics is focused solely on solving the collision integral equation and denoted as

$$\Delta Q = \iiint \left[f(x_i', c_i', t) f(x_j', c_j', t) - f(x_i, c_i, t) f(x_j, c_j, t) \right] \sigma^2 \Omega \cos\varphi \, d\varphi \, dc^3 \tag{2.31}$$

The classic method for solving the Boltzmann-Maxwell equation is by a successive approximation procedure. In general, the Boltzmann equation is treated as a differential operator for the distributions $\zeta(f)$. The result of the successive approximation will be the solution to the unknown distribution functions with increasing order-of-magnitude accuracy. The distribution function is represented by the Enskog infinite series [1].

$$f = f^{(0)}(x_i, c_i, t) + f^{(1)}(x_i, c_i, t) + f^{(2)}(x_i, c_i, t) + \cdots \tag{2.32a}$$

The zero-order term of the series is simply the Maxwell distribution for the null-value collision integral in the dynamic equilibrium state; the distribution function yields an analytic solution as

$$f_i^0(x_i,c_i,t) = n_i(m_i/2\pi\kappa T)^{3/2}\, e^{-m_i(u_{ik}-u_i)^2/\kappa T}$$ (2.32b)

Apply the operator ζ to the series solutions with increased order of accuracy to give;

$$\zeta(f) = \zeta\left[f^{(0)}(x_i,c_i,t) + f^{(1)}(x_i,c_i,t) + f^{(2)}(x_i,c_i,t) + f^{(3)}(x_i,c_i,t) + \cdots\right]$$

$$= \zeta^{(0)}\left[f^{(0)}(x_i,c_i,t)\right] + \zeta^{(1)}\left[f^{(0)}(x_i,c_i,t), f^{(1)}(x_i,c_i,t)\right]$$ (2.33)

$$+ \zeta^{(2)}\left[f^{(0)}(x_i,c_i,t), f^{(1)}(x_i,c_i,t), f^{(3)}(x_i,c_i,t) + \cdots\right]$$

All individual functions of the series are subjected to the condition that each term in the series must be a solution of the Boltzmann equation, $\zeta(f) = 0$. In addition, they must also satisfy the separate and successive equations, such that

$$\zeta^{(0)}\left[f^{(0)}(x_i,c_i,t)\right] = 0$$

$$\zeta^{(1)}\left[f^{(0)}(x_i,c_i,t), f^{(1)}(x_i,c_i,t)\right] = 0$$ (2.34)

$$\zeta^{(2)}\left[f^{(0)}(x_i,c_i,t), f^{(1)}(x_i,c_i,t), f^{(2)}(x_i,c_i,t)\right] = 0$$

After the function $f^{(0)}(x_i,c_i,t)$ is determined, the function $f^{(1)}(x_i,c_i,t)$ will be found successively by the sequence solutions of Equation (2.34). By this means, the resulting distributions are the following series: $f^{(0)}(x_i,c_i,t), f^{(0)}(x_i,c_i,t) + f^{(1)}(x_i,c_i,t), f^{(0)}(x_i,c_i,t) + f^{(1)}(x_i,c_i,t), f^{(2)}(x_i,c_i,t),\ldots$ and will be the successive approximations to $f(x_i,c_i,t)$. The zeroth order approximation, $f^{(0)}(x_i,c_i,t)$ to the approximate series of the distribution, is the Maxwell equilibrium distribution given by Equation (2.18).

Substituting the Enskog series into the collision integral of the Boltzmann equation will acquire the following form;

$$f(x_i,c_i,t)f'(x_i,c_i,t) = f^{(0)}(x_i,c_i,t)f'^{(1)}(x_i,c_i,t) + \varepsilon[f^{(0)}(x_i,c_i,t)f'^{(1)}(x_i,c_i,t)$$

$$+ f^{(1)}(x_i,c_i,t)f'^{(0)}(x_i,c_i,t)] + \ldots$$ (2.35)

The non-vanished collision integrals are contributed by the approximation distribution function beyond the zero-order term, $f^{(0)}(x_i,c_i,t)$, or the Maxwell distribution. Only the distribution from the first-order term $f^{(1)}(x_i,c_i,t)$ is retained, therefore the collision integral of the general transfer equation, Equation (2.28) can be designated as $\Delta Q^{(0,1)}$. The first-order approximation is dependent on the time and location, t and x_i through the variables n, T and u_i;

$$\Delta Q \approx \Delta Q^{(0,1)}(m_i, u_i, n_i, k, T)$$ (2.36)

Beyond the dynamic equilibrium state, the next-order collision term gives [3]

$$\Delta Q^{(0,1)} = f_i^0 \big\{ \partial \log n_i / \partial t + (m_i / \kappa T) U_{ij} \, \partial u_i / \partial t$$

$$+ \big(m_i U_{ij}^2 / 2\kappa T - 3/2 \big) \partial \log T / \partial t + u_{ij} \big[\partial \log n_i / \partial x_i + m_i / \kappa T U_{ij} \big(\partial u_i / \partial x_i \big) \quad (2.37)$$

$$+ \big(m_i U_{ij}^2 / 2\kappa T - 3/2 \big) \partial \log T / \partial x_i \big] \big\}$$

where $U_{ij} = u_i - u_{ij}$; $\partial U_{ij} / \partial x_i = -\partial u_i / \partial x_i$. The partial time derivatives can now be expressed in terms of the spatial derivative through the equations of change.

The approximated collision integral by the Enskog series can even split into the elastic and inelastic components $\Delta Q^{(0,1)} = \Delta Q_{elas}^{(0,1)} + \Delta Q_{inel}^{(0,1)}$ [3], even though the number of inelastic collisions is small compared with the total number of collisions. These integrals delta $\Delta Q^{(0,1)}$ involve detailed derivations by a rather lengthy and tedious procedure, but all the necessary results pertaining to the nonequilibrium dynamics state can be summarized by the following equation [3]:

$$\Delta Q^{(0,1)} = -A \, \partial \log T / dx_i - B \, \partial u_i / \partial x_j + n\Sigma C d\alpha_i \quad (2.38)$$

In short, the perturbed collision terms by the nonequilibrium condition produce gradients of the gas temperature, rate of strain, and species number density that lead to the transport properties of gas in the nonequilibrium state [6]. In Equation (2.38), the linear proportion coefficients, A and B are vectors and the C is a second-order tensor. The approximated nonequilibrium solution will generate additional flux vectors and they are linear functions in terms of the first-order spatial derivatives. The global velocity gradients have a linear relationship between the shear stress and the rate of strain of the gas motion related to the bulk and molecular viscosity coefficients. Any fluid that possess this characteristic is known as the Newtonian fluid. The temperature gradient will lead to an energy transfer flux by the conduction heat transfer through the Fourier's law. Similarly, the gradient of species concentration will generate a very complex diffusion velocity. All these extraneous forces are dissipative in thermodynamic processes, and gas motions under the dynamic nonequilibrium condition are identified as viscous flows.

2.7 NAVIER-STOKES EQUATION

In order to be consistent with the gas kinetic theory, transport properties of the gas mixture for thermal diffusion, molecular viscosity, and thermal conductivity need to be calculated from the Boltzmann equation by the Enskog expansion. It is a landmark achievement by the gas kinetic theory of diluted gas mixtures by describing the transport property of gaseous species with the inter-molecular potential functions [1,3,6].

The contribution to the rate of change is given by Equation (2.28) from the approximated solution to the collision integral of the Boltzmann equation under the collision nonequilibrium condition. Through the dynamic nonequilibrium collisions, the gradients now exist in the velocity of gas motion, in gas temperature, together with the non-uniform species concentration of the gas mixture. These microscopic phenomena lead to shearing stress, conductive heat transfer, and ordinary diffusion in macroscopic quantities. In fact, the molecular diffusion velocity and heat transfer depend on the leading and the last perturbation terms in the rate of change of the collisional process. The shear stress tensor is explicitly related to the second term of Equation (2.38).

In fact, the temperature gradient leads to the linear relationships of the heat transfer by Fourier's law, $q = -k\nabla T$, the rate of strain of the fluid motion generates the shear stress retarding gas motion, $\tau = \lambda\nabla u + \mu\text{def}(u)$, and the gradient of species concentration induces the different diffusion mechanisms. The diffusion phenomenon is rather complex because it includes the ordinary, thermal, pressure, and force diffusions, but for aerodynamic applications, it is limited to the ordinary binary diffusion by Fick's law $\rho_i u_i = D_{ij}\nabla \alpha_i$. More detailed descriptions of the transport properties will be elaborated on in the later topics.

The Navier-Stokes equations for an incompressible medium were first derived by M. Navier in 1827 [7] and incorporated with G.G. Stokes's law for the bulk viscosity to close the relationship between the stresses and rates of strain in 1845 [8]. However, the compressible governing equations presented here are derived from the foundation of Newtonian mechanics through gas kinetic theory. It is the first-order perturbation solution to the Boltzmann equation based on the binary elastic collision model without the presence of an external force field. The resultant Navier-Stokes equations consist of three conservation laws known as the scalar continuity equation which is a fundamental statement for Newtonian mechanics in which the mass and energy are not exchangeable as for quantum mechanics. The conservation of momentum law is the only equation in vector form, which is based on Newton's second law. The conservation of energy law is just the second law of thermodynamics for gas dynamics.

The continuity equation is

$$\partial\rho/\partial t + \nabla\cdot(\rho u) = 0 \tag{2.39a}$$

The conservation of momentum equations

$$\partial\rho u/\partial t + \nabla\cdot(\rho uu - \tau) = 0 \tag{2.39b}$$

where the term uu is the dyadic tensor of second order, which is also the source of turbulence by chaotic interaction between vortices of vastly different sizes. The shear stress tensor is directly related to the rate of the strain for the Newtonian fluid, $\tau = (-p + \lambda\nabla\cdot u)\overline{\overline{I}} + \mu\text{def}(u)$.

The conservation of energy equation;

$$\partial\rho e/\partial t + \nabla\cdot(\rho eu + q + u\cdot\tau) = 0 \tag{2.39c}$$

where the conductive heat transfer is given by Fourier's law, $q = -\kappa\nabla T$. The last term in the divergence operator $u\cdot\tau$ represents the viscous dissipation of the gas motion.

The Navier-Stokes equation is classified as the nonlinear incompletely parabolic partial differential equations system [9]. The closure of the differential equations system requires a well-posed initial value and boundary conditions. A key and fundamental difference in boundary conditions between the Euler and Navier-Stokes equations is the additional non-slip velocity that must be imposed on the medium interface for the second-order partial differential equations system.

In application, the system of equations is usually adopted by the split flux vector form, and solving the inviscid terms (Euler equation) and the viscous terms separately and cyclically [9];

$$\partial U/\partial t + \partial(F_i + F_v)/\partial x + \partial(G_i + G_v)/\partial y + \partial(H_i + H_v)/\partial z = 0 \qquad (2.40)$$

where U denotes the conservative variable $U(\rho, \rho u, \rho v, \rho w, \rho e)$ in the Cartesian frame of reference. However, the system of equations is not closed, because there are only five scalar equations but ten dependent variables, u, v, w, $\rho, p, T, \mu, \lambda, k$ and D_{ij}; a group of auxiliary equations including the equation of state and the definitions of coefficients for the transport properties are needed.

The hierarchy of the governing equations from the theoretical foundation through a series of successive approximations that lead to the most widely adopted governing equations is depicted in Figure 2.3. A graphic depiction delineates a systematic derivation starting from the Liouville theorem via the binary elastic collision model to achieve the Boltzmann equations for aerodynamics on microscopic scales. The translation between the probability description of aerodynamics to the macroscopic global behavior in deterministic variables is carried out by the Maxwell transfer equation and Enskog series expansion.

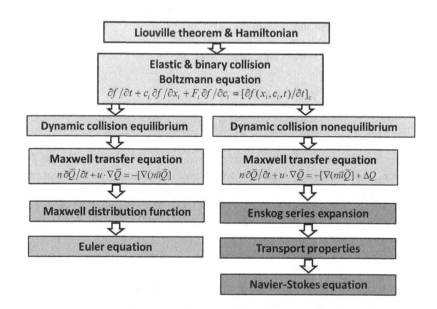

FIGURE 2.3   Hierarchy of the aerodynamic governing equations.

REFERENCE

1. Chapman, S. and Cowling, T.G., *The mathematical theory of non-uniform gases*, Cambridge University Press, New York, 1964.
2. Kennard, E.H., *Kinetic theory of gases, with an introduction to statistical mechanics*, McGraw-Hill, New York, 1938.
3. Clarke, J.F. and McChesney, M. *The dynamics of real gases*, Butterworths & Co, Washington, DC, 1964.
4. Bird, G.A., *Molecular gas dynamics*, Oxford University Press, Clarenden, United Kingdom, 1976.
5. Patterson, G.N., *Molecular flow of gases*, John Wiley & Sons, New York, 1956.
6. Hirchfeld, J.O., Curtiss, C.F., and Bird, R.B., *Molecular theory of gases and liquids*, John Wiley & Sons, New York, 1959.
7. Navier, M., Memoire sur les lois du mouvements des fluids, *Mémoires de l'Académie Royale des Sciences*, Vol. 6, 1827, pp. 389–416
8. Stokes, G.G., On the theories of the internal friction of fluids in motion, *Transactions of the Cambridge Philosophic Society*, Vol. 8, 1845, pp. 287–319.
9. Shang, J.S., An assessment of numerical solutions of the compressible Navier-Stokes equation. *Aircraft Journal*, Vol. 20, No. 5, 1985, pp. 353–370.

3 Inviscid Hypersonic Flows

3.1 SCOPE OF INVISCID FLOW DOMAIN

A hypersonic flow field in a continuum and high Reynolds numbers regime can be divided into an inviscid or shock-dominated external flow region, and a viscous shear layer region near the body surface. All classical theories for inviscid hypersonic flows are accumulating from the crystallized analytic form with exceptional physical insights, ingenuity, and thorough and rigorous analyses. These theories contribute to the basic understanding of hypersonic flow phenomena at the outer limits of extremely high Reynolds numbers and have withheld the test of time. Theoretical developments are derived in the continuum regime, with a focus on the systematic approximations that have been developed from the unique characteristics of hypersonic flow. These theoretical developments describe some of the most important hypersonic flow observations and focus exclusively on the gas dynamics aspect of the phenomenon, in which the effects of internal degrees of freedom of molecules are not considered. Similarly, the detailed treatments of the free molecule flow, rarefied flow, as well as gas in high-enthalpy states are excluded.

Although hypersonic flow phenomena are extremely complex, significant understanding may be instilled from analyses that are done by meticulously chosen simplifications, incisive insights, and limiting processes. Classical knowledge of fundamental inviscid hypersonic phenomena in the continuum domain is derived from the kinetic theory of gas and through the Maxwell distribution function [1,2]. At this level, the effect of rarefaction and the internal structure of molecules are not taken into consideration. Therefore, nonequilibrium thermodynamics, chemical reaction, and radiation are treated in simplified parametric form or as approximations. Despite these simplifications, the fundamental theories are still applicable to a vastly large number of hypersonic flows with relatively little modification.

3.2 HYPERSONIC SIMILITUDES

At a hypersonic speed, the key aerodynamic parameter is the maximum deflection of individual fluid particle paths near the body. The disturbed flow by any finite inclination angles α lead to a shock wave, and the trailing shock wave becomes weak far downstream behind the body and approaches the free stream Mach wave angle $\sin^{-1}(1/M_\infty)$. Since strong shocks and rapid expansion waves are the dominant features of hypersonic flow, it is logical categorizing the distinctive flowfield topology either as a slender or blunt body. For the flow around a slender body, the shock often attaches to the leading edge of the body and the shock wave forms an oblique

DOI: 10.1201/9781003212362-4

envelope with an angle closely related to the body slope and the entire flowfield remains supersonic except a thin boundary over the body. In contrast to that, the shock over a blunt body is usually denoted as a detached bow shock with a continuously varying radius of curvature along the wave. The shock structure around a blunt body always results in large transverse gradients of the flow properties. Immediately downstream to the bow shock and in the stagnation region of the shock envelope, the highly compressed flow field becomes subsonic. The flow past the stagnation point expands and accelerates from this region as it flows downstream. The flow velocity at some distance downstream will exceed the local speed of sound at the sonic line and subsequently becomes supersonic. The rate of intrinsic change from kinetic to thermal energy in the flow field is so dominant over all other processes; as a consequence, the flow medium may be considered inviscid. In other words, the viscous shear stress and heat transfer by conduction are assumed negligible in comparison with the inertia of the gas motion.

The hypersonic flow past slender bodies may be described by small disturbance theory. The magnitude of the disturbance is small only in comparison with the free stream condition, but the theory is essentially nonlinear; unlike at supersonic speeds, linearization approximation is not appropriate. Hypersonic similitude, first described by Tsien [3], refers to the equivalence between inviscid steady hypersonic flows over a slender body and unsteady flow in one less spatial dimension [4]. Hypersonic similitude is inherently connected to small disturbance theory; similitude can be viewed as a direct result of small disturbance theory and conversely the theory is a consequence of applying the similitude concept. A critical point is that hypersonic similitude does not provide solutions to hypersonic flow, but once a solution is found it can be applied to a large group of similar configurations. The benefits are akin to the Prandtl-Glauert similitude for supersonic flow, but the equation of motion in the supersonic regime may be linearized and allows a functional specification independent from the Mach number.

In the classic hypersonic literature, the local flow inclination α is identified by the body thickness ratio τ. In a way, it always defines the maximum inclination of individual particle paths near the body and has the same order of magnitude as the tangent of the induced shock angle; $O(\tau) \sim O(\tan \beta)$. The slender body assumption is that the shock angle $\beta \ll 1$, which underlies the small disturbance theory by requiring the value of τ is small. The parameter is served as a measure of the maximum inclination of the Mach wave in the flow field. From the slender body definition and numerous experimental observations, it concludes the lateral and axial velocity components downstream to the shock are of the order of $u_\infty \tau$ and $u_\infty \tau^2$ respectively. In other word, the perturbations to the crossflow velocity components is one-order-of-magnitude lower than that of the streamwise components. Naturally, it suggests a transformation; $u_x = \tau u_x'$ and $x = \tau^{-1} x'$ to put all terms in governing equation on the same magnitude. The governing equations for the small disturbance equation and the associated boundary condition become

$$\partial \rho / \partial t + \nabla \cdot (\rho u) = -\tau^2 \left(\partial \rho u_x' / \partial x' \right)$$

$$\partial \rho / \partial t + \nabla \cdot (\rho u) + \nabla p / \rho = -\tau^2 \left(\partial \rho u_x' / \partial x' \right) \qquad (3.1a)$$

$$\partial \rho e / \partial t + \nabla \cdot (\rho e u) = -\tau^2 \left(\partial \rho e u_x' / \partial x' \right)$$

And the boundary condition at the free field is unaltered, and on the medium interface, the condition is prescribed as

$$\partial f/\partial t + u \cdot \nabla f = -\tau^2 u_x' \, \partial f/\partial x'; \quad f(x',y,z,t) = 0 \tag{3.1b}$$

It is clearly demonstrated that the error of small disturbance theory in hypersonic flow has an order of magnitude of $O(\tau^2)$, which is more accurate than that of the transonic and supersonic flows. The physical concept of hypersonic similitude is also leading to the equilibrium principle later. In essence, the flow field viewed in the transverse flow plane is independent of each other.

In terms of the quantity of similarity, an independent parameter for similitude is $\sqrt{M_\infty^2 - 1}\tau$ which becomes $\kappa = M_\infty \tau$ at hypersonic Mach number. For slender bodies, the hypersonic parameter is limited to a value around unity, $O(M_\infty \tau) \approx O(1)$. But the parameter $M_\infty \tau$ may have an exceptionally large value at the extremely high Mach number condition or for blunt body flow.

The hypersonic similitude is a similarity rule that expresses equivalence between aerodynamic problems with some intrinsic dissimilarity in other essential respects. Hayes and Probstein point out that the similarity can only accomplish a reduction in the number of independent parameters for the dependent variables [4]. The similitude of small disturbance theory is valid only when the hypersonic similarity parameter $M_\infty \tau$ is around unity and the group of bodies also represents a family of shapes related by an affine transformation that can be described by body-oriented coordinates:

$$f(\tau x, y, z) = 0. \tag{3.2}$$

In the above equation, the body thickness is incorporated into the streamwise coordinate through τ to reduce the number of parameters in the small disturbance formulation. In a uniform steady hypersonic flow with the freestream properties of pressure p_∞, density ρ_∞, and velocity U_∞, the general solution to the inviscid aerodynamic equations satisfying the associated boundary conditions acquires the functional form.

$$p = p(x, y, z; \tau, u_\infty, p_\infty, \rho_\infty) \tag{3.3}$$

When the small-disturbance theory is applicable, the static pressure solution may be written as

$$p = p(\tau x, y, z; u_\infty \tau, p_\infty, \rho_\infty) \tag{3.4}$$

The hypersonic similitude is seen to reduce the independent parameters by one [4]. This result also is applicable to solutions of other dependent variables and gives similar expressions that are derived from the transverse velocity and the axial perturbation velocity components. The pressure coefficient, which plays a crucial role in aerodynamic applications, can be expressed as

$$C_p = \frac{2(p - p_\infty)\tau^2}{\gamma p_\infty (M_\infty \tau)^2} = 2\tau^2 \left[\frac{(p/p_\infty) - 1}{\gamma (M_\infty \tau)^2} \right] \tag{3.5}$$

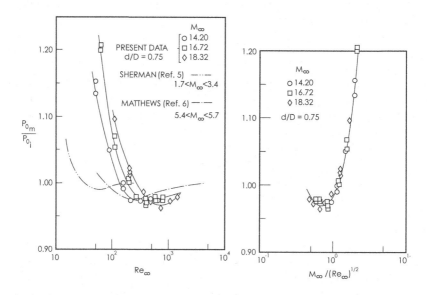

FIGURE 3.1   Similitude parameter for impact pressure probe data.

The thermodynamic property of the gas medium is introduced by the ratio of specific heat $\gamma = c_p / c_v$. This constant single parameter characterizes the complete thermodynamic state of a perfect gas. Additional aerodynamic parameters such as the angle of attack can also be included in the hypersonic small disturbance formulation.

The idea of similitude has been widely adopted in aerodynamics research for flow phenomena or parameters involving multiple dependent variables. By appropriately grouping a large group of dependent variables the specific phenomenon can be represented by a reduced number of variables. A simple example is depicted in Figure 3.1 for collecting the Pitot or impact pressure probe data that depend upon the freestream Mach and Reynolds number in the rarefied gas region. By a suitable grouping of these two independent variables; the data can be evaluated on a single similitude parameter.

In summary, hypersonic similitude considers two flows over affine bodies as equivalence if they have the same solutions from the small-disturbance equations. Succinctly, the similitude reduces the parameters in describing a hypersonic flow, at least by one, when the resultant flow is generated by a small disturbance. The equivalence also applies between unsteady and steady hypersonic slender flow formulation.

3.3 NEWTONIAN FLOW THEORY

Newton's basic approach to high-speed flow is starting from a set of basic laws on a hypothetical physics model and follows by rational deduction to reach the solution according to the adopted physical model. The Newtonian flow theory is the consequence of Newton's impact law which states that when a single particle interacts with a surface, its normal component is completely transferred while the tangential component is unchanged [5]. Therefore, the surface pressure depends only on the orientation of the incident gas particle. Newton deduced the force of impact by a particle on the body as being proportional to the sine squared of the angle of incidence and directed normally to the body. The situation corresponds to the limiting condition at the infinite Mach number and the gas medium is highly rarefied, and only becomes a continuum within the shock layer. This law also introduces the notion of aerodynamic shadow, the region where the direct particle impact is absent, in which the local pressure retains the unperturbed free stream value.

In the context of modern aerodynamic theory, the Sine square law is valid at extremely high Mach numbers and for a very thin shock layer where the resultant shock angle closely approximates the angle of incidence to the body. This asymptotic condition can be met when the specific heat ratio of the gas, γ has a value of unity or the shock layer is infinitesimally thin. The surface pressure corresponding to Newton's theory is

$$p_s - p_\infty = \rho_\infty u_\infty^2 \sin^2 \alpha \tag{3.6}$$

There are many modifications to the Newton's impact law for improving accuracy in practical applications, which gives only the pressure immediately downstream to the shock. Some modifications have been included in the pressure difference across the shock layer due to the flow momentum changes by the curvature. Busemann first includes a centrifugal correction as a rational correction at the limiting condition [4]. In order to take into account gases with different properties, Lees [6] proposed an empirical modification to Newtonian law expanding the applicability of Newton's theory to gases of different specific heat ratios, γ. The modified pressure coefficient is also evaluated at the stagnation point within the shock layer by the Rayleigh Pitot pressure formula, which is on an empirical basis but with incisive insight. This modification becomes unbelievably valuable giving an impressive improvement for practical applications.

$$C_p = C_{p,s} \sin^2 \alpha, \tag{3.7}$$

where, $C_{p,s} = \dfrac{2}{\gamma M_\infty^2} \left\{ \left[\dfrac{(\gamma+1)^2 M_\infty^2}{4\gamma M_\infty^2 - 2(\gamma-1)} \right]^{\gamma/(\gamma-1)} \left[\dfrac{1-\gamma+2\gamma M_\infty^2}{\gamma+1} \right] - 1 \right\}.$

In the limit, as the free stream Mach number approaches infinity, $C_{p,s}$ takes values of 2.0 and 1.839 as γ changes from the value from 1.0 to 1.4. Although this modified formula loses the benefit of independence from the freestream Mach number, it is extremely useful because of an increasing applicable range for gases under different thermodynamic conditions, if they can be characterized by a single parameter γ.

The modified Newtonian law is accurate for hypersonic blunt body pressure coefficient prediction and has been widely used for validating numerical simulations. As an illustration, Figure 3.2 displays the results of the modified Newtonian theory comparing with two numerical solutions of compressible Navier-Stokes equations

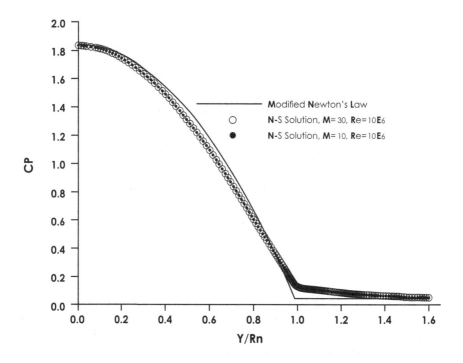

FIGURE 3.2 Comparing Modified Newton Law with Navier-Stokes solutions over the spherical cone.

obtained at Mach numbers of 10 and 30 over a spherical cone at a Reynolds number of one million based on the body nose radius [7]. The numerical simulation imposed the perfect gas law with a specific heat ratio of 1.4. The reasonable agreements with numerical simulations are easily observable and the discrepancy occurs only at the juncture of the hemispherical nose and the conical afterbody where a rapid flow expansion is taken place. In addition, the two high Mach number numerical results are seen to collapse into one over the entire body length to reveal the Mach number independence principle which will be deducted from the limiting hypersonic shock relationships in the following discussion.

3.4 RANKINE-HUGONOIT RELATION

The relationship between states across a shock wave is described by the Rankine-Hugonoit equation in one dimension in 1870 [8]. The equation connects the thermodynamic properties of gas medium with the dynamics of gas motion. The discontinuity in flow properties is an idealization of the extremely high gradient existing in a very thin shock wave within a few mean free paths. The jump condition can be given simply as

$$e_2 - e_1 + (p_2 + p_1)\big[(1/\rho_2) - (1/\rho_1)\big]/2 = 0 \qquad (3.8a)$$

After some rearrangements, The Rankine-Hugonoit relation appears in pressure and density jump across a one-dimensional shock along with a constant specific gas constant $\gamma = c_p/c_v$,

$$\frac{\rho_2}{\rho_1} = \frac{(r+1)/(\gamma-1)+(p_1/p_2)}{1+[(\gamma+1)/(\gamma-1)](p_1/p_2)} \tag{3.8b}$$

For application to the practical problem, it is more convenient to use the pressure ratio across the shock as the basic independent parameter. Then all other thermodynamic variables can be derived from the Rankine-Hugonoit relations. For the widely encountered oblique shock; the tangential component of velocity with respect to the shock front is unaltered, whereas the normal component satisfies the relationship for a normal shock jump. Therefore, the discontinuous condition across the shock can be easily obtained by replacing M_1 by $M_1 \sin \beta$, where β is the shock angle. The basic shock jump conditions are equally valid for supersonic and hypersonic flows that are based on the perfect gas model with a constant value of the specific heat ratio γ;

$$\frac{p_2}{p_1} = 1 + \frac{2}{\gamma+1}\left(M_1^2 \sin^2 \beta - 1\right) \tag{3.9a}$$

$$\frac{\rho_2}{\rho_1} = \frac{(\gamma+1)M_1^2 \sin^2 \beta}{(\gamma-1)M_1^2 \sin^2 \beta + 2} \tag{3.9b}$$

$$\frac{T_2}{T_1} = 1 + \frac{2(\gamma-1)\left(M_1^2 \sin^2 \beta - 1\right)}{(\gamma+1)^2 M_1^2 \sin^2 \beta}\left(\gamma M_1^2 \sin^2 \beta + 1\right) \tag{3.9c}$$

$$\frac{u_2}{u_1} = 1 - \frac{2\left(M_1^2 \sin^2 \beta - 1\right)}{(\gamma+1)M_1^2} \tag{3.9d}$$

$$\frac{v_2}{u_1} = 1 - \frac{2(M_1^2 \sin^2 \beta - 1)\cot \beta}{(\gamma+1)M_1^2} \tag{3.9e}$$

The relationship between the deflection angle α and the shock angle β can be given as

$$\tan \alpha = 2 \cot \beta \frac{M_1^2 \sin^2 \beta - 1}{M_1^2(\gamma + \cos 2\beta) + 2} \tag{3.10}$$

From the graphic presentation between the deflection angle and the oblique shock angle, there are two possible solutions for each deflection angle Figure 3.3. The large value of α yields the so-called stronger shock for which the flow downstream of the shock becomes subsonic. Whereas, the solution of weak shock remains supersonic, except for a small range of the deflection angle α having a value smaller than the maximum value of the α_{max}. In essence, when the deflection α exceeds the α_{max}, a detached curved shock will move upstream from the sharp leading edge. At the center of the detached shock, the shock angle is $\beta = \pi/2$.

In direct contrast to compression waves, the expansion encounters in the hypersonic condition are treated as a simplified Prandtl-Meyer flow. Fortunately, the

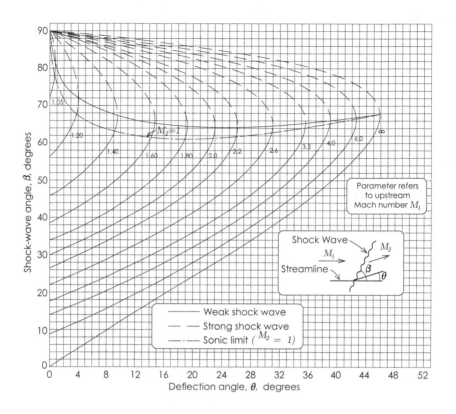

FIGURE 3.3   Relationship between deflection and shave-wave angle, NACA TR 1135.

expansion flow region in hypersonic flow frequently can be approximated as a two-dimensional problem. An explicit integral is available from the Bernoulli equation: $u du + dp/\rho = 0$. In addition, the governing equation for inviscid flow is a partial parabolic differential system and an exact solution can be acquired by the method of characteristics in two-dimensional and axisymmetric flow [9]. The pressure and the angle between the streamlines and the freestream are the basic variables for the method of characteristic. If an explicit integral is possible, then the two-dimensional isentropic flow is equivalent to the Prandtl-Meyer flow. In lacking a simple connection between the speed of sound as a function of the velocity, an implicit connection exists through the equation of state under isoenergetic conditions for isentropic flows; the Prandtl-Meyer angle V, now can be given as [4]:

$$v = \int \left[2(H-h)/c^2 - 1 \right]^{1/2} d(H-h) \Big/ 2(H-h) \tag{3.11a}$$

where the symbols H and h are designated as the stagnation enthalpy and static enthalpy of the gas, and c denotes the local speed of sound. The difference between the stagnation and static enthalpy is the total kinetic energy of the gas motion $\Sigma u^2/2$.

The simplified Prandtl-Meyer angle v acquires the following form and has been a very useful treatment for the expansion region in hypersonic flow.

$$v = -\left(1/\sqrt{\Sigma u^2/2}\right)\int (dp/\rho c) \qquad (3.11b)$$

Equation (3.11b) is integrating along an isentropic path and the Prandtl-Meyer turning angle is defined to be zero in the freestream. The pressure will be given as a function of the freestream condition and the turning angle; $p(vu_\infty, \rho_\infty, p_\infty)$.

3.5 STAND-OFF DISTANCE

A detach shock wave always associated with a blunt body, and a unique phenomenon emerges to know as the bow shock wave stand-off distance from the stagnation point. In the stagnation and adjacent region, the shock wave angle maintains a value near to $\pi/2$ or a normal shock. The density within the shock layer can be considered as a constant if the shock wave with a radius of R_s is similar to the circular body of radius R_b. A constant density formulation for a circular cylinder is derived by the similar approach by Lighthill for a sphere [10]. An analytic solution is obtainable for the constant-density flow in a stream function formulation in the shock layer bounded by the shock and body radii, R_s, R_b [4]. For supersonic flow over a sphere, an approximate solution to the stream function with a small density ratio across bow shock $\varepsilon = \rho_2/\rho_1$ has been found for the stand-off distance. The stream function in spherical coordinates is a quantic equation of R/R_s, and the general solution has not been obtained in analytic form. Only a numerical solution was obtained by Van Dyke using an inverse space marching algorithm [11].

$$\psi = \left(puR_s^2 \sin^2\theta/30\varepsilon^2\right)\left[3(1-\varepsilon^2)(R/R_s)^4 + A(R/R_s)^2 + B(R/R_s)\right] \quad (3.12a)$$

In the solution for the stream function, Equation (3.12a), the Coefficients A and B have been found by invoking the condition that the tangential velocity component must continue across the shock wave $A = -5(1-4\varepsilon)$ and $B = (1-\varepsilon)(1-6\varepsilon)$. The boundary condition of the stream function on the sphere surface is, $\psi = 0$, and leads to a quantic equation of body radius to shock radius R_b/R_s ratio.

$$3(1-\varepsilon)^2(R_b/R_s)^4 - 5(1-4\varepsilon)(R_b/R_s)^2 + 2(1-\varepsilon)(1-6\varepsilon)(R_b/R_s) = 0 \quad (3.12b)$$

A numerical solution is recovered by an inverse method in which a computational marching process from the sphere surface toward the shock front [11]. An approximated solution by solving the boundary condition of a quadratic equation in the small density ratio ε yields [4].

$$\Delta = R_s - R_b = \varepsilon R_b\left[1 - \sqrt{8\varepsilon/3} + 3\varepsilon + O(\varepsilon^{3/2})\right] \qquad (3.12c)$$

The approximate stagnation point detachment distance agrees well with experimental data. In general, the result is equally valid for circular cylinders and spheres.

However, there is an essential difference between the two-dimensional and the axi-symmetric blunt body. For the two-dimensional configuration, the Newtonian solution was divergent, and a reasonable solution for shock wave layer structure and a stand-off distance was only based on the constant density solution [4]. For the axi-symmetric configuration, the Newtonian solution gives a first approximation to the shock layer structure. The leading term of Equation (3.12c) alone reveals that the stand-off distance over a sphere of hypersonic flows is nearly one-sixth of the radius of the sphere in a perfect gas with the specific heat constant of $\gamma = 1.4$.

The verification of numerical results with experimental measurements on the bow shock wave stand-off distance is depicted in Figure 3.4. The comparing freestream Mach numbers span a range from 1.4 to 7.0 and consistently display an incredibly good agreement. Similar results have also been obtained for a hemisphere at the freestream Mach number at 6.8. The calculated and experimental data reveal the stand-off distances of $\Delta/R_b = 0.144$ and $\Delta/R_b = 0.148$ respectively. In particular, the leading term in Equation (3.12c) $\Delta/R_b = \varepsilon$ and an empirical equation $\Delta/R_b = \varepsilon/(1 + \sqrt{2\varepsilon})$ have been widely and frequently used to make a preliminary estimate for practical applications.

The stand-off distance of a detached shock wave over the blunt body is governed by the gas species continuity equation. The total amount of mass flow ingested by the

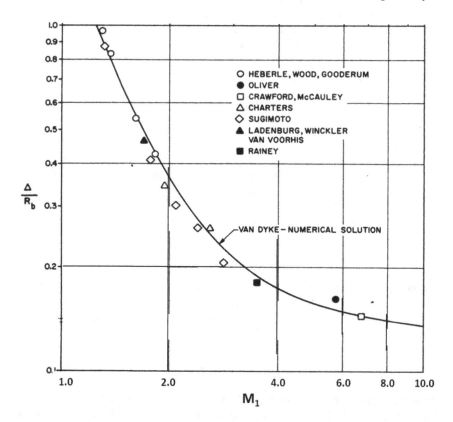

FIGURE 3.4 Comparison of stand-off distance with experimental measurements.

projected frontal area of the bow shock wave must be carried completely within the shock layer. The classic theory accurately reflects a strong dependence of the stand-off distance on the density ratio across the bow shock, in turn, the density ratio across the bow shock wave is actually determining the standoff distance. In a high-enthalpy thermodynamic state, the perfect gas assumption is no longer held for hypersonic flow, and the specific heat ratio will be different from a constant value of 1.4 from the excited internal degree of freedom. The significant change of the stand-off distance in the high-enthalpy thermodynamics state is purely resulting from both the excited internal modes and the changing chemical composition of gas in the high-enthalpy hypersonic regime.

3.6 MACH NUMBER INDEPENDENT PRINCIPLE

As the Mach number of a hypersonic flow is increasing, the flow fields both for the slender bodies with an attach shock and blunt bodies with detach bow shock wave will approach an invariant structure. The unchanging features also include critical aerodynamic parameters for determining the performance of a configuration, such as the pressure coefficient C_p, the lift coefficient C_l, and the wave drag coefficient C_d. The concept of a limiting hypersonic flow is developed by Oswatitisch and Cheryni [12,18]. In order to understand this feature of hypersonic flow, the shock wave jump conditions become the first step to be examined at the limiting hypersonic condition. For a blunt body with a detached shock wave, the shock layer near the stagnation point is bounded mostly by a normal shock. At the infinite Mach number limit, the normal shock relationships in hypersonic flow are easily derived by cognizing that the freestream Mach number is much greater than unity, $M_1 \gg 1.0$ to reach these following asymptotes.

$$\frac{p_2}{p_1} = 1 + \frac{2\gamma}{\gamma+1}\left(M_1^2 - 1\right) \rightarrow \frac{2\gamma}{\gamma+1}M_1^2$$

$$\frac{\rho_2}{\rho_1} = \frac{u_1}{u_2} = \frac{\gamma-1}{\gamma+1+2/M_1^2} \rightarrow \frac{\gamma+1}{\gamma-1}$$

$$\frac{T_2}{T_1} = \frac{\left[2\gamma M_1^2 - (\gamma-1)\right]\left[(\gamma-1)M_1^2 + 2\right]}{(\gamma+1)^2 M_1^2} \rightarrow \frac{2\gamma(\gamma-1)M_1^2}{(\gamma+1)^2} \qquad (3.13a)$$

$$M_2 = \frac{u_2}{a_2} = \left[\frac{(\gamma-1)M_1^2 + 2}{2\gamma M_1^2 - (\gamma-1)}\right]^{\frac{1}{2}} \rightarrow \left[\frac{\gamma-1}{2\gamma}\right]^{\frac{1}{2}}$$

It is observed that the density jump attains a finite limit value which is given by a fixed ratio of the specific heat ratios. The Mach number downstream to the shock wave drops to subsonic in the shock layer and is dependent only on the specific heat ratio of a perfect gas at the extremely high Mach number limit. The pressure and temperature jump across the shock wave however is unbounded to be proportional to the square value of the freestream Mach number.

For the detach shock away from the stagnation region of a blunt body and the shock attach to a slender body with sharp leading edge become an oblique wave. Across the oblique shock, the tangential component of velocity with respect to the shock front is unaltered, whereas the normal component perpendicular to the wave satisfies the relationship for a normal shock jump. Along the wave where the value of shock angle is less than $\pi/2$, $\beta < \pi/2$, the product terms of $M_1^2 \sin^2 \beta$ is still dominant. The limiting discontinuous condition across the shock can be easily obtained From Equations (3.9a)–(3.9c). Similarly, the asymptotic oblique shock relationships in the extremely high Mach number hypersonic stream can be summarized as

$$\frac{p_2}{p_1} \rightarrow \frac{2\gamma}{\gamma+1} M_1^2 \sin^2 \beta$$

$$\frac{p_2}{\rho_1} \rightarrow \frac{\gamma+1}{\gamma-1} \tag{3.13b}$$

$$\frac{T_2}{T_1} \rightarrow \frac{2\gamma(\gamma-1)M_1^2 \sin^2 \beta}{(\gamma+1)^2}$$

The limiting conditions of the shock layer of extremely high Mach number are like that of the normal shock. In order to develop the analysis further, it is convenient to employ the flow deflection angle α when discussing the oblique shock condition. The general relationship between the shock angle β and α, Equation (3.10), is rearranged to give as

$$M_1^2 \sin^2 \beta - 1 = \frac{\gamma+1}{2} \frac{\sin \beta \sin \alpha}{\cos(\beta-\alpha)} M_1^2. \tag{3.14}$$

At an extremely high Mach number and small flow deflection angle, the shock angle is also small, the following approximations are valid: $\sin \alpha \sim \alpha, \sin \beta \sim \beta, \cos(\beta-\alpha) \sim 1$. Utilizing these approximations the relationship between flow deflection and shock angle, Equation (3.14) becomes a quadratic equation in β.

$$\beta^2 - (\gamma+1)\alpha\beta/2 - 1/M_1^2 = 0 \tag{3.15a}$$

The real root of this equation yields

$$\frac{\beta}{\alpha} = \frac{\gamma+1}{4} + \sqrt{\left(\frac{\gamma+1}{4}\right)^2 + \frac{1}{M_1^2\alpha^2}}. \tag{3.15b}$$

From this result, an important asymptotic form of the pressure coefficient is obtained for hypersonic flow over a slender body namely:

$$C_p = \frac{2(p_2-p_1)}{\gamma M_1^2 p_1} \approx 2\alpha\beta \approx 2\alpha^2 \left\{ \frac{\gamma+1}{4} + \sqrt{\left(\frac{\gamma+1}{4}\right)^2 + \frac{1}{M_1^2\alpha^2}} \right\} \tag{3.16a}$$

Again. at the limiting Mach number condition, the pressure coefficient approaches

$$C_p \to (\gamma+1)\alpha^2 \qquad (3.16b)$$

In addition, the important boundary conditions of velocity components u_2 and v_2 immediately downstream of the shock acquire the following form:

$$\frac{u_2}{u_\infty} = 1 - \frac{2\left(M_1^2 \sin^2\beta - 1\right)}{(\gamma+1)M_1^2} \to 1 - \frac{2\sin^2\beta}{\gamma+1}$$

$$\frac{v_2}{u_\infty} = 1 - \frac{2\left(M_1^2 \sin^2\beta - 1\right)\cot\beta}{(\gamma+1)M_1^2} \to \frac{\sin 2\beta}{\gamma+1} \qquad (3.17)$$

Recall $M_1\alpha$ or $M_1\tau$ is the hypersonic similarity parameter κ and is the governing parameter for describing the flow past a slender body whose surface has a small inclination to the flow. From the above formulation, the dependence on the Mach number is seen to diminish as the Mach number increases. At this stage of development, the numerical simulations can be carried out using the asymptotes of the Ranking-Hugonoit relationship by solving the Euler equations to demonstrate the Mach number independent principle.

Equations (3.14) reveal a critical feature of the hypersonic flow structure over a slender body; at the extremely high Mach number condition, the flow deflection angel angle is linearly proportional to the shock wave angle; $\alpha \sim \beta$. From Equation (3.17), the perturbation to the streamwise velocity component is of the order of α^2 and to the transverse component has the order of magnitude of α. Thus, an inclined slender body in a hypersonic stream produces a disturbance that is much greater in a direction normal to the flow than parallel to the flow. This result leads directly to the concept of hypersonic similitude and the equivalence principle in the following discussion.

From the limiting oblique shock relationships for the density ratio, pressure coefficient, Equation (3.13b), and the normalized velocity components, Equation (3.17); all do not depend on the freestream Mach number at the limiting hypersonic flow Mach number. The results suggest a similarity law for the limiting hypersonic flows over geometrically similar shapes with the identical specific heat of a perfect gas model, but different values of freestream pressure, density, and velocity are similar to each other [12].

For high Mach number hypersonic flows, the wave drag dominates over the skin friction and wake drag of the body. Hence the jump condition at the bow shock essentially determines the total aerodynamic force exerted on the body. In Figure 3.5 the data include three groups of drag measurements; two sets of data are for spheres and one set of drag coefficients is for a sharp leading-edge cone-cylinder configuration with a 30° half-cone angle. The Mach number independence principle is confirmed by experimental measurements of the important aerodynamic parameter for spheres and blunt cones. All the experimental measured values are nearly constant when the freestream Mach number approaches the value of seven, or the entire data collection range [12].

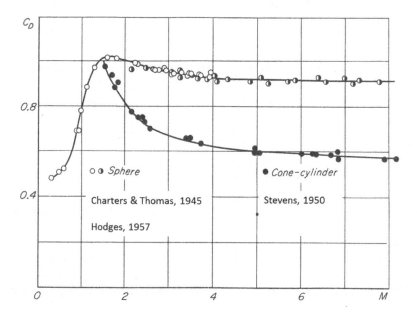

FIGURE 3.5 Experimental data of sphere and cone-cylinder at hypersonic flows.

In essence, the Mach number independent principle provides a clear definition for hypersonic flow in a perfect gas by experimental evidences. As the Mach number increases, the normal and oblique shock jump conditions become only a function of the flow deflection angle and the specific heat ratio. As a consequence, the drag coefficients of a blunt body and a slender body approach a constant asymptote also [12]. By choosing a quantitative measurement for the deviation of the drag coefficient from their respective asymptotes by 1%; the blunt body flow reaches the independent Mach number occurs at the Mach number of 3.4. For the cone-cylinder configuration, the criterion does not reach beyond the measured range of Mach number of seven. The experimental evidence indicates that the hypersonic flow field of a blunt body reaches an invariant structure at a lower Mach number than that of a slender body. The unique aerodynamic behavior is displayed by the graphic presentation, through which the distinctive features of hypersonic flow over the most common shape are quantitatively established.

3.7 TANGENT-WEDGE AND TANGENT-CONE APPROXIMATIONS

The surface pressure can be approximated by using only the local surface inclination angle from Newtonian theory by following the idea that only the momentum normal to the medium interface will be transferred. This practice is the empirical basis for the tangent wedge and tangent cone approximations. In this approach, it is assumed that the local pressure is identical to the pressure on a wedge or cone having the same inclination to the oncoming flow at the same freestream Mach number. However, there do not provide explicit expressions for the pressure on the wedge or cone in terms of the surface inclination. The approximation is justified by the physics that at hypersonic velocities the shock layer is sufficiently thin the normal pressure gradient is negligible. Hence, the surface pressure is identical to that of pressure downstream of the shock.

For a slender wedge, Equations (3.15b) and (3.16a) may be used for the surface pressure to obtain the pressure on the wedge as

$$\frac{p_w}{P_\infty} = 1 + \gamma (M_1\alpha)^2 \left[\sqrt{(\frac{\gamma+1}{4})^2 + \frac{1}{(M_1\alpha)^2}} + \frac{\gamma+1}{4} \right] \qquad (3.18)$$

For a small value of the hypersonic similarity parameter $\kappa = M_\infty \alpha = M_\infty \tau$, the Equation (3.18) can be expanded in the power series to get

$$\frac{p_w}{p_\infty} = 1 + \gamma\kappa + \frac{\gamma(\gamma+1)}{4}\kappa^2 + \frac{\gamma(\gamma+1)^2}{32}\kappa^3 + O(\kappa^5) \qquad (3.19)$$

The first three terms of the above power series are identical to the Busemann expansion series for the pressure downstream of an oblique shock [4]. There is also a special case of the tangent wedge solution for the case that the hypersonic parameter $k = M_\infty \alpha$ is large, but the approximation is rarely used in practical applications because additional information is needed on the specific heat ratio in the shock layer. The Tangent-wedge can lead to an error when the centrifugal force induces by the longitudinal curvature of the wedge and by the pressure gradient generates by the convergence and divergence of the streamline across the shock layer.

By the hypersonic axially symmetric slender body theory, the pressure downstream to the shock with a centrifugal force correction in the shock layer is required. The correction term of pressure difference across the shock layer has been approximated as

$$p_s - p_b = \gamma p_\infty M_\infty^2 (r_b/R_b) \qquad (3.20)$$

The resulting approximation for the body of revolution is the tangent cone method and the pressure on a slender body of revolution is given as [4]

$$\frac{p_c}{p_\infty} = 1 + \frac{\gamma\kappa^2}{1 - \varepsilon/4} - \frac{\gamma M_\infty^2}{2} \frac{r_b}{R_s}, \qquad (3.21)$$

where ε is the density ratio across the shock, r_b and R_s are the corresponding radii of curvature of the body and the conic shock. According to the approximation of the tangent cone method, the pressure at any surface point of a slender body is the same as Taylor-Maccoll's result at the identical Mach number for a circular cone of half–angle equal to the deflection angle [13]. The tangent wedge and tangent cone approximations predict the surface pressure well but provide no detail of the flow field. Both the tangent wedge and tangent cone approximations can also be applied to obtain the pressure distribution for any inclined surfaces. It shall be remarked that in the limits of $\gamma \rightarrow 1$ and $M_\infty \rightarrow \infty$, these approximations degenerate to the Newtonian flow theory as the root of the original concept.

3.8 EQUIVALENCE PRINCIPLE OR LAW OF PLANE CROSS SECTION

The equivalent principle is rest on the observations that any hypersonic slender body moves through the gas; the normal gradients of the gas quantities are much smaller than that parallel to the unperturbed free stream. As the consequence, the resulting gas motion is mostly confined in the crossflow planes. The hypersonic equivalence principle [3,4] or the law of plane cross section of Il'yushin [14] is essentially the statement that the steady hypersonic flow over a slender body is equivalent to an unsteady flow in one less spatial dimension. The physics is that the main effect of a slender body moving in a hypersonic stream is pushing the disturbed air away from the body. As a consequence, the gradients of flow variables are much greater in the normal direction than parallel to the gas flow. For which a steady gas flow closely relates to the unsteady flow in a plane normal to the direction of gas motion. In this sense, the flow field structures in the cross-flow plane are essential independent from each other but to appear as temporal sequence. A pictorial depiction is given in Figure 3.6.

The equivalence is valid for the entire flow field of a hypersonic slender body to the second order accuracy of arbitrary angle of attack $\angle \alpha$, which is small in comparison with the body thickness $\angle \alpha < \tau$. A rigorous and detailed proof was provided in 1960 by Sychev [15]. The derivation is built on two premises; first, the maximum transverse dimension of the body or its span width dimension is much smaller than the body length, $\tau = b / l \ll 1.0$. Second, the hypersonic parameter shall be greater than unity $M_\infty \tau \geq 1.0$, which is the upper limit of the hypersonic small perturbation.

The Euler equations for a hypersonic slender body at a small angle of attack are expressed on the three-dimensional cylindrical polar coordinates. The governing equations consist of continuity, momentum, and energy equation in terms of entropy or (p/ρ^γ) and include the boundary conditions on the body and shock surfaces. The following derivation demonstrates that the governing equation in three-dimension

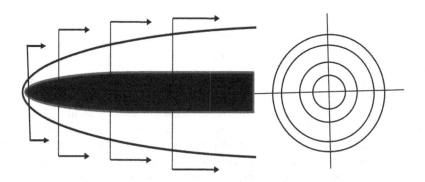

FIGURE 3.6   Shock wave pattern projecting on cross-flow plane.

space can be transformed into an equivalent unsteady two-dimensional formulation by omitting terms of a higher order than τ^2.

In Figure 3.7, the sketch of the flow field over a hypersonic slender body with a slightly blunt nose and the orthogonal coordinates are depicted. The Euler equations in the three-dimensional cylindrical polar coordinate system, (x, r, ϕ) consisting of continuity, momentum, and energy equation in terms of entropy or (p/ρ^γ) can be given as the following.

The continuity equation is expressed in three scalar velocity components (u, v, w) along the axial, radial, and polar angle coordinates (x, r, ϕ),

$$\frac{\partial}{\partial r}(\rho v r) + \frac{\partial}{\partial \phi}(\rho w) + \frac{\partial}{\partial x}(\rho u r) = 0 \qquad (3.22a)$$

The conservation law for momentum is described by the three scalar velocity components in three orthogonal spatial coordinates,

$$u\frac{\partial u}{\partial x} + v\frac{\partial u}{\partial r} + \frac{w}{r}\frac{\partial u}{\partial \phi} + \frac{1}{\rho}\frac{\partial p}{\partial x} = 0$$

$$u\frac{\partial v}{\partial x} + v\frac{\partial v}{\partial r} + \frac{w}{r}\frac{\partial v}{\partial \phi} - \frac{w^2}{r} + \frac{1}{\rho}\frac{\partial p}{\partial r} = 0 \qquad (3.22b)$$

$$u\frac{\partial w}{\partial x} + v\frac{\partial w}{\partial r} + \frac{w}{r}\frac{\partial w}{\partial \phi} + \frac{vw}{r} + \frac{1}{\rho r}\frac{\partial p}{\partial \phi} = 0$$

The conservation law for energy is a simplified consequence for the inviscid isentropic flows, in which the entropy is invariant along a streamline. The pressure, density, and entropy are related as $p/\rho^\gamma = \text{const.} \exp(S/c_v)$.

$$u\frac{\partial}{\partial x}\left(\frac{p}{\rho^\gamma}\right) + v\frac{\partial}{\partial r}\left(\frac{p}{\rho^\gamma}\right) + \frac{w}{r}\frac{\partial}{\partial \phi}\left(\frac{p}{\rho^\gamma}\right) = 0 \qquad (3.22c)$$

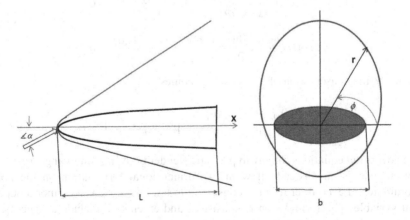

FIGURE 3.7 Sketch of the slender body in cylindrical polar coordinates.

The boundary condition on the impermeable body surface requires the vanished normal component of velocity.

$$U(u, v, w) \cdot n_b = 0 \tag{3.22d}$$

where n_b is the unit outward normal of the body surface. On the shock surface, the tangential velocity is unaltered, but the velocity component normal to the discontinuous control surface must satisfy the continuity condition.

$$\mid U(u, v, w)_s - U(u)_\infty \mid = D\left(1 - \rho_\infty/\rho_s\right) \tag{3.22e}$$

In Equation (3.22e), the normal component of the shock velocity is $D = -U_\infty \cdot n_s$, and n_s is the unit outward normal of the shock envelope.

In an order to conduct an order-of-magnitude analysis, all the evaluated terms need to bring in the same order, this requirement is easily accomplished; by normalizing the axial velocity component by $u_\infty \cos \alpha$, the two transverse velocity components, as well as the normal shock propagation velocity by $u_\infty \sin \alpha$. The density is made non-dimensional by ρ_∞ and the pressure by $\rho_\infty u_\infty \sin^2 \alpha$ respectively the order of magnitude of unity. For the cylindrical polar coordinates, the axial and radius coordinates are scaled by the body length and the body thickness. Substitute the non-dimensional variable into Equations (3.22a)–(3.22e) and omit the higher order than τ^2, the three-dimensional Euler equation in spherical polar coordinate reduces to the following form.

The continuity equation now appears as

$$\tau \cot \alpha \frac{\partial \rho}{\partial x} + \frac{1}{r}\left[\frac{\partial}{\partial r}(\rho v r) + \frac{\partial}{\partial \phi}(\rho w)\right] = 0, \tag{3.23a}$$

The axial momentum equation drops out by virtue of the order of magnitude argument, and only the momentum equations in radius and polar coordinates retain

$$\tau \cot \alpha \frac{\partial v}{\partial x} + v \frac{\partial v}{\partial r} + \frac{w}{r}\frac{\partial v}{\partial \phi} - \frac{w^2}{r} + \frac{1}{\rho}\frac{\partial p}{\partial r} = 0$$

$$\tau \cot \alpha \frac{\partial w}{\partial x} + v \frac{\partial w}{\partial r} + \frac{w}{r}\frac{\partial w}{\partial \phi} + \frac{wv}{r} + \frac{1}{\rho r}\frac{\partial p}{\partial \phi} = 0 \tag{3.23b}$$

Similarly, the conservation of energy law becomes

$$\tau \cot \alpha \frac{\partial}{\partial x}\left(\frac{p}{\rho^\gamma}\right) + v \frac{\partial}{\partial r}\left(\frac{p}{\rho^\gamma}\right) + \frac{w}{r}\frac{\partial}{\partial \phi}\left(\frac{p}{\rho^\gamma}\right) = 0. \tag{3.23c}$$

The governing equations for any hypersonic slender body at a small angle of attack now appear as an unsteady flow in two-dimensional formulation on the polar coordinates; the axial term in the conservation laws behaves as a temporal dependent variable. The formulation is consistent and extends to include all necessary boundary conditions. On the body surface, the radial velocity component acquires the following result.

$$v_b = \frac{w_b}{r_b}\frac{\partial r_b}{\partial \phi} + u_b \tau \cot \alpha \frac{\partial r_b}{\partial x}.$$ (3.24a)

On the shock surface, the velocity components in radius and polar angle coordinates become.

$$v_s = \cos\phi + \frac{D}{|\nabla s|}\frac{2}{\gamma+1}\left(\frac{1}{M_\infty^2 \sin^2\alpha} - 1\right)$$

(3.24b)

$$w_s = \sin\phi + \frac{D}{|\nabla s|}\frac{2}{\gamma+1}\left(\frac{1}{M_\infty^2 \sin^2\alpha} - 1\right)\frac{1}{r_s}\frac{\partial r_s}{\partial \phi}$$

The boundary values of density and pressure after the shock envelope are simply defined by the oblique shock jump conditions.

$$\frac{1}{\rho_s} = \frac{\gamma-1}{\gamma+1} + \frac{2}{\gamma+1}\frac{1}{M_\infty^2 \sin^2\alpha}$$

(3.24c)

$$p_s = -\frac{\gamma-1}{\gamma(\gamma+1)M_\infty^2 \sin^2\alpha} + \frac{2}{\gamma+1}D^2$$

The shock velocity in the transformed space is simply a non-dimensional variable.

$$D = -\frac{U(u,v,w)\cdot n_s}{M_\infty \sin\alpha}$$ (3.24d)

It is important to note that for the boundary condition on the body and shock surface surfaces the following relation holds.

$$u\tau \cot\alpha = \tau \cot\alpha + O(\tau^2).$$ (3.25)

Equations (3.23a)–(3.23c) and the associated boundary conditions; Equations (3.24a)–(3.24d) constitute the exact governing equations in the polar coordinate in the crossflow plane of the original three-dimensional system; Equations (3.22a)–(3.22e). Sychev [15] has demonstrated that to the accuracy of τ^2 the equivalent principle or the law of plane section indeed transforms the body axial axis x to a time-like coordinate for the hypersonic equation of motion including the necessary boundary conditions. Therefore, the law of plane sections or the hypersonic equivalent principle is firmly established on the theoretical ground that starts from an ingenuous physical insight.

3.9 BLAST WAVE THEORY

The explosive compression wave on the blunt body in hypersonic flow generates a blast wave-like behavior to the gas medium. The similarity of wave motion arising from the equivalence principle has illustrated the analogy between steady flow past a blunt-nosed body of revolution or a two-dimensional slab with an instantaneous energy release in the form of a concentrated explosion in the plane perpendicular to the axis of motion. For the body of revolution, the concentric blast waves in the crossflow plane appear as a point discharge, whereas for the blunt-nosed slab the parallel wave behaves as the result of a line charge. Figure 3.5 only illustrates the shock structures in the plane perpendicular to an axisymmetric body. For a two-dimensional configuration, the projected shock waves in the cross-flow planes are appearing as parallel lines extending to infinity along the spanwise direction. This analogy establishes the relationship between an unsteady flow and a steady motion with one more spatial dimension. Based on this analogy, blast-wave results have been successfully used to estimate the pressure distribution for blunted-nosed bodies at hypersonic speeds.

A group of self-similar solutions for unsteady violent gas explosions has been developed analytically by Sedov [16] and a numerical solution was obtained by Taylor [17] for a spherical explosion. The latter is conducted at the same period as the former for nuclear explosion investigation but delayed from publication for security reason. The classic result for the explosion of a line charge applies to axisymmetric hypersonic steady flow through the Lagrangian formulation [16,18]. On the Lagrangian frame, the gas motion is studied by following an identical group of gas particles, instead of observing the gas motion in a fixed control volume in space. The continuity and conservations of momentum and energy are written as functions of the radius distance r for gas particles normal to the initial axis of symmetry; the dependent variables are pressure p, density ρ, and velocity $u = \partial r / \partial t$. All variables are defined by the Lagrangian coordinates t and $m = \rho r^n / n$;

$$\frac{\partial r}{\partial m} = \frac{1}{\rho r^{n-1}}$$

$$\frac{\partial u}{\partial t} = -r^{n-1} \frac{\partial p}{\partial m} \tag{3.26}$$

$$\frac{\partial \left(p / \rho^\gamma \right)}{\partial t} = 0$$

where in Equation (3.26), $n = 1$ is designated for two-dimension configuration and $n = 2$ for body of revolution.

The key connection of the blast-wave theory is between the energy released by the explosion and the wave drag of the blunt nose [16,18]

$$E = D = \int_0^t p u_n (2\pi r)^{n-1} dt = \int_0^r p (2\pi r)^{n-1} dr \tag{3.27}$$

The similarity solution of the blast wave for wave velocity, density, and pressure is obtained in the spatial and temporal independent variables r and t [16]. The relationship between the streamwise coordinates x and t are established by the equivalence principle; $t = x/u_\infty$. For the cylindrical blast wave, the analytic results for the pressure and the radial coordinate of the wave from the similar solution give as

$$p = \frac{\gamma^{[2(\gamma-1)/(2-\gamma)]}}{2^{[(4-\gamma)/(2-\gamma)]}} \, p_\infty \left(\frac{E}{\rho_\infty}\right)^{1/2} t^{-1} \tag{3.28a}$$

$$r = \left(\frac{E}{\rho_\infty}\right)^{1/4} t^{1/2} \tag{3.28b}$$

The formulas of pressure distribution and shock shape for the blunt cylinder and blunt slab in hypersonic flow have been obtained by Lukasiewicz [19]. The accuracy of the blast-wave theory in predicting equivalent steady flows has been tested by comparing it with experimental observation with mixed results. The pressure distribution on a hemisphere cylinder at a Mach number of 7.7 by the blast wave analogy is presented in Figure 3.8 as an illustration. The pressure distributions are derived as a function of the drag coefficient by two consecutive approximations. The approximated results are reasonably supported by experimental data.

One must be mindful of the validity of the blast-wave analogy; it only holds in regions where the small perturbation equations are valid. Thus the approximation will fail in the nose region and in the entropy layer. However, the elegance of the analytic results of the blast wave cannot be overlooked in that it theoretically obtains the general trends observed with experimental data [19,20].

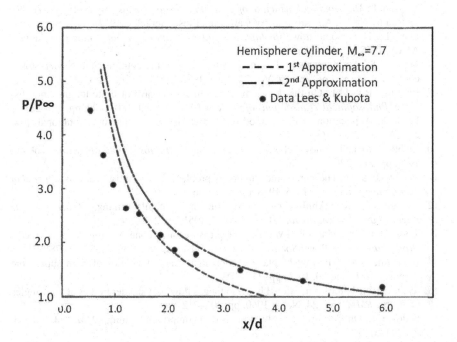

FIGURE 3.8   Pressure distributing on hemisphere cylinder by blast wave analogy.

In concluding our discussion for inviscid hypersonic flows, it will be a serious omission if all the numerous outstanding efforts including the method of characteristics have not been discussed because some of these accomplishments are spreading over a wide area of applications. An exceptional and thorough collection of the classic hypersonic flows theories is exemplified by the illuminating book by Hayes and Probstein which is strongly recommended [4]. It is truly regrettable that the characteristics method is not included but derived from the principle of the domain of dependence for the hyperbolic partial differential equation system. Historically, Ferri [9] first applied the method of characteristic for rotational flow, which is an essential feature of the flow field downstream of a curved shock. In application to hypersonic flows, special provisions are required in posing the initial data line. Specific information can be found in textbook by Zucrow and Hoffmann [21].

REFERENCES

1. Chapman, S. and Cowling, T.G., *The mathematical theory of non-uniform gases*, Cambridge University Press, Cambridge, 1964, pp. 134–150.
2. Clarke, J.F. and McChesney, M., *The dynamics of real gases*, Butterworths & Co, Washington, DC, 1964.
3. Tsien, H.S., Similarity laws of hypersonic flows, *Journal of Mathematical Physics,* Vol. 25, 1946, pp. 247–251.
4. Hayes, W.D. and Probstein, R.F., *Hypersonic flow theory*, Academic Press, New York, 1959.
5. Newton, I., *Mathematical principal of natural philosophy transl.* By Motte, A., (1729), revised by Cajori, A., University of California Press, Berkeley, 1934.
6. Lees, L., Hypersonic flow, *5th International Aeronautical Conference*, Los Angles, AIAA, 1955, pp. 241–276.
7. Shang, J.S., An assessment of numerical solutions of the compressible Navier-Stokes equations, *Journal of Aircraft,* Vol. 22, No 5, 1985, pp. 353–370.
8. Rankine, R.W.J.M., On the thermodynamic theory of wave of finite longitudinal distance, *Philosophical Transactions of the Royal Society*, Vol. 160, 1870, pp. 277–288.
9. Ferri, A. Application of the method of characteristics to supersonic rotational flow, NACA TN 841, 1946.
10. Lighthill, M.J., The flow behind a stationary shock, *Philosophical Magazine*, Vol. 40, 1949, pp. 214–220.
11. Van Dyke, M.D., The supersonic blunt body problem-review and extensions, *Journal of Aero/Space Science*, Vol. 25, 1958, pp. 485–496.
12. Oswatitisch, K., Ahnlichkeitsgesetze fur hyperschallstromung, *Zeitschrift für Angewandte Mathematik und Physik,* Vol. 2, 1951, pp. 249–264.
13. Taylor, G.I. and Maccoll, J.W., The air pressure on a cone moving at high speeds, *Proceedings of the Royal Society of London, Series A,* Vol. 139, 1933, pp. 278–231.
14. Il'yushin, A.A., The law of plane section in the aerodynamics of high supersonic speeds, *PMM*, Vol. 20, 1956, pp. 733–755.
15. Sychev, V.V., Three-dimensional hypersonic gas flow past slender bodies at high angles of attack, *PMM*, Vol. 24, No. 2, 1960, pp. 205–212.
16. Sedov, L.I., On Certain unsteady motions of a compressible fluid, *PMM*, Vol. 9, 1945, pp. 293–311.

17. Taylor, G.I., The formation of a blast wave by a very intense explosion, *Proceedings of the Royal Society of London, Series A*, Vol. 201, 1950, pp. 159–168.
18. Chernyi, G.G., *Introduction to hypersonic flow*, Academic Press, New York, 1961.
19. Lukasiewicz, J., Blast-hypersonic flow analogy-theory and application, *Journal of the American Rocket Society,* Vol. 32, No. 9, 1962, pp. 1341–1346.
20. Anderson, J.D., *Hypersonic and high-temperature gas dynamics*, McGraw-Hill, New York, 1989.
21. Zucrow, M.J. and Hoffmann, J.D., *Gas dynamics, vol. 2: Multidimensional Flows*, John Wiley & Sons, New York, 1977.

4 Hypersonic Viscous Flows

4.1 COMPRESSIBLE BOUNDARY-LAYER FORMULATION

The viscous flows govern completely by the compressible Navier-Stokes equations, which is a system of incompletely parabolic partial differential equations system and constitutes an initial values and boundary condition problem [1]. The classical Prandtl boundary layer approximation is also applicable to compressible hypersonic viscous flows [2,3]. The boundary-layer concept postulates that all effects of transport phenomena are confined within a thin layer over the solid surface. The velocity components normal to the external velocity are nearly negligible in comparison to the component parallel to the external flow, and the boundary thickness is much smaller than the radii of the longitudinal and transverse curvature of the attached surface. Hence, the optimal coordinates for analyzing the boundary-layer equation are either the Cartesian or the axial symmetric frames of reference.

The drastic differences between viscous from inviscid flows are the non-slip velocity boundary condition on the medium interface, and the presence of transport properties originating from nonequilibrium collision dynamics. The transport properties of the gas play an important role in hypersonic flow in two aspects. First, they create a thin subsonic inner layer immediately adjacent to any solid surface by the non-slip velocity condition leading to a high shear stress domain. Second, conductive heat transfer at the fluid-solid interface can substantially alter the energy balance incurring thermal management concerns. Furthermore, a sufficiently strong adverse pressure gradient in the streamwise direction will unavoidably trigger flow separation and consequently induce strong viscous-inviscid interaction. Finally, the formidable challenge of laminar-turbulent transition is always incurred by hydrodynamic instability at high Reynolds number conditions or by external disturbances. At present only an incomplete understanding of the transition phenomena is reached but must be addressed.

The shock wave and viscous process in hypersonic flows decelerate the gas flow producing extremely high temperatures. The elevated temperature increases the boundary layer thickness more than a lower speed flow of the identical freestream Reynolds number. The outward deflection of the streamline from the thickening boundary layer also enhances the interaction with the external flow. Meanwhile, the vorticity generated by the curved shock envelope will affect the growth of the boundary layer. For this reason, viscous-inviscid interaction becomes another unique feature of hypersonic viscous flows. However, the most pronounced difference in hypersonic flow from the lower speed flow is the much higher rate of heat transfer by the high-temperature gradient within the boundary layer. Special attention must

DOI: 10.1201/9781003212362-5

be paid to the development of the conservation of energy equation. At the thermodynamic nonequilibrium state the high temperature or correctly the high-enthalpy environment, dissociation and ionization will occur. These types of complications by gas internal mode excitations are in general not treatable by perfect gas without additional modifications.

A commonly simplified approximation to a hypersonic boundary layer is considering the gas to be a binary mixture so the diffusion coefficient is practically independent of composition. The governing equations of the boundary layer in practical applications are mostly adopted for either two-dimensional or axisymmetric flows at the steady state. The species conservation, continuity, momentum, and energy equations in scalar variables are

$$\rho u \frac{\partial c_i}{\partial x} + \rho v \frac{\partial c_i}{\partial y} - \frac{\partial}{\partial y}\left(\rho D_{12}\frac{\partial c_i}{\partial y}\right) = \dot{w}_i \tag{4.1a}$$

In Equation (4.1a), the symbols c_i and $\dot{w}_i$ represent the mass fraction of species i, and $\dot{w}_i$ is the net production/depletion rate of the i species. For a binary mixture, the coefficient of thermal diffusion D_{12} is determined from the gas kinetic theory [4].

Summing up all gas species of the concentration equation yields the continuity equation, in which the provision for axisymmetric configuration is also inserted. In the following equations, r is the radius from the axis of symmetry and the index j denotes two-dimensional, $j = 0$ or axisymmetric flow, $j = 1$.

$$\frac{\partial \rho u}{\partial x} + \frac{1}{r^j}\frac{\partial \rho v r^j}{\partial y} = 0 \tag{4.1b}$$

The momentum equation is degenerated into the following form by dropping all the higher-order terms scaled by the boundary-layer thickness to the transverse coordinates and velocity component. For the hypersonic laminar boundary layer, the negligible term is proportional to the product of the square of the freestream Mach number and the boundary layer thickness, which is inversely proportional to the square root value of the Reynolds number based on a characteristic length.

$$\rho u \frac{\partial u}{\partial x} + \rho v \frac{\partial u}{\partial y} = -\frac{\partial p}{\partial x} + \frac{1}{r^j}\frac{\partial}{\partial y}\left[r^j\left(\mu \frac{\partial u}{\partial y}\right)\right]$$

$$\frac{\partial p}{\partial y} = O\left(\gamma M_\infty^2 \delta\right) \tag{4.1c}$$

The conservation energy equation is customarily given by the total enthalpy H and two similarity parameters: the Prandtl and Lewis numbers for simplification.

$$\rho u \frac{\partial H}{\partial x} + \rho v \frac{\partial H}{\partial y} = \frac{1}{r^j}\frac{\partial}{\partial y}\left\{r^j\left(\frac{\mu}{\mathrm{Pr}}\frac{\partial H}{\partial y}\right) + \frac{\partial}{\partial y}r^j\left[\mu\left(1-\frac{1}{\mathrm{Pr}}\right)\frac{\partial}{\partial y}\left(\frac{u^2}{2}\right)\right]\right.$$

$$\left. + \frac{\partial}{\partial y}r^j\left[\rho D_{12}\left(1-\frac{1}{Le}\right)\sum\left(h_i-h_i^0\right)\frac{\partial z_i}{\partial y}\right]\right\} \tag{4.1d}$$

In Equation (4.1d), the total enthalpy is defined as

$$H = h + \frac{u^2}{2} = \sum z_i \left(h_i - h_i^o \right) + \frac{u^2}{2} \tag{4.2a}$$

where h_i^o is the standard heat of formation for the species i. Two similarity parameters are the Prandtl number and the Lewis number. The former is well known as $\mathrm{Pr} = C_p \, \mu / \kappa$ by relating the constant pressure specific heat with molecular viscosity and thermal conductivity. The Lewis number is just the ratio between the energy transport due to diffusion and thermal conduction $Le = \rho D_{12} C_p / \kappa$. The binary gas also introduces an additional component of heat transfer by the species recombination process at the solid surface.

$$q_w = -\sum \frac{1}{r^j} \left[k \frac{d\left(r^j T \right)}{dy} + D_{12} \rho h_i^o \frac{d\left(r^j z_i \right)}{dy} \right] \tag{4.2b}$$

The compressible boundary layer equations are valid for hypersonic flow with two special provisions. First, standard gas kinetic theory excludes the internal structure of molecules, and thus, the internal degrees of excitation must be resolved by approximations [3,4]. The connection between molecular structure and thermodynamic behavior of high-enthalpy gas is described by statistical mechanics and augmented by quantum physics [5]. The quantum physical-chemical phenomena of nonequilibrium hypersonic flows are extremely complex and take place at the microscopic scales, and the physical-based modeling must be devised by a sparsely available validation database. In addition, the traditional approximation of neglecting the wall-normal component of the pressure gradient, see Equation (4.1c), may not hold when the freestream Mach number is extremely high, and the boundary-layer thickness is no longer small relative to the streamwise length scale. Under this circumstance, the externally generated wave can also penetrate the boundary layer and produce a pressure gradient across the layer [7].

Figure 4.1 presents the static pressure variation across the turbulent boundary layer at a Mach number of 9.37 with a set accompanying experimental data. The calculated pressure distribution within the boundary layer agrees very well with the measurement. Generally, the pressure decreases from the solid surface and reaches the maximum value near the outer edge of the laminar sublayer, then approaches the external stream asymptotically. This behavior is expected from both the laminar and turbulent boundary layers over the entire Mach number regions without viscous-inviscid interaction.

To solve the parabolic partial differential equations numerically, only the upstream, solid surface and the external flow boundary conditions are required to completely specify the necessary boundary conditions. The boundary condition at the exit plane must leave unspecified to avoid the ill-posed problem by overspecification [1,7]. However, the boundary-layer approximation does not stand alone as the solution to a complete hypersonic problem because it requires information from the inviscid external flow. To bypass solving this system of nonlinear partial differential equations, a compressibility transformation can reduce the system of equations to the incompressible form to be benefitted from a huge collection of classic

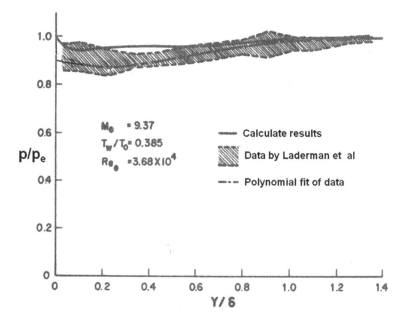

FIGURE 4.1   Static pressure profile of hypersonic turbulent boundary layer.

knowledge [6,8]. The compressibility transformation was introduced by Stewartson and Illingworth reducing the compressible boundary layer equations to incompressible form and can be manipulated to acquire similar solutions analogs to the classic equation of Falkner-Skan [3,8]. In short, the effective approach is to reduce the partial differential equations system to an ordinary differential equation by a similar formulation.

4.2 SELF-SIMILAR SOLUTIONS

The advantages of self-similar solutions can also be exploited for hypersonic flows, in much the same manner as for subsonic and supersonic flows, where a large group of boundary-layer problems can be analyzed in detail by a coordinate transformation to yield self-similar solutions. However, phenomena specific to hypersonic flow must be addressed; in the derivation process, the specific heat of the gas is assumed to be a constant. The transformation by Levy [9] incorporates the density variation of Howarth and Dorodnitsy, and Lees and Probstein [10] combined with the two-dimensional and axisymmetric relations by Manger. The resultant boundary-layer equations are transformed into a form reducible to the classic Falkner-Skan equation.

The compressible to impressible boundary layer coordinate transformation is given as.

$$\xi(x) = \int \rho_w \mu_w u_e r^{2j} dx; \quad \eta(x,y) = \frac{u_e r^j}{\sqrt{2\xi}} \int \rho dy \qquad (4.3a)$$

The dependent variables such as the streamwise velocity, stagnation or total enthalpy, and species mass fraction are defined by the dimensionless quantities.

$$f = \int \frac{u}{u_e} d\eta, \quad \frac{\partial f}{\partial \eta} = f' = \frac{u}{u_e}$$

$$g = \frac{H}{H_e}, \quad \frac{\partial g}{\partial \eta} = g'$$

$$z_i = \frac{c_i}{c_{ie}}, \quad \frac{\partial z_i}{\partial \eta} = z_i'$$

(4.3b)

After the transform, the boundary-layer equations from the (x, y) to the (ξ, η) coordinate by Equation (4.3a), the continuity equation is satisfied automatically by including the stream function during the derivation. The continuity equation gives only the solution of the transverse velocity component as

$$\rho v = -r^{-j} \left(\frac{d\xi}{dx} \frac{\partial \sqrt{2\xi f}}{\partial \xi} + \sqrt{2\xi} \frac{\partial \eta}{\partial x} f' \right)$$

(4.4)

According to the basic assumption of the boundary-layer theory, the transverse velocity component is in the order of magnitude of the inverse square root of the Reynolds number $O\left(1 / \sqrt{\mathrm{Re}_x}\right)$ and is eliminated by the boundary-layer approximation.

The momentum and energy equation of the compressible boundary-layer equations becomes

$$\frac{\partial}{\partial \eta} (cf'') + fg' + 2\frac{d \ln u_e}{d \ln \xi} \left[\frac{\rho e}{\rho} - f'^2 \right] = 2\xi \left(f' \frac{\partial^2 f}{\partial \xi \partial \eta} - \frac{\partial f}{\partial \xi} f'' \right)$$

$$\frac{\partial}{\partial \eta} \left(\frac{c}{\mathrm{Pr}} g' \right) + fg' + \frac{\partial}{\partial \eta} \left[\frac{c(Le-1)}{\mathrm{Pr}} \sum \frac{(h_i - h_i^o)}{H_e} z_i' \right] + \frac{u_e^2}{H_e} \frac{\partial}{\partial \eta} \left[c \left(1 - \frac{1}{\mathrm{Pr}} \right) f' f'' \right]$$

$$= 2\xi \left(f' \frac{\partial g}{\partial \xi} - \frac{\partial f}{\partial \xi} g' \right)$$

(4.5)

where the coefficient c is often referred to as the Chapman-Rubesin parameter [11] and is defined as $c = \rho\mu / \rho_w \mu_w$.

The self-similarity solution of the boundary layer equation is achievable by reducing the partial differential to ordinary equation, including the boundary conditions. These requirements are easily met by the nonslip velocity condition on the solid surface. For the energy equation either the constant surface temperature $g(\theta) = \mathrm{const}$ or an adiabatic wall $g'(\theta) = 0$ is sufficient. At the outer edge of the boundary layer both the non-dimensional velocity and enthalpy need to approach the external flow condition asymptotically. The requirements for the differential equations are a little more complicated; first, the external flow is required to be either constant like that over a flat plate or is a function of streamwise coordinate rising in power of m, $u_e\left(x^m\right)$.

The Chapman-Rubeson parameter, Prandtl ($Pr = c_p\mu / k$) and Lewis numbers ($Le = \rho D_{12}c_p / k$) also require being a function of η only, or simply a constant. Finally, the gradient of mass fraction z_i' to be a function of η or simply vanished. The simplest self-similar boundary-layer equations acquire the following form:

$$f''' + fg' + 2\frac{d\ln u_e}{d\ln \xi}\left[\frac{\rho_e}{\rho} - f'^2\right] = 0$$

$$g'' + fg' = \frac{u_e^2}{H_e}\frac{\partial}{\partial \eta}\left[c\left(1 - \frac{1}{Pr}\right)f' f''\right]$$

(4.6a)

A group of self-similar solutions of the compressible laminar boundary layer has been investigated by Cohen and Reshotko [12]. In their formulation, the enthalpy is defined as $s = H / H_e - 1$. The resultant equations are nearly identical to Equation (4.6a).

$$f''' + ff'' = \beta\left(f'^2 - 1 - s\right)$$

$$s'' + Pr\, fs' = (1 - Pr)\left[\frac{(\gamma - 1)M_e^2}{1 + (\gamma - 1)M_e^2 / 2}\right]\left(f' f'' + f''^2\right)$$

(4.6b)

In their formulation, the Prandtl number is assumed to be a constant, viscous coefficient is assumed to be linearly proportional to the temperature, and with a constant surface temperature. Like all possible similar solutions, the pressure gradient must be generated by an external flow to behave as a power of the streamwise distance $u_e \sim cx^m$; the rate of change in the momentum equation reduces to a constant $\beta = 2m / (m + 1)$. When the pressure gradient parameter vanishes, $\beta = 0$, which corresponds to the solution over a flat plate or an axisymmetric flow. If the boundary layer encounters an adverse pressure gradient, this parameter assumes a negative value $\beta < 0$. The numerical solutions also reveal the upper and lower branches of the similar solution separate the attached and separated boundary layer. When the boundary layer encounters favorable streamwise pressure, then the self-similar solution is described by $\beta > 0$. At the stagnation point, the value of β is unity for two-dimensional flows and β equals to 0.5 for axisymmetric configurations.

Applying the compressibility transformation has expanded the self-similar boundary layer approximation from the incompressible to compressible flow domain. The clear evidence is revealed by the numerical solution produced by compressible similar boundary layer equations by setting the value of $\beta = 0$, $S_w = 0$, and Prandtl number to unity over a flat plate. By which the momentum and energy equation is decoupled and the boundary condition for the energy equation is satisfied as an adiabatic wall. In Figure 4.2, the compressible self-similar solution [12] duplicates the first incompressible boundary layer solution by Blasius in 1908. The classic result has been fully validated by a large group of experimental data spanning a very wide Reynolds number range of around one million.

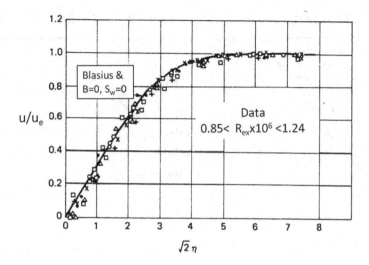

FIGURE 4.2 Compressible similar solution duplicating the classic Blasius solution.

A vast series of solutions have been successfully generated by a similar boundary layer equation over the complete range of the pressure gradient and surface temperature conditions [12]. The investigated pressure gradient conditions include the strong favorable pressure gradient flow at $\beta = 2.0$ until the velocity profile involves boundary layer separation and beyond by the adverse pressure gradient, $\beta < 0$ to recover the lower branch solution. The surface enthalpy also covers the extremely hot surface, the surface enthalpy reaches twice the value of the freestream stagnation enthalpy $s = 1.0$, and the extremely cold surface on which the surface enthalpy is zero $s = -1.0$. Only a cold wall condition close to the stagnation condition $s = -0.40$ are displayed in Figure 4.3 and cover the selected full two pressure gradient conditions at $\beta = -0.235$ and $\beta \leq 2.0$. The reverse velocity profile has appeared as the lower-branch solution in adverse pressure conditions.

4.3 SIMILARITY NUMBERS AND PARAMETERS

The boundary-layers over a constant streamwise pressure gradient flow condition, or most likely over a flat plate, or over a cylinder the governing equation is further reducing to become.

$$f''' + fg' = 0$$

$$g'' + fg' + \frac{u_e^2}{H_e} \frac{\partial}{\partial} \left[c \left(1 - \frac{1}{Pr} \right) f' \, f'' \right] = 0 \tag{4.7}$$

At hypersonic flow conditions and for a perfect gas, the ratio u_e^2 / H_e approaches a value of two; $u_e^2 / H_e = 2$. The coupling between the momentum and energy equation

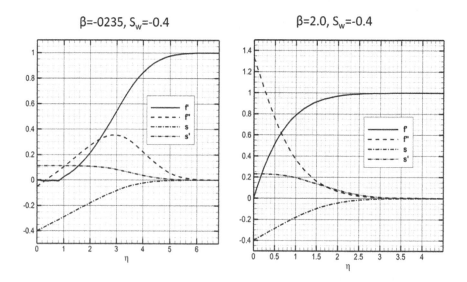

FIGURE 4.3   Typical compressible similar solutions of the compressible boundary layer.

reveals intriguing results. When the Prandtl and Lewis number assumes to be unity at the zero-pressure gradient condition over an adiabatic wall, Buseman's general solution for the energy equation gives $g(\eta) = 1.0$ [2,3]. In similar conditions, the surface temperature is constant, $g(0) = const$. The Crocco's relationship with the enthalpy reveals $g(\eta) = f'(\eta)[1 - g(0)] + g(0)$ [13]. Under this condition, $g' = f''$, the momentum and energy equation of the boundary layer is identical and can be completely uncoupled.

$$f''' + ff'' = 0$$
$$g'' + fg' = 0$$
(4.8)

In general, these simplified conditions may not all satisfy simultaneously, and a physical based explanation is highly valuable for basic understanding. For example, the Prandtl number $Pr = c_p\mu / k$, reflects the relative importance of the viscous dissipation versus conductive heat transfer, and the Schmidt number indicates the relative effectiveness of viscous in contrast to thermal diffusion, $S_c = \mu / \rho D_{12}$. When both the similarity numbers become unity, the boundary-layer thickness and thermal layer thickness are identical for flow over a flat plate. When both the similarity numbers are less than unity, then the thermal layer thickness is greater than the boundary thickness. The third most used similarity number is the Lewis Number and often is given in literature as the Lewis-Semenov number which is defined as $Le = \rho D_{12}c_p / k$ and is the ratio among the Prandtl and Schmidt numbers; $Le = Pr / Sc$. By assuming the Lewis number equal to unity, is equivalent to assume the energy flux is directly proportional to the enthalpy gradient. Under this condition, the heat transfer is independent from the detailed mechanism and how the energy is distributed by different modes of heat transfer. Finally, the Eckert number is defined as $E_c = u_e^2 / h_e$. When the Eckert number becomes large in hypersonic flow, it has been approximated as $Ec = (\gamma - 1)M_e^2$. In this case, the diffusion and conductive heat transfer are secondary for determining the temperature profile in the boundary layer.

The most interesting finding of adopting the similarity numbers for analyzing the compressible boundary layer is that the variation of the Chapman-Rubesin parameter controls the skin friction and the heat transfer rate for the perfect gas [11]. A remarkable agreement of the skin friction coefficient and the heat transfer has been found by the variable Prandtl number and Chapman-Rubesin parameter for perfect gas with constant specific heat.

The skin friction coefficient and heat transfer rate for the compressible laminar boundary layer of the perfect gas under the zero-pressure gradient flow condition are derived as a function of the Reynolds number based on the running length from the leading edge of the flat plate.

The friction skin coefficient is well-known as the classic incompressible counterpart.

$$C_f \sqrt{Re_x} = 0.664 \sqrt{\frac{\rho_r \mu_r}{\rho_e \mu_e}} \tag{4.9a}$$

The heat transfer rate under zero-pressure gradient flows is described by the Stanton number and the Nusselt number, $Nu = St \cdot Re_x$. The connection between the skin friction coefficient and the heat transfer rate is the recognized in Reynold's analogy.

Figure 4.4 depicts the computational simulation of a hypersonic boundary layer over a flat plate that transits from laminar to turbulent conditions, the numerical results are fully supported by experimental observations to show validation of the Reynolds analogy in predicting skin friction coefficient and the Stanton number.

$$St = \frac{-\dot{q}}{\rho_e u_e (h_r - h_b)} = \frac{1}{2} C_f (Pr)_r^{-2/3} \tag{4.9b}$$

$$\frac{Nu}{\sqrt{Re_x}} = 0.322 \sqrt{\frac{\rho_r \mu_r}{\rho_e \mu_e}} (Pr)_r^{1/3} \tag{4.9c}$$

Equations (4.9a)–(4.9c) are equally applicable to axisymmetric flows just by multiplying the right-hand side of the formulas by the Mangler factor, $\sqrt{3}$. The reference conditions in these equations are calculated by the reference enthalpy method which is purely empirical [2].

The reference temperature or the reference enthalpy method is an approximation procedure for calculating skin friction and heat transfer even in a mild pressure gradient condition. This simplified approximation is reasonable if all temperature-dependent air properties are evaluated at an appropriate reference enthalpy, which

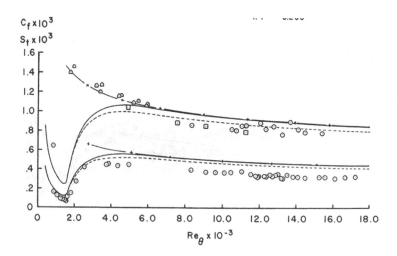

FIGURE 4.4   Verification of the Reynolds analogy.

occupies the extreme limiting condition of the boundary layer. The key parameter is
the altered Chapman-Rubesin parameter as

$$C^* = \frac{\rho^* \mu^*}{\rho_e \mu_e} \approx \left(\frac{T^*}{T_e} \right)^{-1/3} \qquad (4.10a)$$

The reference enthalpy method is developed as a semi-empirical method of correlating skin frication and heat transfer for a compressible laminar boundary layer
over a flat plate. The key association of the aerodynamic properties is the Chapman-
Rubesin parameter $C = \rho\mu / \rho_w\mu_w$, the approximation is thus needed to replace the
local dependence by appropriate reference. Clearly, the reference temperature must
depend on both the Mach number and the viscous dissipation and heat conduction.
The reference temperature or enthalpy was originally given by Eckert for the Prandtl
number of 0.7, Pr = 0.7 [3].

$$h^* = 0.5(h_e + h_w) + 0.22(h_r - h_e), \qquad (4.10b)$$

where the empirical approximation for the recovery enthalpy h_r is defined as

$$h_r = h_e + \frac{\sqrt{Pr^*}}{2} u_e^2 \qquad (4.10c)$$

By the reference enthalpy approximation, the heat transfer rate and skin friction are
connected by the classic Reynolds's analogy, or $St = c_f \, Pr^{-2/3} / 2$. The heat transfer
rate is now approximated as $\dot{q} = \frac{1}{2}\rho_e u_e (h_r - h_w) C_f \left(Pr^* \right)^{-2/3}$. From a wide range of
experimental data and calculated results, the Prandtl number does not change appre-
ciably for a perfect gas, for laminar flows a value of 0.73 has been widely used. But
the Lewis number different form unity is uncertain, for this reason, the reference
enthalpy method has remained as a historical note.

4.4 STAGNATION-POINT HEAT TRANSFER

The steady state compressible boundary layer equations have been transformed from the Cartesian and axisymmetric coordinates by the Levy's transformation using the combination of compressibility and Manger's correlations to a more compact formulation [9]. The compressibility transformation has also been applied successfully by Cohen and Reshotko [12]. This system of equations is a nonlinear partial differential equation and all the explicit dependence of the independent variables ξ is moved to the right-hand side of the momentum and energy equations (4.5). At the stagnation point, the transformed independent variable ξ vanishes then all the terms on the right-hand-side can be eliminated to give

$$(cf'')' + ff'' + 2\frac{d\ln u_e}{d\ln\xi}\left[\frac{p_e}{\rho} - f'^2\right] = 0 \tag{4.11a}$$

$$\left(\frac{c}{\text{Pr}}g'\right)' + fg' - \frac{\partial}{\partial\eta}\left[\frac{c(Le-1)}{\text{Pr}}\Sigma z_{ie}\frac{\left(h_i - h_i^o\right)}{H_e}\frac{\partial z_i}{\partial\eta}\right] - \left(\frac{u_e^2}{h_e}\right)\frac{\partial}{\partial\eta}\left[c\left(1 - \frac{1}{\text{Pr}}\right)ff''\right] = 0 \tag{4.11b}$$

In the stagnation region, the chemical reaction rate is much faster than the characteristic velocity of the flow. The concentration of the binary species can then be uniquely determined by any two independent thermodynamic variables and the species conservation equation, Equation (4.1a) becomes superfluous. The transformed boundary-layer equations can be further reduced.

The flow field in the stagnation region is laminar and the x coordinate is the running length measured from the stagnation point. Thus $r_w = x$ and $u_e = x(du_e / dx)_0$ from the coordinate transformation, we have $\xi \propto x^{2(j+1)}/2(j+1)$ as $x \to 0$. The term related to the streamwise pressure gradient $2\frac{d\ln u_e}{d\ln\xi}\left[\frac{p_e}{\rho} - \left(\frac{\partial f}{\partial\eta}\right)^2\right]$ becomes $\left[\frac{p_e}{\rho} - \left(\frac{\partial f}{\partial\eta}\right)^2\right]/(j+1)$.

A similar boundary layer exists when the dependent variables f and g are only functions of η together with the boundary conditions that $f(0) = 0, \frac{\partial f}{\partial\eta}(0) = 0, g(0) = g_w(\varepsilon)$, or the adiabatic wall condition; $\frac{\partial f}{\partial\eta}(0) = 0$ as well as, f and g approach unity as η tends to infinity. The governing equation for stagnation point heat transfer can be further reduced for a calorically perfect gas; namely $\frac{p_e}{\rho} = \frac{p_e}{p}\frac{T}{T_e} = \frac{p_e}{p}\frac{h}{h_e} = g$ and by letting the Lewis number become unity. Equations (4.11a) and (4.11b) reduce to similar boundary layer equations for the stagnation region by recognizing the external flow is a function of the running distance from the stagnation point. A similar formulation will break down in the stagnation region when the streamwise velocity, u_e^2 / H_e and the pressure across the boundary layer, $(p_s - p_e)/p_s$ are no longer negligible. In fact, for the perfect gas hypersonic flow the ratio between the square of freestream velocity and the total enthalpy becomes $u_e^2 / H_e \approx 2$, and the self-similarity property will be diminished. Theoretically, the breakdown possibility is removed by assuming the Prandtl number to be unity.

$$(cf'')' + ff'' + (1+j)^{-1}\left[\frac{\rho_e}{\rho} - f'^2\right] = 0 \tag{4.12a}$$

$$\left(\frac{c}{Pr}g'\right)' + fg' + \left[\frac{c(Le-1)}{Pr}\Sigma\left(h_i - h_i^o\right)\left(\frac{dz_i}{d\eta}\right)g'\right]' = 0 \tag{4.12b}$$

For hypersonic boundary layers in which the freestream Mach number is high, the ratio of the static and total enthalpy at the outer edge of the shear layer is much lower than unity. A major part of the layer is controlled by the freestream total enthalpy and is isolated from the external flow by a thin transitional outer layer. Because the Prandtl number is assumed to be unity, the viscous dissipation is not included in the energy equation, Equation (4.11b). From numerical solutions of similar boundary layers [12], indeed the effect of the dissipation term is clearly not dominant.

The conductive heat transfer on the stagnation point without chemical reaction is only a function of the enthalpy gradient of the similar boundary layer equation. The dependence of the streamwise velocity gradient or the pressure gradient has also appeared explicitly.

$$\dot{q}_w = \frac{\sqrt{1+j}\,k_w H_e}{c_{pw}}\left[\frac{\rho_w}{\mu_w}\left(\frac{du_e}{dx}\right)_o\right]^{1/2}g'_w \tag{4.13a}$$

Two important classic stagnation point heat transfer results from Equation (4.13a) for two-dimensional and axisymmetric similar laminar boundary layers are developed and correlated by Van Driest [13]. For the two-dimensional body:

$$\dot{q}_w = 0.570\,Pr^{-0.6}\left(\rho_e u_e\right)^{1/2}\sqrt{\frac{du_e}{dx}}\left(h_{aw} - h_w\right) \tag{4.13b}$$

For the axisymmetric body

$$\dot{q}_w = 0.763\,Pr^{-0.6}\left(\rho_e u_e\right)^{1/2}\sqrt{\frac{du_e}{dx}}\left(h_{aw} - h_w\right) \tag{4.13c}$$

The symbol h_{aw} in Equations (4.13b) and (4.13c) designated the enthalpy for the corresponding flow under adiabatic wall conditions.

Two interesting observations can be made from the correlated results; first, the three-dimensional relief effect stands out. This three-dimensional phenomenon causes a reduced boundary-layer thickness for the flows over a sharp cone versus those over a flat plate. The boundary-layer thickness over the cone is thinner by a factor of the square root of 3 than that of the flat plate at the same Reynolds number and is known as the Mangler factor [2,3]. By thinning the boundary-layer thickness through the three-dimensional relief effect in the stagnation region, the temperature gradient normal to the body surface steepens, leading to a higher heat transfer rate to the axisymmetric shape.

The second interesting observation can be made by relating the external velocity gradient term in Equations (4.13a) to the streamwise pressure gradient by Bernoulli's equation to get

$$\frac{du_e}{dx} = \frac{1}{r_b} \sqrt{\frac{2(p_e - p_\infty)}{p_e}} \tag{4.13d}$$

Substitute Equation (4.13d) into (4.13a), the stagnation heat transfer rates of either two-dimensional or axisymmetric configurations are shown to be inversely proportional to the square root of the leading-edge radius.

$$\dot{q}_w \propto \frac{1}{\sqrt{r_b}} \tag{4.13e}$$

From Equation (4.13e) for blunt-body heat transfer in the stagnation region, a piece of critical information for thermal protection is realized. That is, for thermal management of most hypersonic vehicles, increasing the blunt nose radius is an effective means to reduce the peak heat load. This type of fundamental understanding is most effectively derived from analytic approaches.

The catalytic characteristic of an ablating surface contributes a significant effect on the thermal protection of a hypersonic vehicle. The boundary conditions on the surface of a chemical reacting flow including dissociated and ionized air exercise some increasing influence on the conductive heat transfer. One of the extreme conditions of the surface is the non-catalytic wall or the so-called frozen flow where the chemical recombination rates are exceptionally low; the chemical radices reach the surface without recombination. On the other extreme, the surface is catalytic then the species will immediately recombine and release the heat of formation which contributes more heat transfer to the surface. Fay and Riddell extended the applicability of the similarity solution by correlating the tabulated high-temperature properties for stagnation region heat transfer to include equilibrium dissociation. Very useful correction formulas are summarized as follows [14].

For chemical frozen surface

$$\dot{q}_f = 0.763 \, \mathrm{Pr}^{-0.6} \left(\rho_e u_e \right)^{1/2} \left[\left(\frac{du_e}{ds} \right)_o \right]^{1/2} \left(\frac{\rho_w \mu_w}{\rho_e \mu_e} \right)^{0.1} (H_e - h_w) \left[1 + \left(Le^{0.63} - 1 \right) \frac{h_e}{H_e} \right] \tag{4.14a}$$

For full catalytic surface

$$\dot{q}_e = 0.763 \, \mathrm{Pr}^{-0.6} \left(\rho_e u_e \right)^{1/2} \left[\left(\frac{du_e}{ds} \right)_o \right]^{1/2} \left(\frac{\rho_w \mu_w}{\rho_e \mu_e} \right)^{0.1} (H_e - h_w) \left[1 + \left(Le^{0.52} - 1 \right) \frac{h_e}{H_e} \right] \tag{4.14b}$$

There is an anticipated resemblance between Equations (4.14a) and (4.14b); the difference only appears in the term involving the Lewis number. The dependence on the viscosity coefficient at the outer edge of the boundary layer and the uncertainty are mainly incurred by the accuracy in determining its value at the hypersonic condition. For dissociating gas, the ratio of characteristic flow time to reaction rate was also found to have significant variation in heat transfer rate whether the surface is catalytic or not.

The accuracy of similar solutions in the stagnation region has been verified by comparison with shock-tube data by Kemp, Rose, and Detra for the heat transfer distribution over a hemisphere-cylinder [15]. In Figure 4.5, the similarity solution agrees well with the experimental measurement, in general, the discrepancy is confined within the data scattering band. Two different streamwise pressure gradients in the stagnation region have been also investigated the difference only reveals in the region farthest away from the stagnation point. The similarity is expected to fail as the flow leaves the stagnation region, but the increasing experimental data scatter prevents a precise evaluation.

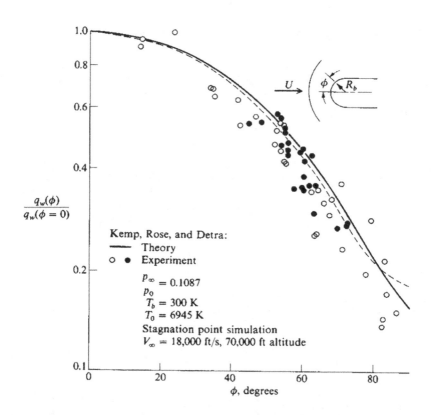

FIGURE 4.5   Verified stagnation point heat transfer with dissociated air.

4.5 LAMINAR-TURBULENT TRANSITION

Laminar-turbulent transition is one of the most challenging topics of fluid dynamics, because it is nonlinear, intermittent, time-dependent, and with fluid dynamic bifurcation and is deeply rooted in hydrodynamic stability [16]. The orderly laminar flow over a flat plate leads to the hydrodynamic instability to appear as the Tollmien-Schlicting wave when the Reynolds number based on the running length increases beyond a critical value varying from 2,000 to 13,000, depending on different initial flow conditions. The amplification of the instability is accentuated by developing a spanwise vorticity and followed by the three-dimensional vortex breakdown or the crossflow instability. Shortly downstream, the intermittent Emmons spot will appear at random locations in the flow path with increasing frequency to become the fully turbulent flow. The process often takes place in a burst and can also bypass the idealized pattern to turbulent by any externally induced perturbations such as free stream disturbance, noise, edge contamination, and surface irregularity including surface temperature variation. The impacts of the cross flow and surface curvature are especially pronounced in practical operations. Especially on a concave surface the flow field is known to incur instability by a longitudinal vortex disturbance known as Taylor-Gortler vortices and makes the transitional boundary layer even more complex.

In the Hypersonic flow regime, a second mode and higher harmonic instability can also be present in the linear instability growth regime [17]. The increased Reynolds number may excite the higher modes without the first mode and grows at a higher rate. The second node once activated displays a higher growth rate than the first. In addition, the higher mode instability is not stabilized by surface cooling, in fact, the upward shift of the frequency and growth rate above Mach number 5.8 are noted from the experimental observations [16].

The freestream disturbances in form of a sound wave or vorticity perturb the basic state of the laminar boundary layer. The process of establishing the initial conditions of the amplitude, frequency, and phase of the disturbance leads to the breakdown of laminar flows and is identified by Morkovin as receptivity [18]. When a large amplitude disturbance is encountered, a laminar flow can transit directly to turbulence. This possibility has been called high-intensity bypass and is not an eigenvalue problem. Morkovin's receptivity theory becomes the mainstay in analyzing the transition phenomenon. The transition of the laminar boundary layer to turbulent is treated as the nonlinear response of an extraordinarily complex stability problem. The broad-spectrum disturbances have an overwhelming amplitude compared to laminar flow perturbations; thus the flow reacts to the disturbance instantly. Nevertheless, the basic assumption of the normal mode analysis is still the foundation for deriving the Orr-Sommerfeld equation. The stability equation solves the eigenvalue problems of the streamwise and spanwise wavenumber, growth rate, as well as the wave frequency and temporal growth rate. A typical and well-established thumb curve for the amplifiable disturbance within the laminar boundary layer with/without crossflow component can provide the critical value of the Reynolds number for transition [16].

All the possible laminar-turbulent transition processes are depicted in Figure 4.6, the five different paths are strongly dependent on the amplitude of the environmental

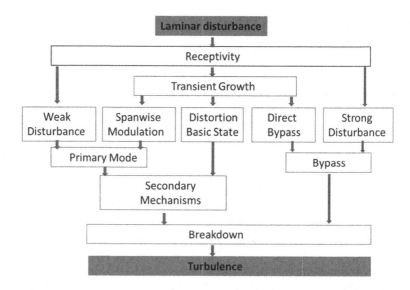

FIGURE 4.6   The processes of laminar-turbulent transition.

disturbance [19]. Rapid transient growth occurs when non-orthogonal and stable instability modes interact and undergo from algebraic to exponential growth. The transition bypass is generally considered to be the process by which the initial growth is not produced by the primary mode of the Orr-Sommerfeld equation. Despite intensive research on the laminar-turbulent process over the years, the understanding of the laminar-turbulent transition is still incomplete.

The temporal approach for systematically solving the Orr-Sommerfeld equation has been achieved by Malik to become a powerful tool for laminar flow control [20]. However, when a large amplitude disturbance is encountered, the predictive approach failed, because the laminar-turbulent transition can lead to turbulent bypass without the amplification of the normal mode, which is not an eigenvalue problem.

4.6 TURBULENT FLOW STRUCTURE

The turbulence phenomenon is completely describable by the Navier-Stokes equations [21]. Although CFD had been subjected to more than sixty years of intensive research and development, turbulence still is the least understood subject and has not been convincingly and successfully simulated. The phenomenon involves a critical condition at which a transition takes place from a subcritical to a supercritical state; known as fluid dynamic bifurcations. Understanding bifurcations by solving directly the Navier-Stokes equations is not necessarily a brute force effort but must be approached by a better understanding based on physics. For example, the vortical breakdown is trackable by the formation of the stagnation point in free space for a vortical motion. The hysteresis of fluid motion is exclusively associated with a delay consequence from an initial condition. The laminar-turbulent transition has arisen from hydrodynamic instability, environmental disturbances or receptivity, and the distribution of intermittencies.

To analyze the most challenging fluid dynamics problem of turbulence must be built on the understanding of basic mechanisms of physics that are generated by the simultaneous and vigorous interactions between large- and small-scale eddies in time and space. The dimensions of vortices of turbulence vary from the Kolmogorov inner scales to large-scale vorticities comparable to the entire flow field. The finest length scale of turbulence is defined as $(v^3 / \varepsilon)^{1/4}$ which involves the kinematic viscosity v and turbulent kinetic energy dissipation rate ε, which is the time-average quantity of the turbulent shear stress and the fluctuating velocity gradient. The inner scaling also knew for the time and velocity of turbulence, $(v / \varepsilon)^{1/2}$ and $(v\varepsilon)^{1/4}$, respectively. For the wall-bounded turbulence, the ever-presented hairpin vortices traveling in packets are always present over the fluid-solid interface.

Since turbulence is highly fluctuating in three dimensions, an energy spectrum becomes necessary for the correlation of the turbulent motion to examine its intensity in different frequencies. A complete statistical decomposition requires knowledge of the probability density function for all variables at all space locations and time, but for practical analysis, a two-point correlation is always applied, $Q_{i,j} = <u'(r,t)u'(r,t)>$. Therefore, the two-point fluctuating velocity correlation by the Fourier transformation is adopted for the velocity spectrum tensor.

In Turbulent flow, the largest eddies contain most of the kinetic energy, whereas the smallest eddies dissipate the energy by viscosity. The rate of compressible turbulent dissipation is controlled by the turbulent transport and production, and is often described as a temporal assemble of turbulent shear stress and rate of fluctuating velocity gradients. The energy spectrum of turbulence is the mean turbulent energy per unit mass as; $(u_i u_j) = \int E(k)dk$. The symbol k designates the wave number. In the energy cascading process, there is an intermediate scales range where neither the direct interacting force nor the significant viscous dissipation is dominated, but a net nonlinear energy transfer is still taking place. This spectrum range is referred to as the inertial range. The viscous dissipation rate is known as $\varepsilon = 2\kappa \int k^2 E(\kappa)d\kappa$, and the turbulent kinetic energy is given as

$$E_{i,j}(t) = \left(\frac{1}{2\pi}\right)^3 \int Q_{i,j}(r)e^{ikr}d^3r \tag{4.15a}$$

Which is the classic quantity describing homogeneous turbulence. It is generally considered that small-scale turbulent motion in any turbulent flows becomes isotropic at a high Reynolds number and therefore the Kolmogorov scales characterize the high wave number region of any turbulent flow. Following a dimensional analysis and assuming there is a universal small-scale spectrum to give [22]

$$\frac{E(k)}{v^{5/4}\varepsilon^{1/4}} = F\left(\frac{kv^{3/4}}{\varepsilon^{1/4}}\right) \tag{4.15b}$$

The energy transfers from the large to the small scales and then by viscous friction dissipated. Kolmogorov's hypotheses led to the universal form of the energy spectrum.

$$E(k) = 1.5\varepsilon^{2/3}k^{-5/3}$$

(4.15c)

According to our understanding of turbulence, the vortical structure of the small-scale fluctuations has a specific continuous energy spectrum that vanishes at high wave numbers by dissipation. The larger eddy extracts energy from the main flow and its structures are highly anisotropic. The wall-bounded turbulent flow is particularly important for turbulence research and engineering applications. The vortical structures in the shear layer are often considered to consist of streamwise vortices or streaks with an intermittent burst of coherent dynamic structures. The turbulent motion thus can be described by vortical dynamics involving interactions between eddies and main flow. The small eddies dissipate energy, tend toward isotropy, and are nearly universal in character. In Figure 4.7, the streamwise turbulent energy spectra for various types of wall-bound flows are assembled by Chapman [23]. It is clearly demonstrated that near the wall or in the inner region of the shear layers, the energy spectra of the small energy-dissipating eddies are clearly universal – it is independent of both Reynolds number and type of flows. This is the guiding principle for subgrid-scale modeling for large eddy simulation (LES).

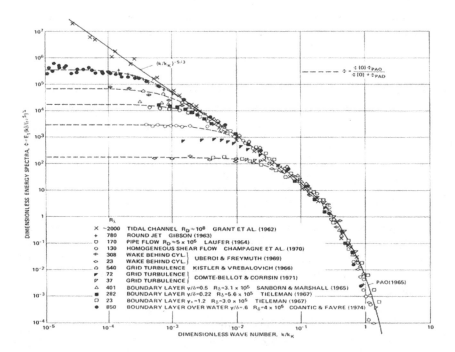

FIGURE 4.7   Streamwise energy spectra for wall-bounded turbulent flows.

From the phonological viewpoint, the classic analytic turbulent boundary-layer structure over a flat plate has been thoroughly investigated by Cole; consisting of the law-of-the-wall and the law-of-the-wake regions [24]. These laws were derived from more than five hundred experimental measurements prior to 1956, and the correlated results are depicted in Figure 4.8. The inner region of the turbulent boundary layer profiles can be accurately described by the logarithm formula when scaled by the friction velocity and a transformed normal distance from the wall, y^+. The friction velocity is the velocity parallel to the wall scaled by the skin coefficient, and the normal distance from the surface is scaled accordingly, as $u_\tau = \sqrt{\tau_w / \rho}$ and $y^+ = (u_\tau / v) y$. A laminar sublayer exists beneath the law-of-the-wall region, and usually has a thickness of $y^+ < 30$. The outer layer or the law-of-the-wake region is highly influenced by the external flow conditions, thus there is not possible to reach a single correlated presentation.

The most recent study of the zero-pressure-gradient flat-plate boundary layer turbulent flows was (ZPGFPBL) conducted by DNS (direct numerical simulation) and presented by the calculated velocity gradients of the kinematic field [25]. Two kinds of vortical structures are observed and known as the hairpin and the vortex clusters. The hairpin formation is the most well-known turbulent coherent structure; these vortex filaments are attached to the solid wall, stretched toward the outer edge of the shear layer, and bent back. The DNS depicts the iso-surface values of the 2nd invariant of velocity gradient tensor Q supported by a 2.098 million grid-point system. On the other hand, the clusters also appear as an intermittent bursting to become an important turbulent energy production from the near-wall sheer. In the wall-bounded domain, the dominant structure carrying the turbulent energy is the streamwise vortex streaks, and the streaks organize the energy dissipation and maintain the energy

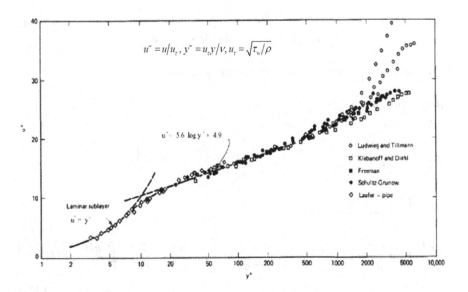

FIGURE 4.8 The phonological structure of turbulent boundary over the plate.

transfer [26]. The effect of streaks diminishes away from the wall. However, it must be remindful that all simulated observations are clearly displayed and subjected to interpretations.

4.7 COMPRESSIBLE TURBULENT BOUNDARY LAYER

The compressible turbulent boundary layer equations are achieved by an ensemble average by decoupling dependent variables into organized motion and turbulent fluctuations. The averaging process eliminates some key characteristics of turbulence such as the frequency and phase relationship of the highly fluctuating motion, which may not be very important for most practical applications. The mass average, due to Favre is based on the observation that the average mass of fluid contained in control volume with mean mass-averaged velocity is constant. This feature is in perfect accord with the known fact that velocity variation in turbulence has a kinematic or volumetric, rather than the dynamic effect on the velocity in a compressible turbulent flow field [27].

The well-known Reynolds average and the Favre or mass-average are defined as

$$\bar{f}(r) = \frac{1}{t}\int f(t,r)dt; \quad \tilde{f} = \frac{\int \rho(t,r)f(t,r)dt}{\int \rho(t,r)dt} \tag{4.16}$$

For simplicity, apply Favre's average to the two-dimensional/axisymmetric Compressible Naiver-Stokes equations like for the laminar boundary-layer flow, and follow Prandtl's boundary-layer approximation to assume

$$u = O(1), v = O(\delta), H = (1)$$
$$\partial/\partial t = O(1), \partial/\partial x = O(1), \partial/\partial y = O(\delta^{-1}) \tag{4.17}$$

The compressible turbulent boundary equations become

$$\frac{\partial r^j \bar{\rho}}{\partial t} + \frac{\partial r^j \overline{\rho u}}{\partial x} + \frac{\partial r^j \overline{\rho v}}{\partial y} = 0$$

$$\bar{\rho}\frac{\partial \overline{u}}{\partial t} + \overline{\rho u}\frac{\partial \overline{u}}{\partial x} + \overline{\rho v}\frac{\partial \overline{u}}{\partial y} = -\frac{\partial \bar{p}}{\partial x} + \frac{1}{r^j}\frac{\partial}{\partial y}\left[r^j\left(\bar{\mu}\frac{\partial \overline{u}}{\partial y} - \bar{\rho}\langle u'v'\rangle\right)\right]$$

$$\frac{\partial \bar{p}}{\partial y} + \frac{\partial(\bar{\rho} < v'v'>)}{\partial y} = 0$$

$$\bar{\rho}\frac{\partial \overline{H}}{\partial t} + \overline{\rho u}\frac{\partial \overline{H}}{\partial x} + \overline{\rho v}\frac{\partial \overline{H}}{\partial y} = -\frac{\partial \bar{p}}{\partial t} + \frac{1}{r^j}\frac{\partial}{\partial y}\left\{r^j\left[\overline{\mu u}\frac{\partial \overline{u}}{\partial y}\left(1 - \frac{1}{\mathrm{Pr}}\right) + \frac{\bar{\mu}}{\mathrm{Pr}}\frac{\partial \overline{H}}{\partial y} - \bar{\rho}\langle H'v'\rangle\right]\right\}$$

$$\tag{4.18a}$$

In deriving Equation (4.18), a few remarks shall be pointed out that within the boundary-layer approximation, the following identities hold true between the Reynolds and the mass-average variables. The cross-correlation terms will be consistently expressed as $\langle f'g' \rangle$.

$$\overline{\rho v} = \overline{\rho}\,\overline{v} + \langle \rho''v'' \rangle = \overline{\rho}\tilde{v}$$

$$\tilde{H} = \overline{H} \qquad\qquad (4.18b)$$

$$\overline{\rho}\langle H'v' \rangle = \overline{\rho}\langle h'v' \rangle = \overline{\rho u}\langle u'v' \rangle$$

The turbulent boundary layer equations now have a non-closure issue, namely the number of equations is less than the unknown dependent variables. The additional Reynolds stress and turbulent heat flux make the fluid medium no longer a Newtonian fluid. The closure problem has been remedied by different turbulent models with vastly different levels of analytic approximations from the zeroth order to a range of second-order formulations.

The first attempt was made by Boussinesq in 1877 by introducing the concept of eddy viscosity via a mixing length; $-\rho\langle u'v' \rangle = \rho \varepsilon_m(\partial u / \partial y)$. In practical application, the closure by the algebraic equation consists of two parts duplicating the viscous sublayer and the velocity defect outer region from the knowledge of the law-of-the-wall and the law-of-the-wake. The outer region of the simplest eddy viscosity model is given as $\varepsilon_m = k_1 y\left[1 - \exp(y / A)\right](\partial u / \partial y)$, and for the outer region, $\varepsilon_m = k_2\left[1 + 5.5(y / \delta)^6\right]^{-1} u_e \delta *$ [29]. The simple eddy viscosity model is one of the very few with modifications including density fluctuation for hypersonic turbulent boundary layer with heat transfer. The comparative studies with experimental data on skin friction coefficient and velocity profiles are presented in Figures 4.9 and 4.10. The laminar-turbulent transition for engineering application is purely an empirical formula [30].

Figure 4.9 displayed the comparison between experimental data of Matting, Chapman, and Nyholm at the freestream Mach number of 2.95 and 4.2. The maximum discrepancy between data and calculations was located immediately downstream to the transition region and the calculated skin-frication coefficient predicts the data by 6%.

The calculated velocity and Mach number profiles with three different eddy viscosity models are compared with data of Cole at the Mach number of 4.55 and are depicted in Figure 4.10. The transverse coordinate is normalized by the boundary-layer momentum thickness. The observable deviation from the experimental data is incurred by the eddy viscosity model without the intermittency correction. The intermittency correction how is not limited to the outer portion of the boundary layer, which might be attributable to the flow history aspect of the solving scheme. The calculated results probably represent the best results by the eddy viscosity model could produce.

The higher-order of turbulence modeling requires more detailed knowledge of the mechanisms of turbulent dynamics. It is particularly critical to understand how the highly ransom and chaotic fluctuations can be sustained, and the distribution of

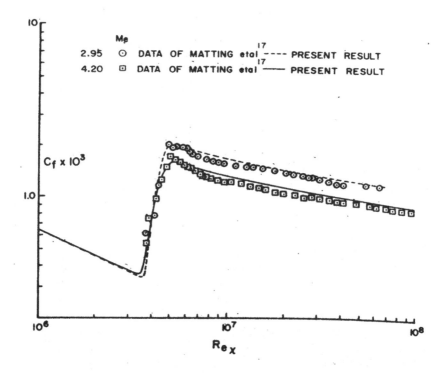

FIGURE 4.9 Comparison of skin-frication distribution with data over hypersonic flows.

energy among interacting vorticities of different frequencies and the phase relationship. Finally, the process describes how the kinetic energy is dissipated by the viscosity of the fluid medium. All these needed knowledges unfortunately must be built on the correlations from the fluctuating components of the fluid dynamic variables. Nevertheless, the transport equation of the Reynolds stress has been derived to show the four basic mechanisms including a triple correction term as [29]:

$$
\frac{D<u_i u_j>}{Dt} = -\left[\left\langle u_i u_j\right\rangle \frac{\partial u_j}{\partial x_k} + <u_j u_k> \frac{\partial u_i}{\partial x_k}\right] - 2v\left\langle \frac{\partial u_i}{\partial x_k}\frac{\partial u_j}{\partial x_k}\right\rangle +
$$
$$
\left\langle \frac{p}{\rho}\left(\frac{\partial u_i}{\partial x_j}+\frac{\partial u_j}{\partial x_i}\right)\right\rangle - \frac{\partial}{\partial x_k}\left[\left\langle u_i u_j u_k\right\rangle + \delta_{ik}\frac{\left\langle u_i p\right\rangle}{\rho} + \delta_{jk}\frac{\left\langle u_j p\right\rangle}{\rho} - \frac{\partial\left\langle u_i u_j\right\rangle}{\partial x_k}\right] \qquad (4.19a)
$$

where the symbol δ_{ij} denotes the Kronecker delta.

In Equation (4.19a), the first group of terms describes the production of the turbulent shear stress, which transfers kinetic energy from the main flow to fluctuation small eddies. The second term is the dissipation rate, and the third term represents the pressure-strain correlation that redistributes energy among all the Reynolds stress components. The last group of terms designates the diffusion transport of turbulent stress. The triple correlation is a form of diffusion of transport turbulent kinetic energy by fluctuating motion but largely remains unknown.

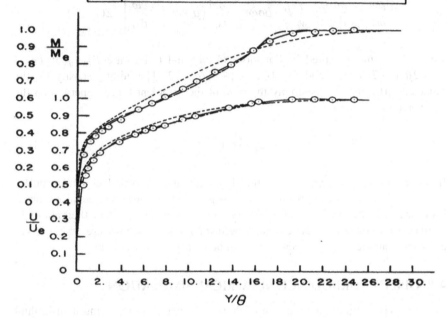

FIGURE 4.10 Comparison of velocity and Mach number profile with data ($M_e = 4.554$).

The two-equation turbulence model is developing on the basic assumption that the Reynolds shear stress is directly proportional to the kinetic energy $k = \left(\overline{u'^2} + \overline{v'^2} + \overline{w'^2}\right)/2$ and inversely proportional to the dissipation rate $\varepsilon = v\langle \partial u_j / \partial x_j \rangle^2$. All the two-equation turbulence models have identical underlying assumption, namely, the turbulent viscosity is isotropic. Under this condition the orientation of the Reynolds stress and the deformation tensor has coincided as $\tau_{ij} = 2\mu_t S_{ij} - 2\rho k \delta_{ij}/3$. The two equation turbulent models often are designated as the $k - \varepsilon$ [31] or $k - \omega$ model [32], but the difference is only focused on how the dissipation term is adopted.

Genetically, the two-equation k-epsilon models can be given as [31]

$$\frac{\partial \rho k}{\partial t} + \frac{\partial \rho u_i k}{\partial x_i} = \frac{\partial}{\partial x_j}\left[\frac{\mu_t}{\sigma_k}\frac{\partial k}{\partial x_j}\right] + 2\mu_t S_{ij}S_{ij} - \rho\varepsilon = 0$$

$$\frac{\partial \rho \varepsilon}{\partial t} + \frac{\partial \rho \varepsilon k}{\partial x_i} = \frac{\partial}{\partial x_j}\left[\frac{\mu_t}{\sigma_\varepsilon}\frac{\partial \varepsilon}{\partial x_j}\right] + 2c_1\left(\frac{\varepsilon}{k}\right)\mu_t S_{ij}S_{ij} - c_2\rho\frac{\varepsilon^2}{k}$$

(4.19b)

The dependent variables of the k-omega model equations like that of the k-epsilon model consist of the turbulent kinetic energy but a specific turbulent dissipation rate relating to the deformation tensor of the fluid motion [33].

$$\frac{\partial \rho k}{\partial t} + \frac{\partial \rho u_i k}{\partial x_i} = P - \beta \rho \omega k + \frac{\partial}{\partial x_i}\left[(\mu + \sigma_k \mu_t)\frac{\partial k}{\partial x_i}\right]$$

$$\frac{\partial \rho \omega}{\partial t} + \frac{\partial \rho \omega_i k}{\partial x_i} = \frac{\gamma}{v_t} P - \beta \rho \omega^2 + \frac{\partial}{\partial x_i}\left[(\mu + \sigma_\omega \mu_t)\frac{\partial \omega}{\partial x_i}\right] + 2(1 - F_1)$$

(4.19c)

where the terms appeared in Equations (4.19b) and (4.19c) are $P = \tau_{ij}\left(\partial u_i / \partial x_j\right)$ $\tau_{ij} = 2\mu_t S_{ij} - 2\rho k \delta_{ij} / 3$ and $S_{ij} = \left(\partial u_i / \partial x_j + \partial_j / \partial x_i\right)/2$. The eddy viscosity for the modeled turbulent flow is simply the ratio of the turbulent kinetic energy and the selected dissipation rate.

$$\mu_t = \frac{k}{\omega} \text{ or } \frac{k}{\varepsilon}$$

(4.19d)

In order to develop these models, tremendous amounts of knowledge of the turbulent structure are mandatory plus impressive insights and ingenuities to achieve a physics-based description of turbulence. These models have demonstrated their ability in engineering applications for internal flow involving adverse pressure gradients but are seldom applied to high-speed turbulent boundary layer simulation.

4.8 DIRECT TURBULENCE NUMERICAL SIMULATION

The most challenging phenomenon of fluid dynamics and the pacing item in fluid dynamics is the turbulence; the genetic physics of the turbulent flow is strictly a chain of events by the unsteady, three-dimensional, nonlinear vortical interactions with a wide range of length scales and frequency spectra. From physical observations, turbulence is generated mostly by the random chaos interactions of vortices with a wide range of length, time, and velocity scales. The highly fluctuating and random motion generates a unique turbulent kinetic energy which is measured by the root-mean-square (rms) value of oscillating velocity components, and the energy will be dissipated into the thermal internal energy by the energy cascade process. The rate of compressible turbulent dissipation is controlled by the turbulent transport and production and is often described as a temporal assemble of turbulent shear stress and rate of fluctuating velocity gradients as $\varepsilon = \left\langle \tau_{ij} \partial u_i' / \partial x_j \right\rangle / \bar{\rho}$.

It is beyond any doubt that the turbulence, laminar-turbulent transition, and all other fluid bifurcation are governed by the time-dependent, three-dimensional Navier-Stokes equations. Based on our understanding of turbulence, the phenomenon is the consequence of chaotic and vigorous vorticity interactions through the gas particles' collisional mechanism. The Navier-Stokes equations are derived from the gas kinetic theory via the binary inelastic gas particle collisional model which accurately describes all phenomena generated by particles' mutual interactions. The direct simulation of turbulence by solving the Navier-Stokes equation (DNS) for basic research was initiated in the later part of the 1990s. The first complex vorticity contours by the DNS of a supersonic ($M = 1.9$) circular jet overlaid on flow dilatation are accomplished [34]. At the atmosphere condition, the Kolmogorov scale has a dimension of around ten millimeters or shorter. For an aircraft simulation in flight by solving directly the Navier-Stokes equations must have a mesh system consisting of more than quintillions of grid points (10^{18}) and a data processing rate of hundreds of Petaflops (10^{15} floating points operations per second) to be satisfactory and practicable (Figure 4.11).

High-performance concurrent computing with multi-computing processors and share/distribution memory have attained remarkable computing efficiency. Recently the Sunway Taihulight supercomputer clocked a data processing rate of 93.01 PFLOPS, running the LINPACK benchmark, and with a mindboggling memory capacity of 1.32 PiB (2^{50} bytes). Within two years, the fastest benchmark data processing rate was claimed by the Summit supercomputer at 122.3 PFLOPS. It is anticipated that the data processing rate of the supercomputer will attain a value over EFLOPS (10^3 PFLOPS) within the next few years [35]. Even to date, the required computational resource by DNS of turbulence for engineering applications is often beyond our reach, so a practical alternative becomes necessary. According to our physical understanding of turbulence, the random fluid properties in vortical structures have a particular form and each has a specific continuous energy spectrum that vanishes at high wave number by dissipation, and the larger eddy extracts energy

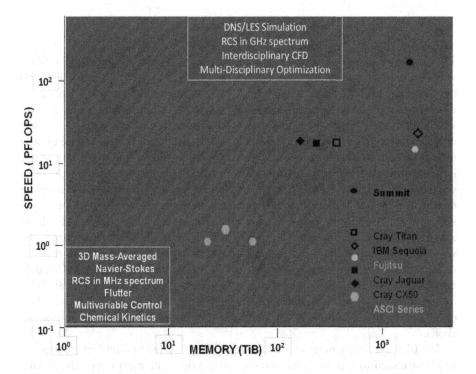

FIGURE 4.11 Performance and requirement for direct numerical simulation.

from the main flow and its structures are highly anisotropic. It is clearly demonstrated that near the wall or in the inner region of the shear layers, the energy spectra of the small energy-dissipating eddies are clearly universal – it is independent of both Reynolds number and type of flows. This is the guiding principle for subgrid-scale modeling for large eddy simulation (LES) [36].

The compressible Navier-Stokes equations using LES formulation are decomposed by the different temporal and spatial scales for the independent variables. The large-scale turbulence is solved by the time-averaged Navier-Stokes equations over the large-scale motion, and the small-scale turbulence is approximated by some dynamic subgrid models. The temporal scale separation for LES formulation is achieved by the mass- or Favre-average for the compressible fluid medium. The small spatial scale fluctuations are eliminated by a low-pass filter $G(x - x')$;

$$\bar{f} = \int G(x - x') f'(x') dx' \qquad (4.20a)$$

After a rather arbitrary separation of the large- and small-scale eddies, the governing equation of LES acquires the following form [36];

$$\partial \bar{\rho} / \partial t + \nabla \cdot \left(\overline{\rho u_i} \right) = 0$$

$$\partial \overline{\rho u_i} / \partial t + \nabla \cdot \left(\overline{\rho u_i u_j} \right) = (1 / \mathrm{Re}) \partial \bar{\tau}_{i,j} / \partial x_i - \partial \tau_{i,j} / \partial x_j \qquad (4.20b)$$

$$\partial \overline{\rho e} / \partial t + \nabla \cdot \left(\overline{\rho u_i e} + \overline{u_i \cdot \tau_{i,j}} - \bar{q}_i + q_i \right] = 0$$

The closure of the above equation systems is provided by the subgrid-scale (SGS) stress tensor which is highly anisotropic; can be given as $\tau_{i,j} = \bar{\rho} \left(\overline{u_i u_j} - \bar{u}_i \bar{u}_j \right)$, and the turbulent heat flux as $q_i = \overline{e u_i} - \overline{e u_i}$ [37]. A range of SGS models have been developed and loosely grouped as the eddy viscosity, dynamic eddy viscosity, and similarity models. The SGS models have also been generally categorized as deterministic and stochastic models. The fundamental difference between these models resides in the fact that the stochastic formulation considers the interaction of the large- and small-scale eddy interactions but not by the deterministic model. The stochastic model also usually consists of two components known as the coherent and incoherent subgrid-scale stress, but the latter contributes little to the large-scale dynamics and the energy transfer [38]. The stochastic model has also been applied for the DNS simulation; the model includes the interaction of the large- and small-scale eddies motion. For this reason, the stochastic model is more realistic in mimicking the filtered direct numerical simulation. Some of the SGS models also use a linear stochastic estimate approach to obtain the local wall-shear stress for the outer flow. For a turbulent boundary layer over a rough wall, the interaction of the inner and outer dynamic structure may be significant to become a model of choice.

The LES approximations have been successfully applied to turbulent combustion and contributed to a better understanding of physics. The most impressive results by LES when compared with experimental observations have also exhibited a close correspondence in the distributions of the standard deviation, skewness, kurtosis, and intermittency of turbulent dynamics.

4.9 VORTICITY AND VELOCITY FORMULATIONS

Since the turbulence is accentuated by vortical dynamics, therefore the dependent variables of Navier-Stokes equations by vorticity and velocity formulation shall be the direct and more advantageous than the traditional conservative variables for studying turbulence. The vorticity-velocity formulation of the three-dimensional, compressible Navier-Stokes equations is derived from the conservation laws of continuity, momentum equations, and energy [35]. The classic definition of vorticity is given as a cross-product of the velocity vector, which describes the infinitesimal rotation of a vector field in three-dimensional Euclidean space. Often the vorticity is also referred to as the circulation density of the flow field.

$$\Omega = \nabla \times u \qquad (4.21a)$$

From the vector identical $\nabla \cdot (\nabla \times u) \equiv 0$, it is immediately recognized that the vortical field is divergence-less of the divergence-free vector; $\nabla \cdot \Omega = 0$. This property becomes highly effective and useful in describing its far-field boundary condition for the vortical field. After some algebraic manipulation and invoking vector identities, the system of vortical dynamic equations for a compressible medium acquires the following form [35]:

$$\partial \Omega / \partial t + \Omega(\nabla \cdot u) + (u \cdot \nabla)\Omega - (\Omega \cdot \nabla)u - \nabla[p - \lambda(\nabla \cdot u)] \times \nabla(1/\rho)$$
$$-\nabla \times [(\nabla v / \rho) \cdot def(u)] - 2\nabla v \times [\nabla(\nabla \cdot u)] + \nabla v \times (\nabla \times \Omega) - v\nabla^2\Omega = 0 \qquad (4.21b)$$

$$\rho \partial(c_v T + u \cdot u / 2) / \partial t + \rho u \cdot \nabla(c_v T + u \cdot u / 2) + \nabla \cdot q + \nabla p \cdot u - \Phi = 0 \qquad (4.21c)$$

where the viscosity dissipation function of the internal energy conservation equation, is well-known as $\Phi = \nabla \cdot (\tau \cdot u) - (\nabla \cdot \tau) \cdot u$.

A few interesting physics of vortical dynamics can be deduced from the vorticity dynamic equations: First, the fourth term of the vorticity dynamics equation (4.21b); the term $(\Omega \cdot \nabla)u$ has no counterpart in the conservation of momentum equation, and it gives the vorticity dynamics a distinguished characteristic that only exists in the three-dimensional flow field. In turbulent research, the $(\Omega \cdot \nabla)u$ commonly referred to as vorticity stretching becomes a critical property of turbulence. As a consequence, the vorticity stretching is identically zero in a two-dimensional field, $(\Omega \cdot \nabla)u \equiv 0$, which directly and indisputably reveals that the turbulence is a genuine three-dimensional physical phenomenon. Second, it has been mentioned before that the divergence of vorticity vanishes by virtue of the vector identity; $\nabla \cdot (\nabla \times u) \equiv 0; \nabla \cdot \Omega = 0$. Third, only the shearing stress component contributes directly by its interactions with the density and viscosity gradients to the vorticity rate of change. The normal stress component, including static pressure, contributes only to the vorticity rate of change through the interaction with the density and the bulk viscosity gradients. For an incompressible medium, the effect of the normal stress component vanishes.

The governing equation system of the compressible medium is rather complex and involves simultaneously the vorticity and velocity components, unlike the Navier-Stokes equations in conservative variables. However, the pressure in the vorticity-velocity formulation is needed to be determined from the momentum conservation law or some other procedure such as the Poisson equation for pressure. For this reason, the direct turbulence simulation via vorticity dynamics equations becomes a genuine path-finding research challenge, but a successful endeavor also offers the incomparable reward for understanding a previously unsolvable fundamental fluid dynamics phenomenon.

It may be noticed that the vorticity dynamic equation and the energy conservation equation is coupled only through the velocity component. In order to link explicitly the two conservation laws with vorticity and energy to gain additional physical insights, the energy equation can be split to get the kinetic energy transport equation as

$$\rho\, \partial(u \cdot u\,/\,2)\,/\,\partial t + \rho u \cdot \nabla(u \cdot u\,/\,2) + \nabla p \cdot u - (\nabla \cdot \tau) \cdot u = 0 \qquad (4.22a)$$

Substitute Equation (4.22a) into the energy conservation law, Equation (4.21c), the conservation energy law in thermodynamic variables yields.

$$T\nabla S + u \times \Omega = \partial u\,/\,\partial t + \nabla h_o + \left[\nabla(\lambda \nabla \cdot u) + \nabla \mu \left(\nabla + \nabla u^T \right) \right] \Big/ \rho + 2\eta\nabla(\nabla \cdot u) - \eta\nabla \times \Omega$$

$$(4.22b)$$

Equation (4.22b) is the generalized Crocco's relationship for compressible flow which provides the additional physical relationship between the vorticity and the thermodynamic properties of the fluid medium. The Navier-Stokes equation in vorticity and velocity formulation is formerly closed by including the equation of state for gas and the constitute relationships of transport property. The governing equations in the vorticity-velocity formulation are completely consistent with the gas kinetic theory and remain to be a nonlinear, second-order partial differential system of vorticity.

For the vorticity dynamics formulation, once the vorticity is known the formal integral relationship between vorticity and velocity gives [39].

$$u(r,t) = -1\,/\,4\pi \left\{ \iiint \left[\Omega \times (r - r_o)\Big/|r_o - r|^3 \right] dV + \iint \left[(u_o \cdot n)(r_o - r)\Big/|r_o - r|^3 \right] ds \right.$$

$$\left. - \iint \left[(u_o \times n) \times (r_o - r)\Big/|r_o - r|^3 \right] ds \right\}$$

$$(4.22c)$$

where the subscript symbol o designates the value at the center of the spherical control volume. The velocity vector can also be determined from the definition of the vorticity via an iterative numerical procedure.

The vorticity dynamic equations for compressible flow appear to be formidable and led to an extremely tedious process for resolving the hydrodynamic pressure term. In the past, the solving procedure has adopted the Poisson projection method to determine the pressure. However, the problem can be substantially simplified to understand the compressibility effect on turbulence by imposing the zero-pressure-gradient-wall-bounded-turbulent (ZPGWBT) condition. Based on the boundary-layer theory, the transversal pressure gradient across the shear layer becomes vanishingly small to the order-of-magnitude error of $O\left(1/\sqrt{\text{Re}_x}\right)$. Thus, the pressure is essentially a constant value over the entire flow field. Equally importantly, the density and temperature are now directly related by the reciprocal relationship; $\rho = p/RT = C/T$. In addition, density and temperature are the only dependent variables with respect to the coordinate perpendicular to the turbulent shear layer. The governing equation becomes.

$$\partial\Omega/\partial t + \Omega(\nabla \cdot u) + (u \cdot \nabla)\Omega - (\Omega \cdot \nabla)u - (R/p)\nabla[\lambda(\nabla \cdot u)] \times \nabla T$$
$$-(R/p)\nabla \times [(\nabla\mu)T \cdot def(u)] - 2\nabla v \times [\nabla(\nabla \cdot u)] + \nabla v \times (\nabla \times \Omega) - v\nabla^2\Omega = 0 \tag{4.23a}$$

$$\partial(c_vT + u \cdot u/2)/\partial t + u \cdot \nabla(c_vT + u \cdot u/2) + (RT/p)(\nabla \cdot q - \Phi) = 0 \tag{4.23b}$$

where the symbol R is the universal gas constant for air; $R = 287 \text{ m}^2/\text{s}^2\text{k}$.

The equations system, Equations (4.23a) and (4.23b), describes the vorticity dynamics in DNS formulation for solving the simplified but still physically meaningful turbulent structure in a compressible medium. Under this circumstance, the governing equations and the solving scheme are significantly simplified. The pressure becomes an input condition instead must be solved by the method of the Poisson projection. In addition, the viscosity coefficients can be approximated as an exclusive function of temperature. The governing system consists of second-order, inhomogeneous, nonlinear partial differential equations. The process still must be conducted in an iterative process to satisfy the boundary condition.

However, numerical direct simulation for incompressible Navier-Stokes equations in vorticity and velocity formulation has been successfully simulated [40]. The flow of an equilibrium zero-pressure-gradient laminar-turbulent transition boundary layer over a flat plate is simulated by imposing a time-harmonic Gaussian excitation over a circular patch and the entire span of the computational domain. The numerical results were obtained by solving the governing equations on staggered grids for vorticity and velocity. The numerical simulation captures the three-dimensional wall-excited multiply turbulent or the Emmons spots grow and multiply in numbers toward downstream. The regeneration mechanism is the hallmark of the spatial-temporal wavefront that is different from any other transient wave. The disturbances are seen to amplify and stretch the vorticity as they propagate downstream, and eventually undergo a transition into full turbulent flow.

Take a step even further by eliminating the tremendous complexity from the compressibility of the fluid medium for analyzing the incompressible turbulent flows. For an incompressible fluid medium, all the terms involving the density gradient and the gradient of the molecular viscosity drop out. At a low speed, the conservation law for internal energy can also be eliminated by virtue of the fact the heat transfer is negligible; the internal energy becomes essentially invariant. For incompressible flow, the divergence of the velocity component also vanishes; $\nabla \cdot u = 0$ the vorticity dynamics equation degenerates into a much simpler form.

$$\partial \Omega / \partial t + (u \cdot \nabla)\Omega - (\Omega \cdot \nabla)u - \eta \nabla^2 \Omega = 0 \tag{4.24a}$$

The classic vorticity equation is well known and has been derived nearly fifty years ago [35,39]. The second-order partial differential equation is homogenous in vorticity. Therefore, if the initial value of vorticity of flow field is null then the rate of change of vorticity must vanish. It leads to the classic theorem that vorticity cannot be created nor destroyed in the interior of a homogeneous fluid and can only be produced only at the media interface boundaries.

For the incompressible flow, the generalized Crocco's relationship, Equation (4.22b) reduces to its simpler and classic form.

$$T\nabla S + u \times \Omega + \eta \nabla \times \Omega = \partial u / \partial t + \nabla h_o \tag{4.24b}$$

From Classic Crocco's relationship, it is observed that only under the condition of constant total enthalpy and the steady state, there is a unique relationship between entropy and vorticity.

For the incompressible flow, a simpler reciprocal relation between the vorticity and velocity can be developed from the definition of the vorticity and the continuity law to give a group of Poisson equations for each velocity component on the Cartesian frame. Therefore, once the vorticity distribution is known, the velocity can be solved by the Poisson equation to complete the coupling between vorticity and velocity.

$$\partial^2 u/\partial x^2 + \partial^2 u/\partial y^2 + \partial^2 u/\partial z^2 = \partial \Omega_y/\partial z - \partial \Omega_z/\partial y$$
$$\partial^2 v/\partial x^2 + \partial^2 v/\partial y^2 + \partial^2 v/\partial z^2 = \partial \Omega_z/\partial x - \partial \Omega_x/\partial z \tag{4.24c}$$
$$\partial^2 w/\partial x^2 + \partial^2 w/\partial y^2 + \partial^2 w/\partial z^2 = \partial \Omega_x/\partial y - \partial \Omega_y/\partial x$$

Another numerical simulation for laminar-turbulent transition also sheds some light on a better understanding of the complex fluid dynamic physics. A hybrid vortex filament scheme simulates the boundary layer flow by combining a finite-volume solution to the full viscous vorticity equation on a thin prismatic mesh adjacent to the wall [41]. The numerical result finds the vortex filaments agglomerate to form a coherent structure and the vortex furrows are the dominant structure that is subject to Klebanoff-type instability [42]. The well-known hairpin vortices in turbulent boundary layers are also found to have no standing as structures. The possibility of elucidating the details of instability in transition and the formation of the intermitting Emmon's spots can be achieved by direct numerical simulations by the Navier-Stokes equations either in conservative form or the vortex dynamic formulation is very promising [35].

Direct numerical simulation results using the vorticity-velocity formulation for incompressible transitional and turbulent flows are depicted in Figure 4.12 [40].

In Figure 4.12, the disturbed streamwise velocity contours downstream of a circular bump at three different instances after the perturbation are projected on the x-z plane. The flow of an equilibrium zero-pressure-gradient laminar-turbulent transition boundary layer over a flat plate is simulated by imposing a time-harmonic Gaussian excitation over a circular patch and the entire span of the computational domain. The numerical results were obtained by solving the governing equations on staggered grids for vorticity and velocity. The numerical simulation captures the three-dimensional wall-excited multiply turbulent spots that grow and multiply in numbers downstream. The regeneration mechanism is the hallmark of the spatial-temporal wavefront that is different from any other transient wave. The disturbances are seen to amplify and stretch the vorticity as they propagate downstream, and eventually undergo a transition into full turbulent flow.

There is no rigorous mathematical theorem concerning the proper initial value and boundary conditions to ensure the existence and uniqueness of the solution to Navier-Stokes equations written in vorticity dynamic formulation. However, our experience in solving Navier-Stokes equations [28] has shown that the imposed conditions must take into account the physical meaningfulness of the problem and the

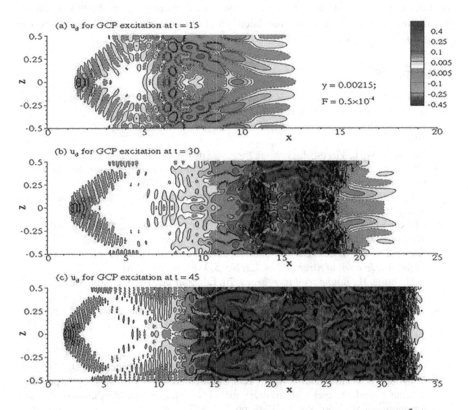

FIGURE 4.12 Laminar-turbulent transition by a bump at a frequency of 5×10^{-5}.

mathematical nature of the system of equations. This physical requirement of a mathematical formulation is referred to as the well-posed problem. A well-posed problem can be achieved only if the boundary conditions are properly specified; the solutions to the differential equation vary continuously with respect to perturbations to the initial value and boundary conditions. Successful numerical simulations must possess this property, namely, that slight perturbation of the initial and boundary conditions yield a correspondingly small change to the numerical result.

In essence for simulating turbulence by solving the vorticity dynamics equations, the initial value and the inflow conditions require a complete specification of the velocity, vorticity, and thermodynamic properties for the simulation. These initial conditions are exclusively determined by whether the simulation includes the laminar-turbulent transition or not. For the initial laminar inflow conditions, the specific value can be found from the classic boundary-layer theory. Otherwise, it may be derived from Taylor's hypothesis for turbulent flow, in that the convective velocity is assumed to be independent of both frequency and wave number but equal to the local mean flow [43]. In fact, the hypothesis asserts that the convective velocity fluctuation and the convective fluid structure are statistically independent. The approximation has been found to be applicable to the divergence-free domain but failed in the high shear region. Therefore, the placement of the inflow condition also needs to be carefully considered. In particular, the turbulence is assumed to be homogeneous in the streamwise direction, which has been adopted by nearly all DNS applied to the Navier-Stokes equations.

REFERENCES

1. Shang, J.S., Computational fluid dynamics application to aerospace science, *The Aeronautical Journal*, Vol. 113, No. 1148, 2009, pp. 619–632.
2. Hayes, W.D. and Probstein, R.F., *Hypersonic flow theory*, Academic Press, New York, 1959.
3. Dorrance, W.H., *Viscous hypersonic flow*, McGraw-Hill, New York, 1962.
4. Chapman, S. and Cowling, T.G., *The mathematical theory of non-uniform gases*, Cambridge University Press, Cambridge, 1964, pp. 134–150.
5. Hirschfelder, J.O., Curtiss, C.F., and Bird, R., *Molecular theory of gases and liquids*, Wiley & Sons, New York, 1959.
6. Clarke, J.F. and McChesney, M., *The dynamics of real gases*, Butterworths & Co, Washington, DC, 1964.
7. Shang, J.S., An assessment of numerical solutions of the compressible Navier-Stokes equations, *Journal of Aircraft*, Vol. 22, No. 5, 1985, pp. 353–370.
8. Schlichting, H., *Boundary-layer theory*, 7th Ed., translated by J. Kestin, McGraw-Hill, New York, 1979.
9. Levy, S., Effect of large temperature changes upon laminar boundary layers with variable free-stream velocity, *Journal of the Aeronautical Sciences*, Vol. 21, 1954, pp. 459–474.
10. Lees, L. and Probstein, R.F., Hypersonic viscous flow over a flat plate, Rept. No. 195, Department of Aerospace Engineering, Princeton University, Princeton, NJ, 1953.
11. Chapman, D.R. and Rubesin, M.W., Temperature and velocity profiles in compressible laminar boundary layer with arbitrary different surface temperatures, *Journal of the Aeronautical Sciences*, Vol. 16, 1949, pp. 547–565.

12. Cohen, C.B. and Reshotko, E., Similar solutions for compressible laminar boundary layer with heat transfer and pressure gradient, NACA Tech. Note 3325, Washington, DC, 1955.

13. Van Driest, E.R., Investigation of the laminar boundary layer in compressible fluid using the Crocco method, NACA Tech. No. 2597, 1952.

14. Fay, J.A. and Riddell, F.R., Theory of stagnation point heat transfer in dissociated air, *Journal of the Aeronautical Sciences*, Vol. 25, 1958, pp. 73–85 & 121.

15. Kemp, N.H., Rose, R.H., and Detra, R.W., Laminar heat transfer around blunt bodies in dissociated air, *Journal of Aerospace Sciences*, Vol. 26, No.7, 1959, pp.421–430.

16. Reshotko, E., Boundary-layer stability and transition, *Annual Reviews of Fluid Mechanics*, Annual Reviews, Inc., Palo Alto, pp. Vol. 8, 1976, 311–349.

17. Mack, L.M., Linear stability theory and the problem of supersonic boundary-layer transition, *AIAA Journal*, Vol. 13, 1975, pp. 278–289.

18. Morkovin, M.V., On the many faces of transition. In: Sinclair Wells, C. (Ed.), *Viscous drag reduction*, Plenum, New York, 1968.

19. Saric, W.S., Reed, H.C., and Kerschen, E.J., Boundary-layer receptivity to freestream disturbance, *Annual Review of Fluid Mechanics*, Vol. 34, 2002, pp. 291–319.

20. Malik, M.R., A black box compressible stability analysis code for transition prediction in three-dimensional boundary layer, 1982, NASA CR 165925.

21. Bradshaw, P., Turbulence: The chief outstanding difficulty of our subject, *Experiment in Fluid*, Vol. 16, 1994, pp. 203–216.

22. Kraichnan, R.H., The structure of isotropic turbulence at very high Reynolds number, *Journal of Fluid Mechanics*, Vol. 5, 1959, p. 497.

23. Chapman, D.R., Computational aerodynamics development and outlook, *AIAA Journal*, Vol. 17, 1979, pp. 1293–1313.

24. Coles, D., The law of the wake of the turbulent boundary layer, *Journal of Fluid Mechanics*, Vol. 1, 1956, pp. 191–226.

25. Wu, X. and Moin, P., Direct numerical simulation of turbulence in a nominally zero-pressure-gradient flat plate boundary layer, *Journal of Fluid Mechanics*, Vol. 630, 2009, pp. 5–41.

26. Jimenez, J., Near-wall turbulence, *Physics of Fluid*, Vol. 65, 2013, pp. 1–28.

27. Favre, A., A equations des gaz turbulents compressible, *Journal of de Mecanique*, Vol. 4, No.3, 1965, pp. 361–390.

28. Shang, J.S., An assessment of numerical solutions of the compressible Navier-Stokes equations, *Journal of Aircraft*, Vol. 22, No. 5, 1985, pp. 353–370.

29. Cebeci, T. and Smith, A.M.O., *Analysis of turbulent boundary layers*, Academic Press, New York, 1974.

30. Shang, J.S., Hankey, W.L., and Dwoyer, D.L., Numerical analysis of eddy viscosity models in supersonic turbulent boundary layers, *AIAA Journal*, Vol. 11, 1973, pp. 1677–1683.

31. Hanjalic, K. and Launder, B.E., *A Reynolds stress model of turbulence and its applications to thin shear flows, Journal of Fluid Mechanics*, Vol. 52, No. 4, 1972, pp. 609–638.

32. Wilcox, D.C., Formulation of the k-Omega turbulence model revisited, *AIAA Journal*, Vol. 46, 2008, pp. 2823–2838.

33. Menter, F.R., Two-equation eddy viscosity turbulence model for engineering applications, *AIAA Journal*, Vol. 32, 1994, pp. 1598–1605.

34. Moin, P. and Mahesh, K., Direct numerical simulation; a tool in turbulent research, *Annual Review of Fluid Mechanics*, Vol. 30, 1998, pp. 539–578.

35. Shang, J.S., Landmarks and new frontiers of computational fluid dynamics, *Journal of Advances in Aerodynamics*, Vol. 1, No. 5, 2019, pp. 1–36.

36. Piomelli, U., Large-eddy simulation: achievement and challenges, *Progress in Aerospace Sciences*, Vol. 35, 1999, pp. 335–362.

37. Smagorinsky, J.S., General circulation experiments with the primitive equations, *Monthly Weather Review*, Vol. 29, 1963, pp. 511–546.
38. De Stefano, G., Goldstein, D.E., and Vasilyev, O.V., On the role of subgrid-scale coherent models of large scale eddy simulation, *Journal of Fluid Mechanics*, Vol. 525, 2005, pp. 263–274.
39. Gatski, T.B., Review of incompressible fluid flow computation using the vorticity-velocity formulation, *Applied Numerical Mathematics*, Vol. 7, 1991, pp. 227–237.
40. Bhaunik, S. and Segapta, T.K., A new velocity-vorticity formulation for direct numerical simulation of three-dimensional transitional and turbulent flows, *Journal of Computational Physics*, Vol. 284, 2015, pp. 230–260.
41. Bernard, P.S., Vortex dynamics in transitional and turbulent boundary layer, AIAA Journal, Vol. 51, 2013, pp. 1828–1842.
42. Klebanoff, P.S., Tidstrom, K.D., and Sargent, L.M., The three-dimensional nature of boundary-layer instability, *Journal of Fluid Mechanics*, Vol. 12, 1962, pp. 1–24.
43. Taylor, G.I., The spectrum of turbulence, *Proceedings of the Royal Society of London Series A*, Vol. 164, 1938, p. 478.

5 Viscous-Inviscid Interaction

5.1 COMPUTATIONAL SIMULATIONS

Prior to the 1970s, the viscous-inviscid interactions are investigated by experimental observations and analytic means. Only a limited computational capability has existed and several attempts have been made to investigate the viscous-inviscid interaction by coupling the pressure gradients of the external flow field to the boundary-layer equations. The shortcoming of these modeling and simulation tools becomes glaring [1]. Especially, the separated boundary-layer solution is recovered by a trial-and-error method; the final solution is the numerical result that passed the saddle point at the flow separation point. In this connection, the triple deck theory by Stewartson [2] has demonstrated that the singular point of flow separation of the interacting boundary layer is indeed removable. The scaling law by matching asymptotic expansions of the inner and outer layers has been successfully incorporated into the interacting laminar boundary-layer theory. The analytic approach to viscous-inviscid interaction is mathematically elegant and rigorous but subject to severe limitations for practical application. Finally, it's realized that the capability of solving the compressible Navier-Stokes equations for inviscid-viscous interaction is essential. For computational simulation, the issues of piecewise shock jump discontinuity in the flow field, the numerical algorism in the multi-dimensional temporal-spatial domain, and computational efficiency are paramount.

The shock wave capturing ability of any numerical simulation for viscous-inviscid interaction is essential because shock formation is always present in the flow field. A conceptual innovation for solving the inviscid blunt body problem by a time-marching technique was demonstrated by Moretti and Abett [3]. By solving the time-dependent Euler equations, the governing differential system becomes hyperbolic partial differential equations. The mixed subsonic and supersonic flow domains can be solved simultaneously. In fact, the time-marching technique becomes the standard procedure for computational fluid dynamics (CFD) regardless of whether the solution has a steady state asymptote or not.

The technique of vector flux splitting of the flux vector actually germinates the idea by splitting the Navier-Stokes equations into inviscid and viscous parts solve separately and recombine to produce the complete solution. The formulation has been given by Equation (2.40).

$$\partial U/\partial t + \partial(F_i + F_v)/\partial x + \partial(G_i + G_v)/\partial y + \partial(H_i + H_v)/\partial z = 0 \qquad (5.1a)$$

DOI: 10.1201/9781003212362-6

It is recognized that the hyperbolic partial differential equation possesses all real eigenvalues and the flux vector can be split according to their signs to honor the zone of dependence. The piecewise discontinuity of shock waves is treated as the Riemann problem over the entire computational domain according to their domain of influence. The directional signal propagation is constructed according to the phase velocity from the permissible database. Steger and Warming have shown a systematic relationship between the real eigenvalue and eigenvector for the split flux formulation [4]. They also demonstrate the Euler equations, together with the equation of state for a perfect gas, possessing the homogeneous function of degree one property, namely

$$F_i = (\partial F_i/\partial U)U; \; G_i = (\partial G_i/\partial U)U; \; H_i = (\partial H_i/\partial U)U \qquad (5.1b)$$

where the $\partial F_i/\partial U$, $\partial G_i/\partial U$, and $\partial H_i/\partial U$ are the Jacobian matrices of the flux vector or the coefficient matrices. The governing equations in split flux vector form become

$$\partial U/\partial t + \partial\left[\left(F_i^+ + F_i^-\right)+ F_v\right]/\partial x + \partial\left[\left(G_i^+ + G_i^-\right)+ G_v\right]/\partial y + \partial\left[\left(H_i^+ + H_i^-\right)+ H_v\right]/\partial z = 0$$
$$(5.1c)$$

The split flux vectors containing all inviscid terms are formed according to the signs of the eigenvalue λ of the coefficient matrices, $\partial F_i/\partial U$, $\partial G_i/\partial U$, and $\partial H_i/\partial U$ are as follows:

$$F_i = F_i^+ + F_i^- = \left(S_x\lambda_x^+ S_x^{-1} + S_x\lambda_x^- S_x^{-1}\right)U$$

$$G_i = G_i^+ + G_i^- = \left(S_y\lambda_y^+ S_y^{-1} + S_y\lambda_y^- S_y^{-1}\right)U \qquad (5.1d)$$

$$H_i = H_i^+ + H_i^- = \left(S_z\lambda_z^+ S_z^{-1} + S_z\lambda_z^- S_z^{-1}\right)U$$

In Equations (5.1d), the similarity and its inverse matrices for diagonalization of coefficient matrices are designated as S_x and S_x^{-1}, and the positive and negative eigenvalues are denoted as λ_x^+ and λ_x^-. The solving procedure for the split equation is by applying one-side differencing approximation to achieve the approximate Riemann problem. The basic approach is to remedy the difficulties that the split inviscid flux components are not differentiable at singular sonic points and on the shock surface. The continuous viscous terms are solved simultaneously by spatially central scheme. An incisive summary for using the approximated Riemann approximations can be found from the work of Roe [5].

When the shock wave is treated as the Riemann problem, the Gibbs phenomenon across the discontinuity now can be alleviated by a series of limiters. Figure 5.1 depicts the effectiveness of thinning the thickness of the captured shock wave. When the computing grids are perfectly aligning with the shock wave, the TVD (Total vanish diminishing) limiter can catch the shock wave by three grid points. Nevertheless, the numerical results are still unable to duplicate the physics for the shock wave thickness; because the thickness of the shock wave is of the order of a few mean free paths.

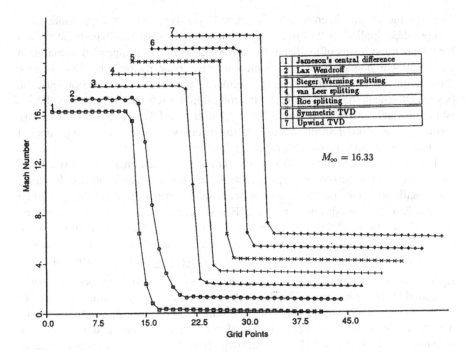

1	Jameson's central difference
2	Lax Wendroff
3	Steger Warming splitting
4	van Leer splitting
5	Roe splitting
6	Symmetric TVD
7	Upwind TVD

$M_\infty = 16.33$

FIGURE 5.1   Effect of limiter on captured shock definition.

For simulating a realistic multi-dimensional flow field all numerical algorithms solve the three-dimensional problem with three one-dimensional problems consecutively or in a cyclic sequence. For the explicit scheme, the solution is advancing uncoupled to each other such as the alternative direction explicit (ADE) methods, fractional step method, or the MacCormack scheme [6]. All explicit schemes are carried out by time-split techniques such as

$$U^{n+2} = L_x(\Delta t/2)\,L_y(\Delta t/2)\,L_z(\Delta t)\,L_y(\Delta t/2)\,L_x(\Delta t/2)\,U^n \qquad (5.2a)$$

where the difference operators aligning with each respective coordinate are designated as L_x, L_y, and L_z. The solution is second-order accurate if the mixed derivatives are commutative. MacCormack's scheme is exclusively utilized for simulating viscous-inviscid interactions at the very beginning.

When CFD ventures into increasingly complex fluid phenomena, more efficient and stable implicit schemes are required. The ADI (Alternating direction implicit) algorithm also spits a three-dimensional problem into three one-dimensional problems by succession approximations in time as

$$[\partial U/\partial t + \partial F(U)/\partial x]^* = -[\partial G(U^n)/\partial y + \partial H(U^n)/\partial z]$$

$$[\partial U/\partial t + \partial G(U)/\partial y]^{**} = -[\partial F(U^*)/\partial x + \partial H(U^n)/\partial z] \qquad (5.2b)$$

$$[\partial U/\partial t + \partial H(U)/\partial z]^{n+1} = [\partial F(U^*)/\partial x + \partial G(U^{**})/\partial y]$$

The ADI method has been effectively applied to all types of partial equations [7,8], except when applied to the time-dependent, three-dimensional hyperbolic system for which some forms of artificial dissipative terms must be appended to maintain computational stability [9]. This shortcoming has been removed by finite-volume formulation with an iterative solving scheme. In the subsequent developments, the basic ADI scheme has evolved into the strong implicit scheme (SIP) and the diminishing residual return (DRR) formulations [10]. Both ADE and ADI algorithms are formerly second-order accurate plus splitting errors, which affect the computational efficiency by requiring a higher grid-point density.

The heart of any physics-based simulation is to eliminate the artifacts of numerical error incurred by approximation through truncation and to be computationally efficient. In the Fourier space, the numerical error can be quantified accurately by the wave number of the Fourier components; the wavenumber ω, $(\omega = 2\pi/\lambda = 2\pi f/v_p \ \ 0 \leq \omega \leq \pi)$ is the spatial frequency of a wave relating to the phase velocity; it gives the number of waves per unit length and related directly to the phase velocity of a Fourier component. The dissipative error is determined by the L_2 norm, or simply the mean-square-root error, $\sqrt{\Sigma(\Delta u^2)}$ and the dispersive error is measured by the phase error. The desirable high-resolution algorithms are therefore assessed by these criteria. A series of outstanding and remarkable accomplishments have been accomplished for the definite-difference, finite-volume, and finite-element formulations. In particular, they are starting from Harten's reconstructive process, the Essential non-oscillatory (ENO) method, the Compact-difference approximation, Spectral-like local grid refinement approach, and the recent Discontinuous Galerkin scheme. More elaboration and directly usable details can be found in reference [11].

The characteristic and advantage of the high-resolution schemes over the convenient numerical procedures are displayed in Figure 5.2. On the left-hand side of the presentation, the numerical results reveal that the compact differencing scheme CD404 (fourth-order approximation with low-pass filter) is the last numerical solution departing from the known exact solution at the increasing wave number. On the right-hand side of the graph, the grid-point spacing of the high-resolution algorithms

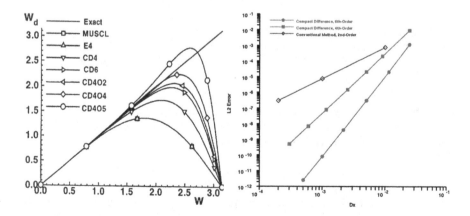

FIGURE 5.2 Dispersive and dissipative error of high-resolution algorithms.

shows significantly less demand for required grid spacing at the identical L_2 norm error by the second order scheme.

5.2 LEADING EDGE MACH WAVE INTERACTION

Viscous-inviscid interaction is an intrinsic component of hypersonic flows and appears at the leading edge of the hypersonic body. Such interactions may be divided into two distinct categories. One is the vorticity interaction associated with a bow shock over a blunted body, which arises from the variation in entropy across streamlines and thus may be considered a streamlined interaction. However, the vorticity interaction is not easily singled out because the phenomenon is developing from the continuous shedding of vorticity from the curved shock wave. The other phenomenon unique to hypersonic flow is known as the pressure interaction or Mach wave interaction which occurs conspicuously at the sharp leading edge of a control surface [12].

In hypersonic flow, the presence of a boundary layer creates added flow deflection to the oncoming freestream. According to the boundary-layer theory, the rate of the growth of the boundary layer thickness is singular at the sharp leading edge. Regardless of the predictions of the asymptotic theory, the rate of growth of boundary layer displacement is significant at the leading edge. This physical phenomenon has been routinely observed in hypersonic flow over a sharp leading edge even when the surface is immersed into the hypersonic stream at zero incidences. The Schlieren image shown in Figure 5.3, generated by a sharp leading edge wedge at zero angle incidence and Mach 5.5 and Reynolds number 1.61×10^6/m clearly shows the induced leading edge shock structure even over the plat surface – this feature is absent in low supersonic or subsonic flows.

The hypersonic pressure interaction is induced by the longitudinal curvature of the boundary displacement thickness. According to classical boundary-layer theory,

FIGURE 5.3 Schlieren photograph of hypersonic leading edge viscous-inviscid interactions.

the equivalent body in the inviscid flow is the original body thickness plus the displacement thickness of the attached boundary layer. The local hypersonic similarity parameter is then related to the sum of the body inclination angle and displacement thickness growth rate:

$$\kappa = M_\infty \left(\alpha_b + \frac{d\delta^*}{dx} \right).$$

(5.3a)

For the case of a flat plate, the boundary-layer thickness can be approximated as

$$\delta^2 \sim \left(\frac{\mu_e x}{\rho_e u_\infty} \right) \sim \left(\frac{\mu_e}{\mu_\infty} \right) \left(\frac{\rho_\infty}{\rho_e} \right) \left(\frac{\mu_\infty x}{\rho_\infty u_\infty} \right).$$

(5.3b)

For a linear viscosity-temperature relation, $\mu_\infty / \mu = c\,T/T_\infty$, because the pressure gradient vanishes across a shear layer, the density ratio across the boundary layer is just the reciprocal of the temperature ratio which is proportional to $(\gamma - 1)M_\infty^2/2$. By considering the boundary-layer displacement thickness can be approximated by the boundary-layer thickness, it yields

$$\frac{\delta^*}{x} \sim \frac{\delta}{x} \sim \frac{\gamma - 1}{2} \frac{M_\infty^2 \sqrt{C_\infty}}{\sqrt{Re_x}}.$$

(5.3c)

At the leading edge of a sharp flat plate, the hypersonic similarity parameter becomes

$$\kappa = M_\infty \left(\frac{\delta}{x} \right) = \frac{M_\infty^3 \sqrt{C_\infty}}{\sqrt{Re_x}} = \bar{\chi}$$

(5.3d)

Traditionally the pressure interaction is classified into strong and weak strengths according to whether the value of κ is greater than or smaller than three. Asymptotic results for large and small hypersonic similarity parameters may be obtained by substituting the hypersonic similarity parameter into the tangent wedge approximation and expanding from the surface to freestream pressure ratio. From Equation (3.19), by dropping all the higher order terms, the induced pressure from a weak interaction becomes

$$\frac{p}{p_\infty} \approx 1 + \frac{\gamma - 1}{2} \bar{\chi} + \frac{\gamma(\gamma + 1)}{4} \bar{\chi}^2$$

(5.3e)

At the relatively weak induced pressure by the Mach wave interaction, the laminar boundary-layer growth remains as the square root power of its running length.

For the strong pressure interaction, the induced pressure is given approximately by the tangent wedge approximation again as

$$\frac{p}{p_\infty} \approx \frac{\gamma - 1}{2} \bar{\chi}$$

(5.3f)

An important point for the strong pressure interaction, the boundary layer now grows like the three-quarter power of the boundary-layer running length. And yet the induced pressure still retains as a function of the square root of the streamwise distance. The most widely applied pressure distributions for the leading-edge pressure interaction are the second-order formulation on an insulated flat plate with $p_r = 1$ and $\gamma = 1.4$ [12],

$$\frac{p_w}{p_\infty} = 0.514\,\bar{\chi} + 0.759 \quad \text{Strong}$$

$$\frac{p_w}{p_\infty} = 1 + 0.31\bar{\chi} + 0.05\bar{\chi}^2 \; \text{Weak}$$

(5.3g)

The comparison of the analytic results of pressure interaction with the experimental data by Bertram is given in Figure 5.4 for several conditions over an insulated flat plate [13]. The hypersonic flows are tested at a Mach number range from 5.8 to 9.6 at the Prandtl number of 0.725 and a specific heat ratio of 1.4.

The pressure interaction is also displayed in numerical simulations solving the fully compressible Navier-Stokes equations. In Figure 5.5, two solutions at Mach number 5 and 10 and Reynolds number based on the running length of 1.5×10^5 are included. The surface pressure distributions are obtained for an isothermal wall. The sharp rise of the surface pressure, which increases with the Mach number, is followed by a rapid expansion and is characteristic of the pressure interaction. At the

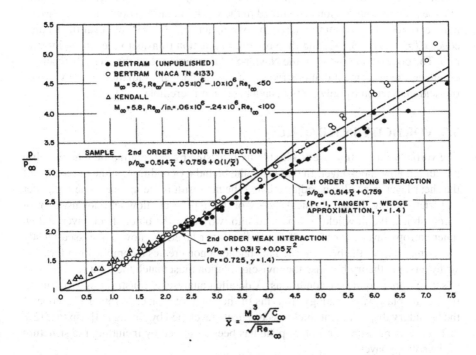

FIGURE 5.4 Comparison of analytic results with experimental data.

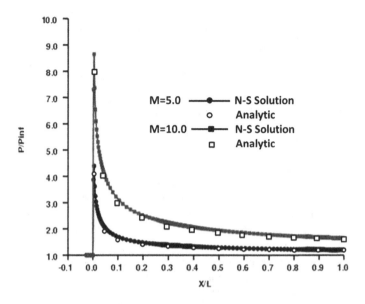

FIGURE 5.5 Comparison of analytic and computation results.

Mach of five the minimum interaction parameter is $\bar{\chi} = 1.130$, over nearly the entire plat the pressure interaction is confined in the weak interaction domain. At the Mach number of ten, the minimum interaction parameter is $\bar{\chi} = 1.414$ but near the leading edge, $x/l = 0.2, \bar{\chi} = 3.162$, and the pressure interaction transits to the strong interaction. The comparison between the Navier-Stokes solutions and the analytic results is excellent; the agreement demonstrates that the Numerical simulation of the viscous-inviscid interaction by solving the Navier solution is credible.

5.3 VORTICITY INTERACTION

The vorticity interaction over a blunt body is arising by the rotating flow field within the curved bow shock layer. In that layer, the boundary condition on the outer edge of the shear layer must also be adjusted, and the net result is the increased heat transfer and friction drag. In the classical theory of the hypersonic flowfield downstream of a bow shock, the flowfield is separated into two parts; the outer shock layer and an inner entropy layer near the surface. In this inner layer, the flow has passed mostly the normal part of the bow shock. This phenomenon creates a complication for classic hypersonic theory because the inviscid solution must match the high entropy core to satisfy inner boundary conditions. A detailed analysis is constructing a perturbation to the blast wave analogy in terms of the angle of attack, or rather the slope of the boundary displacement thickness has been developed by Cheng [14]. Up to a 12% difference in pressure for a blunt plate has been observed by including the structure of the entropy layer.

Because the entropy layer originates from the normal portion of the bow shock, for some distance downstream of the blunt nose, the boundary layer grows within the entropy layer and eventually contains the entropy layer far downstream. The vorticity of the external flow over the boundary layer exerts influence to the thin layer structure, and the external boundary condition of the boundary layer ceases to be an inviscid surface streamline. In the framework of the traditional boundary theory, the effect of the entropy layer is treated as a vorticity interaction and a vorticity interaction parameter is defined at the edge of the equivalent body in the equivalent inviscid flow. A relationship between velocity and stream function has been obtained by Hayes [12], to replace the usual boundary condition $f_\eta \rightarrow 1.0$ at the outer edge of the boundary. The vorticity interaction parameter is defined as $\varsigma = \Omega_{inv}/(u_{inv}/\delta)$ and Ω_{inv} is the vorticity of the inviscid stream $\Omega = \nabla \times u$; if the parameter $\varsigma \ll 1.0$ the vorticity interaction is negligible.

Under the strong vorticity interaction condition, $\varsigma > 1.0$ the shear stress in the shock layer is of the same order of magnitude within the boundary layer. The concept of the boundary layer breaks down; therefore the entire shock layer must be considered a totally viscous flow. Nevertheless, the investigation of the vorticity interaction shows the rigorous and thorough altitude in classic hypersonic flow research which shall set an example to be followed.

5.4 SHOCK-BOUNDARY-LAYER INTERACTION

The maturation of numerical simulation in solving the compressible Navier-Stokes equations leads to an explosive growth in understanding the challenging viscous-inviscid interacting phenomena. However, the computational simulations are severely limited by computer memory size and data processing speed, some of the simulations may be viewed as only able to duplicate the essential physics, but it's the beginning of a new research avenue. In the first decade of CFD development, the aerodynamics research is limited mostly to the canonical viscous-inviscid interactions over a compression ramp, shock impingement including separated flow, and three-dimensional corner flows with shock-shock interaction. These research activities have been sustained for more the 50 years and still carrying on to date [15].

The first numerical simulation of compressible laminar flow over a two-dimensional compression corner is produced by Carter [16], the inviscid-viscous interaction is automatically accomplishes by the direct simulation without matching the external flow with boundary-layer. The first ever counterpart of turbulent flow is conducted by the computational simulation on a really coarse (64×22) grid mesh system over the entire domain of $(30.5\,cm \times 182.9\,cm)$ [17]. As a consequence, a uniform distributed complete mesh spaces are highly stretched in both streamwise normal to the ramp surface. The turbulence closure is achieved with the two-layer mixing length model of Cebice-Smith model, equally important, a relaxation modification is also implemented to the Reynolds stress model to mimic the nearly frozen state along the streamline in a highly accelerated or decelerated state. Bradshaw suggests an empirical correction with time scale corresponding to a streamwise distance roughly $10\,\delta$ in the outer part of the turbulent boundary layer [18]. In order to resolve the viscous sublayer, the finest mesh spacing immediately next the surface is 0.003 cm, which corresponds to the law-of-the-wall variable $y^+ = 1.0$.

The solving scheme adopts MacCormack's alternation-direction-explicit numerical scheme, and the convergence criterion is preset until the consecutive calculations indicate no significant change (0.1%).

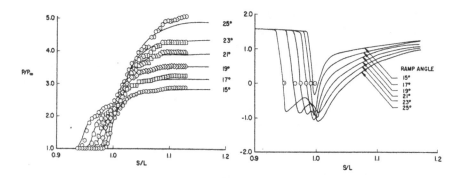

FIGURE 5.6   Surface pressure and skin friction coefficient distributions over ramps.

A series of calculations has been accomplished over six ramp angles from the incipient separate flow of 15°–25°. The surface pressure and skin friction coefficient distribution over the ramp configurations are depicted in Figure 5.6. The pressure distributions reach excellent accordance with the experimental data. Excellent predictions of the pressure propagation upstream of the ramp corner have been consistently achieved. For the higher compression ramp angles, a pressure plateau is presented in direct contrast to the lower ramp angle pressure distributions. Only significant discrepancies between data and the computed results appear in the downstream corner region for the high ramp angle. In general, the computed results predicted accurately the point of separation with experimental observations with a different technique [17]. One also notes a systematic evolution of the skin friction coefficient from a single-minimum to a dual-minima pattern as the separation domain increases substantially. This behavior is known for the investigations of separated laminar interactions, it also has been revealed by the lower branch solutions from the self-similar solutions under adverse pressure conditions (Figure 5.7).

The complete viscous-inviscid interacting flowfield, including the flow separation phenomenon, is presented together with a density interferogram [17]. The clearly discernable turbulent boundary layer enters the interaction undisturbed. The normal velocity component first adjusts when approaching the ramp by a pressure rise. The strong adverse pressure gradient by coalescence of compression waves into the separation shock initiates the boundary layer separation. The recirculation flow beneath the dividing streamline covers the entire corner region. Further downstream of the corner, a second coalescing compression wave at the boundary-layer reattachment point also becomes detectable. The reattachment shock eventually intersects the leading separation shock. Far downstream of the corner region, the converged shock wave approach asymptotically to the Rankine-Hugoniot condition for an oblique shock solution with a 25° deflection angle.

According to experimental observations, substantial variations exist between experimental measurements for shock-wave-boundary-layer investigations performed under supposedly similar conditions in differential facilities. It is generally agreed that the discrepancies are primarily due to the three-dimensional effects. Meanwhile, the shock wave-boundary interaction by shock impingement may be considered to be a variant of the compression ramp configuration [19,20]. The global behaviors of the viscous-inviscid interaction bear out the commonly accepted assertion by the surface pressure distributions from these two types of viscous-inviscid interactions in Figure 5.8. Although the computed incident shock yields the accurate incident shock-wave angle, the numerically smearing shock creates difficulty in the determination of an accurate coordinate reference point. Thus, the point of flow separation is used as the comparing coordinate origin [21].

In spite of the shortcoming in depicting a sharp wave definition by numerical simulations and the computational simulations are only supported by a coarse grid system of (64×30) for the $(20\delta \times 5\delta)$ entire two-dimensional simulation domain, the numerical result still displays a subtle flow field structure difference. For the nearly identical surface pressure distributions between the compressing ramp and shock impingement interactions, the latter displayed a more complex flowfield structure in the intensively interacting zone [21]. A consistent appearance of an expansion zone wedges between the incident and reattachment shocks, due to the thinning of the displaced shear layer thickness suppressed by the incident shock. This observation has been confirmed later by an outstanding and studious review by Gaitonde [22]. The distinguishable differences between the computational and measured density profiles of the two types of viscous-inviscid interactions are easily noted in Figure 5.8.

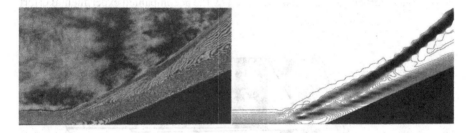

FIGURE 5.7 Turbulent viscous-inviscid structure over the 25° compression ramp.

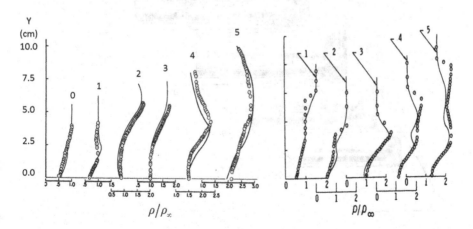

FIGURE 5.8 Density profiles of compression ramp and impinging shock interaction.

5.5 THREE-DIMENSIONAL CORNER FLOWS

Taking a step beyond the idealized two-dimensional viscous-inviscid interacting phenomena, basic research for physically realistic interactions induced by corner configurations was also initiated in the 1970s. The first three-dimensional laminar corner flow computation is conducted for a 15° wedge mounting on a sharp leading edge plate, at the hypersonic Mach number of 12.5 and a Reynolds number of 1.21×10^6 based on the plate length of 31.75 cm [23]. The numerical simulations are made possible by invoking the hypersonic equivalent principle or the law of plane-cross-section [12], which implies that the inviscid-dominated flow is nearly invariant in the streamwise direction. The salient characteristic permits only a few stream-wise planes to be included; the maximum three-dimensional grid-point distribution is $(8 \times 32 \times 36)$ with a total number of grid points of 9,216 on an analytically transformed rectangular coordinate system. The guiding hypersonic principle is clearly demonstrated by the overall computational flowfield structure presented in Figure 5.9.

The triple-point shock wave is forming at the intersection of the induced horizontal shock wave and the stronger oblique shock over the vertical wedge. The triple-point is the intersecting point of the two shock waves and Mach stem, as well as a slipstream. The triple-point shock wave is aware by the reflecting shock wave of typical separated corner flow [24] and is verified by the accompanied experimental effort

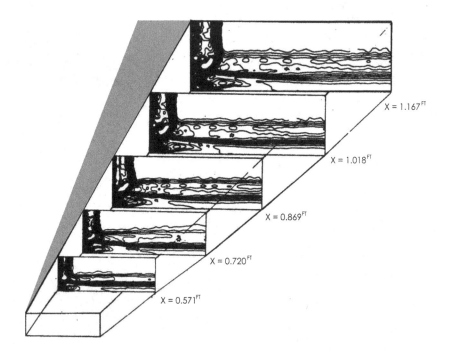

$X = 1.167^{FT}$

$X = 1.018^{FT}$

$X = 0.869^{FT}$

$X = 0.720^{FT}$

$X = 0.571^{FT}$

FIGURE 5.9   Hypersonic flowfield structure of a hypersonic axial corner flow.

[25]. The computational results also reveal the flow separation in the corner region and experimental investigation indicates similar observation from the surface oil film pattern that shows an extensively separated flow region which indeed occurred. The numerical result duplicates the outer separation and inner reattachment limiting surface streamlines but fails to capture the footprint of the secondary vortex.

The shear layer plays a relatively passive role in the formation of the shock envelope over the corner configuration. However, the inner shear layer produces the most intricate aerodynamic phenomenon. Drastic flowfield changes beneath the intersection of the wedge and the induced shock waves to form the triple point shock wave that must satisfy the pressure continuity and parallel velocities across the slipstream conditions. The boundary layer thickness changes gradually from the corner outward, which signifies the return to a two-dimensional flow field farther away from the corner.

Figure 5.10 depicts the most striking and previously unknown features of the corner flow which is found in the vicinity of the corner region under the triple-point shock. The cross-flow component in this region retains a relatively high speed and eventually impinges on the solid surface away from the corner juncture. The evidence is generated by the numerical results in the form of the conical Mach number contours on the spherical surface. The high-energy stream impacts and scavenges the solid surface at a small distance away from the corner junction. As the consequence, an order of magnitude higher heat transfer rate locates in the impinging area on both the vertical wedge and horizontal plate surface. The numerical result is given in the insert of the normalized heat transfer rate over the flat plate. The highly concentrated hot spot near the corner junction offers a convincing explanation for the severe heat-damaged ramjet pylon on the X-15 flight test vehicle [26].

A three-dimensional interacting turbulent flow along a symmetric corner has also been performed. The corner is constructed by two $9.48°$ wedges at a Mach number of three, with the Reynolds number spanning a range from 0.4×10^6 to 60.0×10^6 for

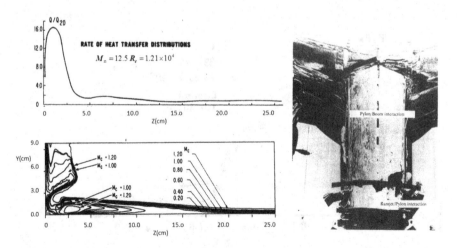

FIGURE 5.10 Crossflow Mach number contours and heat transfer distribution of corner flow.

which experimental data existed [27,28]. The data displays a rather minor Reynolds number dependence over the tested condition based on the evidence from the oil film pattern. The fully turbulent condition is reached for the Reynolds number as low as 1.1×10^6 and is adopted for the computational condition. The flowfield investigated contains a supersonic leading edge with the subsequent development of laminar, transitional, turbulent flow along the streamwise corner.

For computational simulation, the turbulent model is an eddy viscosity formation with the modified length scale by Gessner [29]. The computational physical space traces a truncated frustum of a pyramid with dimensions of $12.7 \times 15.3 \times 15.3\, cm^3$ by a $(17 \times 23 \times 23)$ mesh system of a total of 8993 grid points. The grid spacing is highly stretched to set the minimum value next to the wedge surfaces by the law-of-the-wall variable y^+ and z^+ around unity. The solving scheme utilizes MacCormack's ADE scheme. It is amazing to recall that the two time-level dependent variables and seven spatial coordinates transformed metrics required computer core storage of 615,600 octal words. To fit the available computer memory, a data manager has implemented and reduced the maximum core memory from 615,600 to 235,000 octal words.

Figure 5.11 presents surface pressure distributions for laminar and turbulent interacting flows within a symmetric corner. The agreement between the experimental measurements and computational results is surprisingly good and the maximum deviation between data and calculation is encountered at the location where the turbulent interaction emanates [27]. From the experimental data, the pressure rise toward the corner junction from the turbulent interaction also exhibits a consistently mild dependence on the Reynolds numbers [28]. The computed surface pressure distributions clearly display different characteristics between the laminar and turbulent interacting flows. The outward propagation of the laminar interaction zone extends further than the turbulent counterpart and produces a less intensively spanwise pressure gradient toward the symmetric corner. However, both interactions generate nearly identical pressure plateaus at the corner juncture.

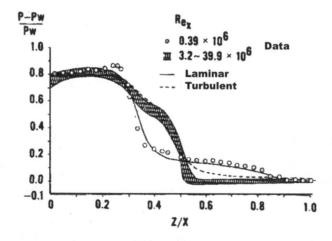

FIGURE 5.11   Surface pressure distributions of laminar and turbulent corner flow.

Limited by the coarse grid-point density for numerical resolution the computational results only duplicate all the essential experimental observations, especially for the envelope of the limiting surface streamline for three-dimensional flow separation. However, the numerical simulations capture the distinguished triple-point shock wave structures between the laminar and turbulent flows as exhibited by Figure 5.12. The embedded triple-shock wave formation within the corner region, including the slip surfaces is verified by experimental measurements. The Mach stem of the triple-point shock wave for the turbulent flow is terminated by a clearly discernable lambda formation from the laminar flow counterpart. For practical applications, the impingement heating of the symmetric corner configuration has minimized the local heating load in contrast to the asymmetrical corner configuration. Finally, the path-finding turbulent corner flow simulation appears to offer promise for investigating flow fields of high-speed inlet and fuselage-wing junctions.

Over the years, impressive progresses have been made in experimental measurement techniques and in high-resolution algorithm innovations for physics-based computational simulation. The accumulated knowledge in basic science such as aerodynamic topology, bifurcation, and turbulence modeling is further augmented by the leaping and bounding high-performance computing technology. A much better understanding of the physics of three-dimensional shock-wave-turbulent-boundary-layer interaction generating by the asymmetrical double fin geometries is achieved for multiple combinations of 7°, 11°, and 15° fins [30]. The simulated viscous-inviscid interaction flow condition is characterized by a Mach number of 3.85 and a Reynolds number of $87.5 \times 10^6 / m$. The solving scheme adopts Roe's third-order flux-difference split scheme with limiter to enforce monotonicity for inviscid terms [31], and the viscous terms are resolved by an ADI algorithm. The turbulent closure is achieved by the $k - \varepsilon$ two-equation models [32]. The numerical simulations are supported by different grid-point distributions according to the arrangement of different

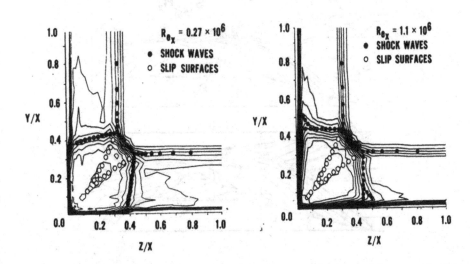

FIGURE 5.12 Triple-point shock wave structures in laminar and turbulent corner interactions.

configurations of fins; the densest grid point density is $(109 \times 88 \times 197)$ system with a total of 1.9×10^6 grid points.

The flow field structure is best described by the vortical structure that impinges on the surface of the fins and realigns with media interfaces [30]. The footprints of the vortical flow projects on the side wall are recognized as the limiting surface stream-line, which actually represents the surface shear map. The trajectory is constructed by a continuous vector field and must obey the topological rule to understand the sep-arated flow field [33]. For an asymmetric corner formed by mounting a combination of a 7° and 15° fins on a plate, the computational (on top) and experimental results are depicted in Figure 5.13. The overall affinity between experimental and compu-tational results is remarkable. Multiple singular points for flow separation (saddle point) and reattachment (nodal point) are unmistakably identifiable by the divergent and convergent limiting streamlines or shear lines. There are some minor discrepan-cies between results, but is only the consequence of different discriminating scaling between the two different simulation techniques. Nevertheless, the rule of topology is completely complied with to ensure the correct physics. Once the most difficult flow field features are determined, the understanding of the viscous-inviscid interaction within the asymmetric corner can be obtained.

Numerous and other viscous-inviscid interactions have been conducted for flow fields that have intrinsic values for basic aerodynamic such as generating by shock-on-shock [34], vortical flow upstream to blunt fins [35], and cross-flow jet injection [36] will be deferred later as needs. The all-inclusive survey and review for addi-tional viscous-inviscid interactions can be found in the works by Dolling [15] and Gaitonde [22].

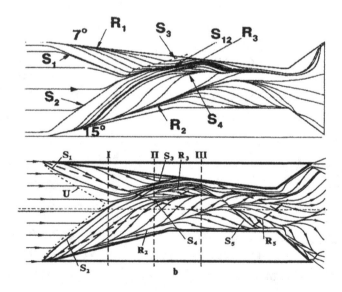

FIGURE 5.13   Surface limiting streamline patterns of experiment and computation.

5.6 RESONANCE AND BIFURCATION

A category of the intrinsic and less understood aerodynamic phenomenon is the self-sustain viscous-inviscid interactions, which involve hydrodynamic stability, aerodynamic bifurcation, and perturbation feedback amplification. And yet, these unsteady flow fields are encountering overall speed regimes [37]. For example, supersonic flow over an open cavity induces pressure oscillation by the inherently unstable shear layer structure. The fluctuating external flow induces periodic mass addition and expulsion from the cavity with discrete frequencies by a selective amplification process.

In 1880, Rayleigh showed when velocity profiles containing inflection points are unstable and later confirmed all shear layers are unstable but only at low frequencies ($\lambda/\delta > 2\pi$) [38]. In the cavity, a resonant situation arises when a forcing function excites the shear layer in the frequency range where amplification is possible. The forward traveling wave of the pressure pulsation propagates around the half speed of the external stream and the rearward traveling wave is known to be at the acoustic speed. The disturbance will grow until a limit cycle is reached by viscous dissipation. Under this condition, pressure resonants appear and give commensurable waves with distinctive discrete frequencies. The selected wave pattern and the comparison of spectral analysis results of discrete resonant wave frequencies from computational Navier-Stokes equations have been duplicated by computational simulation [39]. The pressure oscillation is presented at regular intervals of 5.5 seconds, and the data is collected over a period of over 900 seconds and processed through a spectral analysis to determine discrete frequencies. The numerical simulations duplicate the first two primary modes but are unable to resolve the high-order modes due to the shortfall of high wave number resolution.

The analytic results also reveal that the unstable shear layer with inflection points in the velocity profile only for low frequencies $f < u_e/4\pi\delta$. The open cavity will not resonate for the cavity length shorter than $2\pi\delta$, and the Rayleigh instability will not occur above the Mach number above 2.5 [39]. Therefore, the mixing enhancing cavity is mostly applied to the internal flow of the air-breath propulsion system and is widely implemented for hypersonic scramjet [40].

The spike-tipped buzzing phenomenon has been detected on an ablated reentry configuration which is the consequence of a wide variety of shear layer impingement viscous-inviscid interactions [37,41,42]. The oscillatory flows also occur in supersonic inlet, hypersonic drag reducing blunt body, and on ablating reentry vehicles. For the spike-tip configuration the unstable shock structure constitutes of a strong detached bow shock at the spike tip then collapses into a conical shock that intersects the bow shock at the juncture between the spike and the after-body. A side-by-side comparison of experimental observations and the calculated density contours by solving the three-dimensional Navier-Stokes equation is presented in Figure 5.14. The investigated flow conditions are characterized by a Mach number of 3.0 and a Reynolds number of 7.87×10^6 / m, and the basic configuration consists of a hemispherical spike cylinder connected to a truncated conical after-body [42].

The sequence of oscillatory density contours is validated by the accompanying schlieren photographs. The disturbance propagates downstream and upstream through the subsonic-separated flow region to reinforce the oscillatory dynamics by

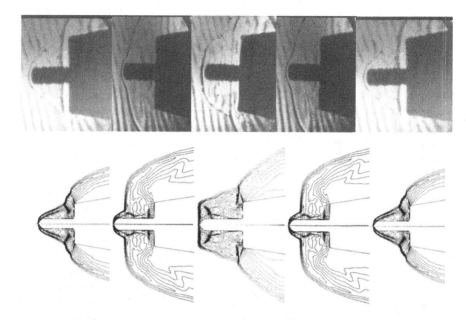

FIGURE 5.14 Comparison of oscillatory shock structure over the spike-tipped body.

reintroducing the disturbance to the free shear layer to close the feedback loop. Only an appropriate spike length permits the in-phase feedback. From the Root Mean Square (RMS) pressure correlation and power spectral analyses, two major oscillatory modes are found. The primary mode is associated with the longitudinal shock oscillations, and the secondary or the odd mode wave is related to the rotational wave [42]. The odd mode wave has about half the frequency of the oscillatory motion and actually modulated the primary wave.

The spike length is the controlling factor for stable and unstable shock formation. When the spike length is less than or greater than a critical value according to the separated shear-layer condition; the oscillating motion ceases. At the investigated condition, the critical spike length for the buzzing phenomenon is within the values of 12.75 and 45.72 mm, in which violent shock oscillation takes place. The hysteresis produces a shift of the primary mode frequency from 2.8 to 3.5 kHz, but the frequency switches at different spike lengths by increasing and decreasing movements. The observation gives a glimpse of the complex viscous-inviscid interaction in unsteady flows to show the initial condition comes to play an important role.

A detailed confirmation between experimental data and computational simulations in wave pattern is attained at a spike length of 38.1 mm [42]. The secondary or the odd mode does not appear outside the hysteresis region and the shock wave oscillations are perfectly symmetric. The numerical results demonstrate that the self-sustained oscillatory motion is maintained by the selective amplification of disturbances in the separated shear layer [41].

The high-speed blunt body wave drag reduction [43] using counterflow jet and bow shock interaction is achieved by altering a single bow shock into a multiple shock formation. From the Rankine-Hugonoit condition, the entropy increment across a shock is proportional to the cubic power of the pressure jump; $\nabla S \sim \nabla p^3$. For a given total pressure jump, the advantage of a multi-shock increment over a single shock is clearly perceivable. The aerodynamic bifurcation also exists in the interacting flowfield to have steady and oscillatory dynamic bi-states. The critical point of the bifurcation occurs in a range of the stagnation pressure ratio between the freestream and the jet from 0.85 to 1.05 in a hypersonic flow. When the ratio of the jet injecting stagnation pressure versus the freestream is lower than a critical value, the interacting flow is unsteady and the interacting zone elongates upstream with increasing jet stagnation pressure. The investigated oscillatory flow has a dominant frequency of 100 Hz and a secondary discrete frequency of 440 Hz by spectral analysis from the measured drag coefficient. The shock-jet interaction [44] generates a complex flowfield and has successfully stimulated by solving of the time-dependent, mass-averaged Navier-Stokes equations.

In Figure 5.15, the instantaneous streamline of a counterflow jet issues from the stagnation point of a hemispherical cylinder is given. The freestream has a Mach number of 6.0 with a stagnation pressure of 344.9 kPa. Under this condition, the injection stagnation pressure is below the critical point of bifurcation and the flowfield is unsteady. The flowfield is composed of a secondary separated flow ring at the jet exit, but mainly with a rapidly expanding counterflow jet from the stagnation region and reversing its direction downstream by the Mach disk as a free shear layer. A part of the shear layer is entrained to form a toroidal recirculation zone beneath the dividing stream surface, and the rest of the free shear layer flows over

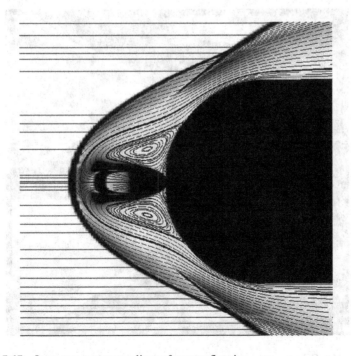

FIGURE 5.15 Instantaneous streamlines of counterflow jet.

the recirculating ring and impinges on the blunt fore-body. The impinging shear layer induces a series of compression waves coalescing into a reattachment shock. The counterflow jet interaction replaces a single shock over the hemispherical cylinder with a triple shock structure. In spite of the added reverse thrust by the counterflow jet, the reduced wave drag from the multiple shocks is still overwhelming.

Figure 5.16 presents a validation of computational simulations with schlieren photographs for the counterflow jet interaction at Mach number of 6 at the stagnation pressure of $p_o = 689.0$ kPa, and stagnation temperature of 610 k. The presentation on the top row gives the bifurcation phenomenon in subcritical state of the viscous-inviscid interaction ($p_{jo}/p_o = 0.75$), where the photograph of the oscillatory flow-field becomes blurred by the multiple exposure and the computational result only captures an instantaneous oscillatory state. The bifurcation at the supercritical state ($p_{jo}/p_o = 1.05$) returns to a steady state like that prior to jet injection, and the computational results duplicate the experimental observation.

The bifurcation is the result of the breakdown of the subsonic feedback loop between the Mac disk and the unstable shear layer. At the subcritical state, the toroidal recirculating zone and the immediate downstream subsonic zone to the Mach disk are completely embedded by the supersonic reverse jet; the feedback loop of disturbance is closed. At the supercritical state, the supersonic jet separates the two zones and cut off the signal feedback loop; the oscillatory motion ceased. The physics is depicted in Figure 5.17 by the Mach number contour, the supersonic flow region of the counterflow jet injection is presented by the same hue of color.

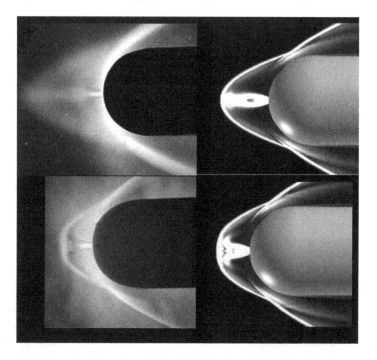

FIGURE 5.16　Comparing computational simulations with schlieren photographs.

Computations correctly recover the oscillatory flow structure for all subcritical states; calculated drag coefficients for these cases shows a scattering band that is totally absent from the supercritical cases. At the bifurcation point, the drag also attains its minimum. Computed drag coefficients generally underpredict the data by ~11%, but the impressive drag reduction by counterflow jet approaches 55% at the Mach number of six. However, there is some discrepancy with experimental data in the precise critical state (Figure 5.18).

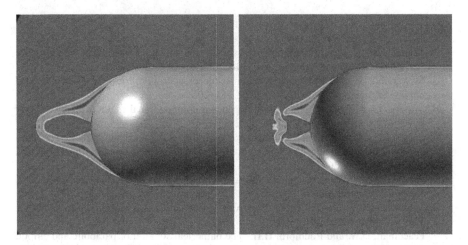

FIGURE 5.17   Mach number contour of the counterflow jet interaction.

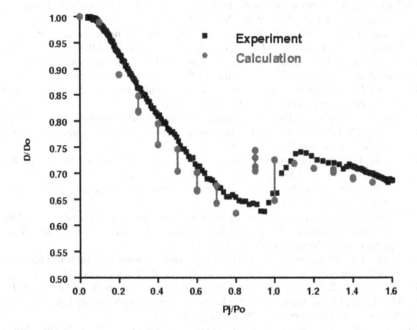

FIGURE 5.18   Comparison of drag coefficient by computation and data.

Despite significant advances in experimental and computational methods, the classical hypersonic theory continues to be highly relevant in obtaining a fundamental understanding of trends and providing a framework for observations. This article has provided a brief glimpse of the many elegant results that facilitate the analysis of pressures and heat transfer rates, for local and integrated analysis. These conclusions, based on advanced mathematical theory and limiting techniques, have enabled rapid analysis of extremely complex phenomena for practical problems.

REFERENCES

1. Shang, J.S., Computational fluid dynamics application to aerospace science, *The Aeronautical Journal*, Vol. 113, No. 1148, 2009, pp. 619–632.
2. Stewartson, K. and Williams, P.G., Self-induced separation, *Proceeding of the Royal Society of London A*, Vol. 312, 1969, pp. 181–206.
3. Moretti, G. and Abett, M., A time-dependent computational method for blunt body flows, *AIAA Journal,* Vol. 4, 1966, pp. 2136–2141.
4. Steger, J.L. and Warming, R.F., Flux vector splitting of the Inviscid Gasdynamics equations with application to finite difference methods, *Journal of Computational Physics*, Vol. 40, 1981, pp. 263–293.
5. Roe, P.L., Approximate riemann solvers, parameter vectors and difference schemes, *Journal of Computational Physics*, Vol. 43, 1981, pp. 357–372.
6. MacCormack, R.W., The effect of viscosity in hypervelocity impact cratering, AIAA 1969-354, Cincinnati OH, 1969.
7. Peaceman, D.W. and Rachford, H.H., The numerical solution of parabolic and elliptic differential equations, *Japan Journal of Industrial and Applied Mathematics*, Vol. 3, 1955, pp. 28–41.
8. Richtmyer, R.D. and Morton K.W., *Differential methods for initial-value problem*, 2nd Ed. Interscience Publishers, Wiley, New York, 1967.
9. Beam, R.M. and Warming R.F., An implicit factored scheme for the compressible Navier-Stokes equations, *AIAA Journal*, Vol. 16, 1978, pp. 393–401.
10. Anderson, D., Tannehill, J, Pletcher, R., Munipolli, R., and Shankar, V., *Computational fluid dynamics and heat transfer*, 4th Ed. CRC Press, Boca Raton, FL, 2021.
11. Shang, J.S., Landmarks and new frontiers of computational fluid dynamics, *Journal of Advances in Aerodynamics*, Vol. 1, No. 5, 2019, pp. 1–36.
12. Hayes, W.D. and Probstein, R.F., *Hypersonic flow theory*, Academic Press, New York, 1959.
13. Bertram, M.H., Boundary-layer displacement effects in air at Mach numbers of 6.8 and 9.6, NACA Tech. Note No. 4133, 1957.
14. Cheng, H.K., On the structure of vortical layers in supersonic and hypersonic flows, *Journal of Aerospace Sciences*, Vol. 27, 1960, pp. 155–156.
15. Dolling, D.S., Fifty years of shock/boundary Interaction: What next, AIAA Journal, Vol. 39, 2001, pp. 1517–1531.
16. Carter, J.E., Numerical solution of the supersonic laminar flow over a two-dimensional compression corner, In: Citro, R., et al. (eds.), *Lecture Notes in Physics*, Vol. 19, Springer-Verlag, New York, 1973, pp. 69–78.
17. Shang, J.S. and Hankey, W.L., Numerical solution for supersonic turbulent flow over a compression ramp, *AIAA Journal,* Vol. 13, No. 10, 1975, pp. 1365–1374.
18. Bradshaw, P., Effects of streamline curvature on turbulent flow, AGARD-AG-169, 1973.
19. Green, J.E., Interaction between shock waves and turbulent boundary layer, *Progress in Aerospace Science*, Vol. 11, 1970, pp. 233–340.
20. Law, C.H., Supersonic turbulent boundary-layer separation, *AIAA Journal,* Vol. 12, 1974, pp. 794–797.

21. Shang, J.S., Hankey, W.L., and Law C. H., Numerical simulation of shock wave: Turbulent boundary-layer interaction, *AIAA Journal*, Vol. 14, 1976, pp. 1451–1457.

22. Gaitonde, D.V., Progress in shock wave/boundary interactions, *Progress in Aerospace Science*, Vol. 72, 2015, pp. 80–99.

23. Shang, J.S. and Hankey, W.L., Numerical simulation of the Navier-Stokes equations for three-dimensional corner, *AIAA Journal*, Vol. 15, 1977, pp. 1575–1582.

24. Hornung, H., Regular and Mach reflect of shock waves, *Annual Review of Fluid Mechanics*, Vol. 8, 1986, pp. 33–58.

25. Cooper, J.R. and Hankey W.L., Flowfield measurements in an asymmetric axial corner at M=12.5, *AIAA Journal*, Vol. 12, 1974, pp. 1353–1357.

26. Shang, J.S., Hankey, W.L., and Petty, J.S., Three-dimensional supersonic interacting turbulent flow along a corner, *AIAA Journal*, Vol. 17, 1979, pp. 706–713.

27. Korkegi, R.H., Survey of viscous interaction associated with high Mach number flight, *AIAA Journal*, Vol. 9, 1971, pp. 771–784.

28. West, J.E. and Korkegi, R.H., Supersonic interaction in the corner of intersecting wedges and high Reynolds numbers, *AIAA Journal*, Vol. 10, 1972, pp. 652–656.

29. Gessner, F.B., The origin of secondary flow in the turbulent flow along a corner, *Journal of Fluid Mechanics*, Vol. 58, 1973, pp. 1–25.

30. Gaitonde, D.V., Shang, J.S., Zheltovodov, A.A., and Maksimov, A.I., Investigation of 3-D turbulent interactions caused by asymmetric cross-shock configuration, *AIAA Journal*, Vol. 37, No. 12, 1999, pp. 1602–1608.

31. Roe, P.L., Approximate Riemann Solvers, parameter vectors, and difference schemes, *Journal of Computational Physics*, Vol. 43, 1981, pp. 357–372.

32. Jones, W.P. and Launder, B.E, The prediction of laminarization with two-equation model of turbulence, *International Journal of Heat and Mass Transfer*, Vol. 15, 1972, p. 301314.

33. Tobak, M. and Peak, D.J., Topology of three-dimensional separated flows, *Annual Review of Fluid Mechanics*, Vol. 14, 1982, pp. 61–85.

34. Edney, B.E., Effects of shock of shock impingement on heat transfer around blunt bodies, *AIAA Journal*, Vol. 6, 1968, pp. 15–21.

35. Master, D.L. and Shang, J.S., A numerical study of three-dimensional separated flows around a Sweptback Blunt Fin, AIAA 88-0125, Reno NV, 1988.

36. Shang, J.S., McMaster, D., Scaggs, N., and Buck, M., Interaction of jet in hypersonic cross stream, *AIAA Journal*, Vol. 27, No 3, 1989, pp. 323–329.

37. Rockwell, D. and Nandascher, E., Self-sustained oscillations of impinging free shear layer, *Annual Review of Fluid Mechanics*, Vol. 11. 1979, pp. 67–94.

38. Michalke, A., On the inviscid instability of the hyperbolic tangent velocity profile, *Journal of Fluid Mechanics*, Vol. 19, 1964, pp. 543–556.

39. Hankey, W.L. and Shang, J.S., Analyses of pressure oscillations in an open cavity, *AIAA Journal*, Vol. 18, 1980, pp. 892–898.

40. Seleznev, R.K., Surzhikov, S.T., and Shang, J.S., A review of the scramjet experimental data base, *Journal of Progress in Aerospace Science*, Vol. 106, 2019, pp. 43–70.

41. Widhopf, G.F. and Victoria, K.L., Numerical solution of the unsteady Navier-Stokes Equations for the oscillatory flow over concave body, In: Citro, R., et al. (eds.), *Lecture Notes in Physics*. No. 35, Springer-Verlag, New York, 1974, pp. 431–444.

42. Shang, J.S., Hankey, W.L, and Smith, R.E., *AIAA Journal*, Vol. 20, 1982, pp. 25–26.

43. Shang, J.S., Plasma injection for hypersonic blunt: Body drag reduction, *AIAA Journal*, Vol. 40, No. 6, 2002, pp. 1178–1186.

44. Shang, J.S., Hayes, J., Wurtzler, K., and Strang, W., Jet-spike bifurcation in high-speed flows, *AIAA Journal*, Vol. 39, No. 6, 2001, pp. 1159–1165.

Section II

High-Enthalpy Hypersonic Flow

6 Quantum Transition

6.1 HEISENBERG UNCERTAINTY PRINCIPLE

Historically, the quantum concept was from its inception in 1901 by Planck to predict how electromagnetic energy was distributed amongst the electromagnetic wave frequencies [1]. The energy distribution has a quantum restriction, namely the energy is dependent upon its natural frequency of oscillation. Sufficient experimental evidence has demonstrated that discrete energy levels are prevalent for all dynamic systems. Thereby, the quantum theory describes the energy transfer is not a continuous process as that of classic physics but through discrete quanta, and the size of quantum energy depends on the natural frequency of wave motions. Planck discarded the ancient maxim that governs the energy distribution by considering the energy of an atom has no fixed value but depends on the frequency of oscillatory motion [2].

$$\varepsilon = h\nu \qquad (6.1a)$$

where the Planck constant in the SI unit is 6.6252×10^{-27} erg-s and the symbol ν denotes the fundamental frequency. From the spectroscopic measurement of the hydrogen atom, a regular relationship between the frequency and the atomic line of hydrogen in the visible region of the spectrum was discovered. This finding leads to a discrete energy level between quantum states from the first Bohr orbit or the ground state is given by $n = 1$. The quantum transition of energy between the ground state and n level greater than the ground state is [1].

$$\varepsilon_n - \varepsilon_1 = 2\pi^2 Z^2 e^4 m / h^2 \left(1 - 1/n^2\right) \qquad (6.1b)$$

The coefficient $2\pi^2 Z^2 e^4 m / h^2$ is known as the Rydberg constant has a value of 109,737 cm^{-1} and the symbol Z is the number of elementary charges in an atom.

In 1924, de Broglie follows the analogy of the wave-particle quality of photons considering that electrons must have wave properties on purely theoretical ground. He demonstrated that the Bohr orbits of hydrogen could also be defined that the circumference of an orbit could be defined the momentum of electron by the ratio of the Planck constant and the whole number of wavelengths; $p = h/\lambda$. After a series of experimental confirmations, it established beyond any doubt that all particles have dynamic process wave properties.

Following the wave property of matter, a second basic principle is developed by the indeterminacy or uncertainty that is introduced by the experimental measurement. It emerges through the motion on the atomic scales in that the quantum character of the object and the measurement process must be accepted. This uncertainty gives

DOI: 10.1201/9781003212362-8

in terms of the position of the elementary particle is evaluated from the principle of physical optics as $\Delta q \approx f\lambda/D$ where D is the diameter of the microscopic object. The uncertainty of the component of the momentum of particle dynamics is approximately as

$$\Delta p \approx (h\nu/c)(D/2f)$$

$$= (f\lambda/D)(h\nu/c)(D/2f) \qquad (6.1c)$$

$$= h/2$$

It means that the uncertainties are irreducible either by the measuring process or the same quantum law of the system being measured. These minimum uncertainties become the Heisenberg uncertainty principle.

$$\Delta q \Delta p \eqsim h \qquad (6.1d)$$

In Equation (6.1d) the quantities q and p are conjugate quantities according to Hamilton's canonical equation. Namely, if q is the coordinate, p is the momentum along that coordinate. If q is the energy, then p will be the time. The uncertainty is the classic limit of quantum mechanics.

6.2 QUANTUM STATES OF AN ATOM

Bohr's theory of atomic hydrogen provided the physical foundation for defining the allowable energy states of an atom. From the spectroscopic observations, the results conclude not only those spectral lines representable in series form, but all spectral lines arise from the allowed quantum transition. A sketch of the spectral series hydrogen atom is presented in Figure 6.1.

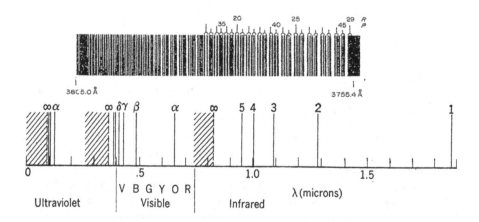

FIGURE 6.1   Rotational spectrum of nitrogen and sketch of spectral series of the atom.

These spectral lines display a series of quantum transitions at discrete wavelengths which is simply the reciprocation of the emitting or absorbing frequency by the electromagnetic waves during the quantum transition. The spectroscopic lines have also been recorded for all atoms and revealed their atomic structure.

According to the Bohr's theory of atom and molecule, the electrons of a molecule is orbiting in a Coulomb field and describing by the quantum number n. The orbits introduce another quantum number l relating to the eccentricity of the orbital shape as the azimuthal quantum number. A general three-dimension motion of electron introduces the third quantum number m_l. From the electronic states of an atom, two more quantum numbers add to accounting for the angular momentum of the electron s as it spins around its axis. Another quantum number of atom m_s is associated with the orientation of an externally applied magnetic field. The dynamic states of an atom under the influence of a magnetic field in rotational and spinning motions are designated by quantum numbers of n, l, m_i and s. Physically, the principal quantum number n is a discrete integer number which is originated from Bohr orbits in radial coordinates. The quantum number, l denotes the azimuthal quantum number related to angular momentum. The magnetic quantum number is related to the orientations of electrons under influence of an external magnetic field. The behavior of electronic motion reveals that the reaction of angular momentum vector in a field of magnetic force is characterized as $m_i = \pm l, \pm(l-1), \ldots$. Finally, the spin quantum number of atoms is designated by s, and it has a fixed value for a single-electron atom of ½, and the component of spin quantum number in an external magnetic field is identified as m_s [3].

In principle, one can construct the eigenfunction corresponding to the energy states of an atom that occupies a definite set of orbits, but the energy state cannot specify which electron occupies a particular orbit. In fact, there are no distinct orbits of the electron, which perhaps shall be described as the shells as electron clouds about the nuclei [4,5]. The assessment has been affirmed by the photographic representation of the electron probability density distribution in a plane that is vertical for several energy quantum states. Modern quantum mechanics shows that a finite probability the electron will be found in the region is unpredictable by the classic theories [1].

Four different elements or subshells exist for each set of energy levels or shells K, L, M, and N: they are designated as S, P, D, and F [5]. The electron configurations of gaseous atoms have been given as [2].

In each shell, the energy levels are further distributed into subshells or elements. In general, the transition of energy level is from the higher s subshell to the lowest p subshell. In most atoms, the energy of the various clouds or subshells lies in the

TABLE 6.1

Electron configurations of gaseous atoms

Shell	K	L		M			N			
Subshell/element	1s	2s	2p	3s	3p	3d	4s	4p	4d	4f

order of *1S, 2S, 2P, 3S, 4S, 4P, 6S, 4F, 5D, 7S, 6D*, etc. Where the *1s* orbit has the lowest energy.

The important atomic structure in the high enthalpy environment is the weakly ionized air and the ground state datum is critical for analyses. For example, *3p* means the quantum numbers $n = 3$ and $l = 1$. The notation of $1S^2$ denotes those two electrons are in the identical *1s* subshell. And $4P^3$ means three electrons are in the *4p* subshell. The general energy level specification of a configuration is *1S, 2P* for the various subshells with the exponent indicating the number of electrons in the L shell, namely, $(1S)^2(2S)^2(2P)^6$. Therefore, the ground state of a nitrogen atom is described as S^2P^3 and for atomic oxygen is S^2P^4. The energy of the next level of the energy state is the energy necessary to remove an electron completely from the field of the nucleus and is the potential for ionization.

The term symbol of slightly different energy levels in a quantum state is indicated by the upper left index 1, 2 or 3 on subshell or element; $^2S, ^2P, ^2D, \ldots$ and the quantum number l for doublets terms is indicated as $^2S_{1/2}, ^2P_{1/2}, ^2P_{3/2}, ^2D_{5/2}, \ldots$ For examples; the lowest orbits of one electron of a hydrogen atom are in the 1S orbit and the ground state is $^2S_{1/2}$. The lowest electron configuration of the nitrogen atom has three electrons $^2P^3$ which give the terms $^4S, ^2D, ^2P$, the ground state is therefore $^4S_{3/2}$. The inverted triplet ground states of oxygen are given as $3p_2, 3p_1, 3p_0$.

The Pauli Exclusion Principle governs the basic rule of quantum mechanics, namely that no two electrons of the same atom can have the same four quanta numbers of the same atom. For example, the unexcited nitrogen and oxygen atoms have the electron configuration of $(1s)^2(2s)^2(2p)^3$ and $(1s)^2(2s)^2(2p)^4$ respectively.

A summarization for the quantum states of atom that describe to the energy state of an atom is best provided by the quantum numbers. There are a total of five quantum number for an atom, the most important quantum number for specifying the energy values of the atom is n, its integer value in theory without theoretic bound. The azimuthal or orbital quantum number l is related to the revolving angular momentum of the electron, again it is an integer limiting to $(n - 1)$ values. The magnetic quantum number m_l is connected to the orientation of an externally applied magnetic field. It has positive and negative integers. The spin quantum number s is related to the angular momentum of the electron by the verified hypothesis. The electron spin produces a magnetic field that is constant in magnitude and orientation and it has a fixed value of ½. In an externally applied magnetic field, the component of the spin quantum number has two possible values of $\pm 1/2$.

6.3 QUANTUM STATES OF A MOLECULE

The molecular quantum number can be described similarly to that of the atom and the interaction between electronic angular momenta and the nuclear angular momenta is important. From the molecular spectroscopic result, the electron spin is also important; both the spin and orbital magnetic moments of the electron and their interaction must be considered [6]. The molecular orientation is also analyzed by quantum mechanics in which the rotational momentum has different directions. Thus, the difference in the quantum number of molecules and atoms appears as an increased distinguishable state or the degeneracy of the same energy level.

There are five quantum numbers that specify the spin-orbit interaction of molecules [3]. The quantum number Λ denotes the component of electron orbital momentum about the internuclear axis and has values according to the value of the electronic state. The orbital momentum is designated as the Σ state for $\Lambda = 0$, and the electron states are invariant about the symmetric axis. Another electronic state is denoted by Π for $\Lambda = 1$, and the Δ state is characterized by $\Lambda = 2$. In addition, the $\pm$ sign associated with Σ state designates whether a reflection in a plane on the symmetric axis lies, and the subscript g (gerde) and u (ungerde) tell whether the wave function remains invariant or not, upon an inversion at the center of symmetry; $\Sigma_g^+, \Sigma_u^+, \Sigma_g^-$ and Σ_u^-.

The total spin angular momenta of the electrons are designated as S. The commonly adopted notations v and J are the vibrational and rotational quantum numbers, respectively. Finally, the orbital angular momentum about the axes perpendicular to the inter-nuclear axis is identified as K. For example, the most diatomic molecules in the electronic ground state are designated as $^1\Sigma$ for $\Lambda = s = 0$ and with a statistical weight of unity.

There are five quantum numbers for describing the spin-orbit interaction [3], but for investigating the air at the hypersonic high-enthalpy condition, the rotational quantum number of the nuclei J, the vibrational quantum number of the nuclei, and the total spin angular momenta of the electrons are the major concerns. The set of electronic-vibration-rotation states of molecules dominates the molecular behavior at the hypersonic high-enthalpy condition. If there is no coupling between rotation and vibration motions for the idealized diatomic molecule, the quantized energy level of rotation motion is.

$$e_r = J(J+1)h^2/8\pi^2 I \tag{6.2a}$$

where the symbol I designates the momentum of inertia about the internuclear axis, and an extra degeneracy factor or the statistical weight arises by the double degeneracy of the rotational levels.

The quantum energy level of vibration is given as.

$$e_v = (v+1/2)hv \tag{6.2b}$$

To describe the dissociation physics of gas at the upper limit of the vibrational quantum level of electrons, the Morse anharmonic energy level provides a modified description.

$$e_v = (v+1/2)hv - x_e(v+1/2)^2 hv \tag{6.2c}$$

The number of energy levels and the number of possible quantum transitions between molecules are much greater than those for atoms. Therefore, molecular spectra are much more complex than atoms [7]. In addition, the individual spectral lines are tightly packed in a certain region, and they are almost continuous. In order to describe the simultaneous changes in the vibrational and electronic energy states of a molecule, information from vibrational spectroscopy is often included such as the bandwidth of the electronic excited states or stable and unstable molecules.

Another designation of the quantum number for the molecule is consistent with the atom and the transition in a bandwidth is especially noticeable. For example, for an oxygen molecule transition occurring in the Schumann-Runge band ($B \rightarrow X$) the energy states are designated as $B^3\Sigma_u^-$ to $X^3\Sigma_g^-$. For a nitrogen molecule transition of the second positive band, $C \rightarrow B$ is described as $C^3\pi_u$ to $B^3\pi_g$, the transition from the first positive band; $B \rightarrow A$ is given as $B^3\pi_g$ to $A^3\Sigma_u^+$. These processes are also the most often encountered transitions in air ionization. Finally, in the Kaplan forbidden band $A \rightarrow X$, the end state is presented as $X^1\Sigma_g^+$. The intensity of allowable vibrational transitions is governed by the Franck-Condon principle (Figure 6.2).

The molecular quantum transitions have been recorded by radiative heat transfer on the Star Dust sample return probe with computational simulation and experimental measurements [8,9]. These frequencies correspond to a wavelength range from 10,000 to 196 nm. At the earlier stage of reentry, the high intensity of surface radiative flux is concentrated in the upper range from 10^4/cm to 2.0×10^5/cm and shifts toward the middle wave-number range. This lower-wave-number range contains the first positive band of nitrogen (740–870 nm), Schumann–Rung system of oxygen (780–850), and the violet system of CN (386.421.5 nm). The radiative emissions in this spectrum are detected by the measurements using an Echelle spectrograph [9]. The atomic nitrogen N and oxygen O, in fact, have lines and multiples (80–1,400 nm) at the higher wave number range up to 1.25×10^5/cm.

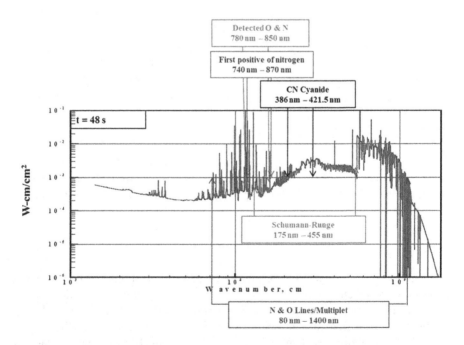

FIGURE 6.2 Quantum numbers associated with emission and absorption wave band.

6.4 SCHRÖDINGER EQUATION

The governing wave mechanics of quantum mechanics is the Hamiltonian of a conservative system for kinetic and potential energy [1]. The second-order partial differential equation for quantum mechanics of a single particle of mass m, and potential function $V(r)$ is the Hamiltonian operator which forms the Schrödinger wave equation.

$$\left(h^2/2m\right)\nabla^2\overline{\varphi} - V(r)\overline{\varphi} + ih\,\partial\overline{\varphi}/\partial t = 0 \qquad (6.3a)$$

It is a type of eigenvalue problem with a spectrum of an entirely different structure. The spectrum consists of continuous and discrete components and the discrete part does not extend to infinity but has a finite point of accumulation [10]. The Schrödinger is linear partial differential equation in $\overline{\varphi}$ and the solutions obey the principle of superposition. The eigen-function can be complex and by the orthogonal theorem the eigen-function and its complex conjugate satisfy the following condition.

$$\int \overline{\varphi}_m^* \overline{\varphi}_n dV = \delta_{mn} \qquad (6.3b)$$

where δ_{mn} is the well-known as the koranic delta function; if $m = n$. $\delta_{mm} = 1$ otherwise $\delta_{mn} = 0$.

In wave mechanics, the motion is generally restricted to a definite region and constituted by a group of waves with different wavelengths and phases without destructive interference. The plausible boundary conditions of the wave functions have required the solution to be single-valued, finite, continuous, and vanished at infinity. These boundary conditions are well-posed and solvable for a certain definite value of potential energy. The types of solutions to the wave equation are known as eigenfunctions. Schrödinger usually imposed the periodic boundary condition for the standing wave phenomena.

For wave mechanics, the equation of wave motion needs to be time-independent to analyze both progressive matter waves and standing waves. The Schrödinger equation is customarily decomposed into time- and space-dependent parts by a defendant variable separation technique. The temporal variation of the eigenfunction is represented by the continuous time-harmonic function.

$$\overline{\varphi}(t,r) = \varphi(r)\exp[-2\pi i(\varepsilon/h)t] \qquad (6.3c)$$

The time-independent Schrödinger equation in spherical polar coordinates (r,θ,ϕ) transforms to the following form with the combined kinetic and potential energy $V(r)$. By Bohr's theory of the atom and molecule, the potential energy of the amplitude equation for the relative motion of the electron and nucleus is given as $V(r) = \varepsilon + Ze^2/2r$, Z is the number of elementary electric charge e. The sum of the internal energy of the molecule includes the contributions by electron rotation and vibration and is characterized by the notation as ε.

$$(1/r^2)\partial(r^2\,\partial\varphi/\partial r)/\partial r + (1/r^2\sin\theta)(\sin\theta\,\partial\varphi/\partial\theta) + (1/r^2\sin^2\theta)\partial^2\varphi/\partial\phi^2 \qquad (6.3d)$$

$$+(2\mu/h^2)[\varepsilon + Ze^2/r]\varphi = 0$$

The eigenfunction φ of the three-dimensional wave equation is sometimes called the probability amplitude function which is required to be single-valued, finite, and continuous for all physically possible values in space. In Schrödinger's equation, the symbol μ denotes the reduced mass of the interacting particles $\mu = (m_1 + m_2)/m_1 m_2$. The eigenfunction of the time-independent Schrödinger equation is separable along the three-dimensional orthogonal coordinates in space, and the separated eigenfunctions are the product of its independent components.

$$\varphi(r) = R(r)\Theta(\theta)\Phi(\phi) \tag{6.3e}$$

$$\left(1/r^2\right)\partial\left[r^2(\partial R/\partial r)\right]/\partial r + \left[(2\mu/h^2)(\varepsilon + Ze^2/r) - l(l+1)/r^2\right]R = 0 \tag{6.3f}$$

$$(1/\sin\theta)\partial(\sin\theta\,\partial\Theta/\partial\theta)/\partial\theta + \left[l(l+1) - m_l^2/\sin^2\theta\right]\Theta = 0 \tag{6.3g}$$

$$(1/\Phi)\partial^2\Phi/\partial\phi^2 = -m_l^2 \tag{6.3h}$$

From a pure mathematic viewpoint, the independent eigenfunctions are of the second-order, ordinary differential equations. An extraordinary feature of these equations, Equation (6.3f)–(6.3h), the separation constants l and m_l are recognized as the azimuthal quantum number and the magnetic quantum number of quantum mechanics. The Equation (6.3f) is known as the radial Schrödinger wave equation and is the equation appropriate to spherically symmetrical interaction potential $V(r)$. The solution to Equation (6.3f) is known in terms of Laguerre polynomials $L_n(r) = e^r\,d^n(r^n e^{-r})/dr^n$ [10,11]. This equation also introduces the principal quantum number n which is the most important quantum number for specifying the energy value of an atom. The solution to Equation (6.3g) is also a series solution, $\Theta(\theta) = \sin\theta^{|m_l|}P_l^{m_l}(\cos\theta)$ which is the associated Legendre polynomials [11].

The quantum solution of translational particle dynamics is the simplest to obtain along any one of the three-dimensional frames of reference. The one-dimensional translational motion in the x-coordinate between two reflecting walls with a separation distance of l can be solved easily without the potential energy $V(r)$ for the Schrödinger wave equation.

$$\partial^2\varphi/\partial x^2 + (2m/h^2)\varepsilon\varphi = 0 \tag{6.4a}$$

The general analytic solution to Equation (6.4a) is the simple trigonometry function $\varphi = a\sin\sqrt{2m\varepsilon/h^x}\,x$. The solution of the eigenfunction is obtained after the satisfied boundary condition of the vanished wave function at the walls of $\varphi(0) = 0$ and $\varphi(l) = 0$.

$$l\sqrt{2m\varepsilon/h^2} = n_x\pi \tag{6.4b}$$

where n_x is the quantum number of an elementary particle in translational motion. From Broglie's hypothesis, the wavelength λ associate associated with the wave of moving mass at the velocity is $mu = \lambda/h$. The length between the perfect reflecting walls must have the value of $l = n_x \lambda / 2$ or $\lambda = 2(n_x/l)$. The kinetic energy of the moving particle is $(mu)^2/2m = (h/\lambda)^2/2m$, thus the discrete translational energy of a particle in a force-free field becomes.

$$\varepsilon_x = h^2 n_x^2 / 8ml^2 \qquad (6.4c)$$

For a three-dimensional translational motion, the translation quantum energy of the Schrödinger wave equation is the sum over three coordinates. On the Cartesian frame, the energy is

$$\varepsilon_t = (h^2/8m)\left[(n_x/l_x)^2 + (n_y/l_y)^2 + (n_z/l_z)^2\right] \qquad (6.4d)$$

The constants n, l, and m_l are resulting from the separation of the dependent variables or the eigenfunctions along the orthogonal coordinates from the three-dimensional Schrödinger's wave equation. These energy eigenvalues obtained during the solving procedure of the wave equation are verified completely with the original Bohr's theory. The only quantum number s is not directly derived from Schrödinger's equation. The energy level is associated with the electron spin, for this reason, is considered as the fine structure energy level of a one-electron atom. In 1925 just before Schrödinger's theory appeared, the quantized energy shifts were considered to be the consequence that electrons possess angular momentum and a magnetic moment. These forces and moments would interact with the force field of the nucleus. The hypothesis proved to be correct to become an intrinsic angular momentum as one of the fundamental properties of the elementary particle [1]. Based on the realization an electron does possess an intrinsic angular momentum or spin. The magnitude of the spin angular momentum is assigned to half-unit quantum numbers to a single possible eigenvalue along an axis. The quantity is known as the Bohr magneton $\pm eh/2m$. From spectroscopic results, the correct energy state that some energy states require half-integer quantum numbers for the observed phenomena. Therefore, the quantum numbers for discrete energy levels elevated to five; n, l, m_l, m_s, and s.

The eigenvalue of the separated wave equation $l(l + 1)$ appears in Equations (6.3f) and (6.3g) are independent of their eigenfunctions. The number of independent wave functions associated with an energy level is called the degeneracy or the statistical weight of that energy level. It is the consequence from the intrinsically complex atomic and molecular structure of any matter. The degeneracy from solving Schrödinger's wave equation actually defines how each energy level connects to the atomic and molecular structure.

Schrödinger's wave equation is equally applicable to obtain the solutions for rotation and vibration excitations of a molecule with simplified physical models. For rotation motion, a simple rigid rotor is adopted for the purpose, and the model assumes the distance between mass center r is held as constant. The quantum wave equation is reduced by limiting the potential energy to only contain the internal energy ε of the molecule. Now, the wave equation acquires the simplified form.

$$(1/r^2 \sin\theta)\partial(\sin\theta\, \partial\varphi/\partial\theta)/\partial\theta + (1/r^2 \sin^2\theta)\partial^2\varphi/\partial\phi^2 + (2\mu\varepsilon/h^2)\varphi = 0 \quad (6.5a)$$

Or

$$(h^2/2I)\left[(1/r^2\sin\theta)\partial(\sin\theta\,\partial\varphi/\partial\theta)/\partial\theta + (1/r^2\sin^2\theta)\partial^2\varphi/\partial\phi^2\right] + \varepsilon\varphi = 0 \quad (6.5b)$$

Apply the separation of variable technique to the wave function into two orthogonal angular displacements as before to acquire the following independent ordinary differential equation and an algebraic equation.

$$(h^2/2I)\left[(1/\sin\theta)d(\sin\theta\,d\Theta/d\theta)/d\theta - (u^2/\sin^2\theta)\Theta\right] + \varepsilon\Theta = 0 \quad (6.5c)$$

$$\Phi = \exp(\pm iu\phi)$$

The solution of the first part of Equation (6.5c) is found and provides the rotation energy with a set of quantum numbers $J(J + 1)$. The momentum of inertia is designated by the symbol I with respect to the axis of mass centers of the rigid rotor.

$$\varepsilon_r = J(J+1)h^2/2I \quad (6.5d)$$

For describing the vibrating electrons dynamics, a simple harmonic oscillator is adopted with two rigid masses connecting by a spring with a constant of f, and the displacing distance from its equilibrium position is denoted by its location x. Thus, the potential energy of the Schrödinger wave equation is $\varepsilon - fx^2/2$. The two connected mass centers are assumed to have the same mass, so the reduced mass is simply $\mu = m$. The Schrödinger wave equation acquires the following form:

$$\partial^2\varphi/\partial x^2 + (2m/h^2)(\varepsilon - fx^2/2)\varphi = 0 \quad (6.6a)$$

The oscillating frequency of the harmonic system is known from classic mechanics as

$$v = \sqrt{f/2m\pi^2} \quad (6.6b)$$

The vibration energy with the vibration quantum number of v becomes. However, the two frequencies may not be synchronized.

$$\varepsilon_v = (v+1/2)hv \quad (6.6c)$$

At this point, the basic quantum solutions of the translation, rotation, and vibration motions have been illustrated. The results from the Schrödinger wave equation duplicate the classic Bohr's theory for the hydrogen atom. In addition, a unique characteristic of quantum mechanics stands out from the classic mechanics, in that a specific eigenvalue can associate with several independent eigenfunctions. The resulting degeneracy from multiple eigenfunctions associated with an energy level provides an incisive understanding of the intriguing internal structure of any elementary particles. All the presented results are obtained by solving Schrödinger's wave equation with simple physical models to demonstrate the feasibility of applying quantum mechanics to particle dynamics with internal atomic and molecular structures. Continuous refinements are still required for a better model for the physical-consistent description of microscopic-scale particles.

In any event, quantum numbers are indisputable conclusions based on many physical observations. From a pure mathematics viewpoint, the quantum numbers are eigenvalues of the partial differential wave equation for each degree of freedom in space. The solutions of the wave equation represent a set of discrete stationary energy states for an atom, characterized by the five quantum numbers n, l, m_l, m_s, and s for three-dimensional motions.

An important rule governing the allowed quantum numbers for an electron in an atom is known as the exclusion principle enunciated by Pauli in 1924. The Pauli principle states that no two electrons can exist in the same quantum state, therefore no two electrons in an atom can have all the four same quantum numbers n, l, m_l, and m_i. Pauli's exclusion principle is universally true to eliminate any ambiguous electronic configuration of an atom. The exclusion principle is especially important for the completeness rather the incompleteness of the electron shells of atoms.

6.5 RELAXATION OF QUANTUM TRANSITION

A major challenge of expanding hypersonic flows beyond the realm of classic gas kinetic theory is the requirement to include the quantum transition in high-enthalpy conditions. When the temperature of compressed air downstream to a bow shock is elevated over 8,000 K, the properties of air differ considerably from the perfect gas, because the excitations of the internal structure of gas particles become significant [12]. At the temperature reaches 1,500 K, the vibrational excitation of oxygen molecules begins to occur, and dissociation decomposes into separated atoms around 3,000k. Meanwhile, chemical reactions generate a certain amount of nitric oxide, which dissociates into oxygen and nitrogen atoms as the temperature is elevated further. Beyond 5,000 K, the oxygen molecules are nearly completely dissociated together with some nitrogen molecules. Ionization of atomic oxygen and nitrogen, and nitrogen and nitric oxide molecule starts at these temperatures. An appreciable amount of free electrons equal number of positively charged ions is also presented when ionization takes place. The internal degrees of freedom of molecules include the translational, rotational, vibrational, and electronic modes in excited states. All these internal excitations are quantized, and the only bridge between the microstate of individual particles and the macroscopic properties of the gas mixture is linked by statistic mechanics through theories of probability and distribution functions. Only in an equilibrium state, the internal energy distributions of a gas molecule or atom have been made easily understood by quantum mechanics through the partition functions.

Quantum transitions dominate the atomic and molecular in the high enthalpy state of gas, the elastic collisions of particles in this state are subject to strict quantum laws and the radiation emitted by the gas is also quantum restricted. The energy transfer or cascading between quanta is noted to follow the ladder-climbing and the big-bang processes [3,13]. These two processes are depicted in Figure 6.3 for the vibrational internal degree of freedom and the different processes can happen in any internal mode. The big-bang quantum jump occurs only when the collision frequency is high and the external energy source supply is an amble, the accumulated energy

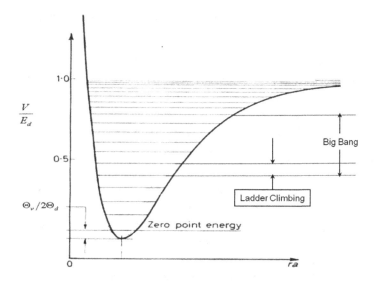

FIGURE 6.3   Possible quantum transition processes in vibration excitation.

by collision permits the transition accomplished over multiple quanta. However, the relaxation of the quantum transition of molecules presents difficulty of describing the discrete transition.

The equilibrium condition either in a chemical reaction or quantum transition subjects to a relaxation phenomenon to a disturbance that displaces it from equilibrium [2]. The rate of restoration follows a first-order kinetic law providing the displacement from the equilibrium is small. A molecule with random velocities will gain or lose energy by collision, and its population at that location will be changed also. The rate of change of number density can be given as.

$$dn_i/dt = c_{ji}n_j - c_{ij}n_i \tag{6.7a}$$

In Equation (6.7a), the notations c_{ij} are constants of collision transition probabilities. At the equilibrium condition, the rate vanishes, $c_{ji}n_j - c_{ij}n_i = 0$. At this condition, the total number of molecules is the sum of the populations in two states $n = (n_1 + n_2)_e = (n_1 + n_2)$. After some algebraic manipulation to get the net rate of change in terms of the initial states, it yields.

$$dn_i/dt = \left(k_{ij} + k_{ji}\right)\left(n_{ie} - n_i\right) \tag{6.7b}$$

The reciprocal of the rate constants of collision probabilities for restoration of equilibrium is called the relaxation time. In short, it is the elapsed time required for a chemical reaction to occur when the interacting system accumulates enough energy to overcome the activation energy barrier. For the quantum transition, it is the elapsed time needed for the colliding partners to accumulate a sufficient amount of energy for a quantum jump.

$$\tau = 1/(c_{ij} + c_{ji}) \tag{6.7c}$$

For all first-order processes, the displacement from the equilibrium state is $\Delta n/\Delta n_e = \exp(-t/\tau)$. The relaxation rate equation now can be written as.

$$dn/dt = (n_e - n)/\tau \tag{6.7d}$$

Therefore, the net rate of energy gain per unit volume is equal to.

$$d\varepsilon/dt = \varepsilon_i (dn_i/dt) + \varepsilon_j (dn_j/dt) \tag{6.7e}$$

Equations (6.7d) and (6.7e) constitute the foundation for analyzing the relaxation phenomena for molecules in quantum transition.

6.6 CONSERVATION EQUATIONS WITH QUANTUM TRANSITION

For molecular nitrogen, oxygen, and nitric monoxide, the characteristic temperatures of rotational excitation at the equilibrium states are 2.86, 2.07, and 2.42 K respectively [3]. Therefore, the rotational mode for air molecules is fully exited in most circumstances and equilibrates with the translational mode; the two air components are analyzed collectively. From quantum mechanics, the energy states of all air internal citations have discrete spectra, except the translational mode which pacts so tightly so is approximated as continuous.

In the region of mostly hypersonic flow immediately downstream to the enveloping bow shock, the vibration energy level of the gas is far lower than that of translational mode. Furthermore, the energy transfer among the internal modes of excitation is dominated by the adjacent quantum levels, the so-called ladder climbing process. The amount of energy transfer between internal degrees of freedom is limited and the vibrational quantum states in hypersonic flow are mostly limited to the first few quanta above the ground level. Under this environment, the anharmonic correction to the quantum vibrational mode usually may not always be necessary.

The governing equations of high-enthalpy hypersonic flows constitute a multiple-disciplinary differential system. The nonlinear incompletely parabolic partial differential system of equations couples the compressible Navier-Stokes equations and the Maxwell equations of electromagnetics in the time domain incorporating quantum mechanics [14,15]. In the formulation, all molecular and ionized components of air occupy different quantum levels and are treated as distinct species. Since there is a huge disparity in time scales between the particle collision rate and the quantum transition, the energy transfer of quantum jumps is treated as the source term in the conservation laws. The quantum effects are implicitly contained in the thermodynamic and kinetic variables and only the quantum energy contributions are only explicitly appearing in the conservation laws of vibration and electronic energy equations, and finally the distribution or the redistribution of all internal energy of the system. The modeling of quantum transition is focused on energy conservation laws.

The energy transfer during the quantum transitions of the internal degrees of freedom between translational, vibrational, and electronic modes are described by the individual conservation energy laws. The rotational mode of internal excitation is exempted from the formulation because only a few collisions are needed for rotational excitation to reach thermal equilibrium with the translational mode.

When air is ionized, the medium becomes an electrically conductive and the electromagnetic force will be automatically presented. The Joule heating, electrostatic forces from space charge separation, and Lorentz force from an externally applied magnetic field need to be included by integrating the Maxwell and Navier-Stokes equations. For now only a specific connection to the quantum transition will be delineated. According to Coulomb's law, the electrostatic force between charged particles q_i and q_j which measured in Coulomb is a collinear force along the unit space vector by a distance of r_{ij} [16]. The electrostatic force by the Coulomb's law is.

$$F_{ij} = q_i q_j / 4\pi\varepsilon_o\, r_{ij} \qquad (6.8a)$$

In Equation (6.8a), the electric permittivity is denoted by ε_o ($\varepsilon_o = 8.85 \times 10^{-12}$ Farad/m). The movement of the charged particles also leads to a conductive electric current $J = \Sigma n_i q_i u_i$. In turn, the current in the electrically conducting medium will generate a differential magnetic field intensity which is governed by the Biot-Savart law of magnetostatics [17]. The definition of magnetic induction is given as $dB_{ij} = (\mu_m J_j / 4\pi)\left[(dl_j \times (1/r_{ij}))\right]$ and μ_m is known as the magnetic permeability and in free space, it has a value of $4\pi \times 10^{-1}$ Henry/n. The electromagnetic force on a current element is.

$$dF_{ij} = J_i dl_i \times dB \qquad (6.8b)$$

Therefore, a moving charge in an externally applied steady and uniform magnetic field with a magnetic flux density B will produce a pushing force normal to B, known as the Lorentz force or acceleration.

$$F_m = qu \times B = J \times B \qquad (6.8c)$$

These remote-action forces will contribute to the conservation of the momentum equation as an external applied force and work done to the system through the conservation of the energy equation.

Joule heating is also known as Ohmic and resistive heating [18]. It is the electromagnetic field does work to overcome the resistive force and dissipates heat in response to resistance. The amount of energy released is proportional to the square of the current. This relationship is known as Joule's first law and the energy was subsequently named after Joule by the symbol J (J = 1 Newton-meter or $J = 10^7$ erg).

The most general and fundamental formula for Joule heating is,

$$q = E \cdot J = I^2/\sigma = \sigma E^2 \qquad (6.8d)$$

For high-enthalpy hypersonic applications, the governing equations are actually an approximation to the classic magneto-hydrodynamics equations for low magnetic Reynolds number [19,20]. In gist, only the electrostatic force and Lorentz acceleration are appending to the momentum and the Joule heating to the conservation of energy equation.

The vibrational energy conservation equations for either the diatomic or polyatomic molecular species are [21,22].

$$\partial \rho_i e_{iv} / \partial t + \nabla \cdot \left[\rho_i (u + u_i) e_{iv} + q_{iv} \right] = Q_{V,\Sigma} \tag{6.9a}$$

In Equation (6.9a), the subscript i denotes each individual vibration-excited species at different quantum levels. Since the vibration internal degree of freedom of the molecule does not contribute to a partial pressure, the work done by pressure is zero and only conductive heat transfer within the control volume is still possible. The source term $Q_{V,\Sigma}$ is the sum of collective energy transfer by the vibration energy from translational-vibrational and cascading form electronic modes.

The electronic energy conservation equation for electrically charged species have been traditionally given as

$$\partial \rho_e e_e / \partial t + \nabla \cdot \left[\rho_e (u + u_i) e_e + u \cdot p_e \overline{\overline{I}} + q_e \right] \tag{6.9b}$$

$$= E \cdot J + \left[\rho_e E + (J \times B) \right] \cdot (u + u_i) + Q_{e,\Sigma}$$

In Equation (6.9b), the symbols $\overline{\overline{I}}$ designate the identity matrix, the electronic partial pressure p_e, and conductive heat transfer q_e respectively. The ionized gas automatically generates an electromagnetic field with an electrical field intensity E (Volt/m) and a magnetic flux density B (Weber/m^2). On the right-hand side of Equation (6.9b), the leading term represents the Joule heating $E \cdot J$ within the control volume and follows by the work performed to the system by the combination of the electrostatic $\rho_e E$ and Lorentz acceleration $J \times B$. The last term is the electron energy source or sinks within the control volume which is the total vibration energy added or removed from the system by the electronic quantum transition.

The conservation of energy law for the internal energy of the system is expanded to include the electromagnetic internal energy that is created by the electrically charged species (electron and ion) and the energy transfer within the control volume by the net amount of the quantum energy transition $(Q_{vt} - Q_{et})$.

$$\partial \rho e / \partial t + \nabla \cdot [\rho e (u + u_i) - \kappa \nabla T + \sum \rho_i u_i h_i + q_{rad} + u \cdot p \overline{\overline{I}} + u \cdot \overline{\overline{\tau}} \tag{6.9c}$$

$$= E \cdot J + \left[\rho_e E + (J \times B) \right] \cdot (u + u_i) + Q_{vt} - Q_{et}$$

The shear stress tensor $\overline{\overline{\tau}} = \lambda_g \nabla \cdot u \overline{\overline{I}} + \mu_g \, \text{def}(u)$ in Equation (6.8c) has two components that are proportional to the rate of strain by the bulk molecular viscosity λ_g and viscosity coefficients μ_g. The rate of strain also consists of two components for the translational motion $\nabla \cdot u$ and the rotational gas motion is described by the deformation tensor of rank two; $\mathrm{def}(u) = \nabla u + (\nabla u)^T$. On the left-hand side of the equation, the divergence bracket contains the conventional Fourier's law for conductive heat transfer, convective heat transfer by different species diffusion velocities, and the radiative heat exchange as a source or a sink. The last two terms give the work done by the pressure and the energy dissipated by the shear stress.

The left-hand side of the equation includes the energy contributions by the electromagnetic field and the energy transfer by quantum jumps. The conservation of energy of the system is now balanced by the energy cascading and transmitting from the translational mode to vibrational and electronic internal excitations of each chemical species, and the electromagnetic field of the ionized medium.

The definition of the internal energy that needs to include the kinetic and thermal energy of charged species is now given as

$$\rho e = \sum_{i \neq e} \rho_i \left(c_{V,i} T + (u \cdot u/2) \right) + \sum_{i \neq e} \rho_e e_{V,i} + \sum_{i \neq e} \rho_i h_i^o + \rho_e \left(c_{V,e} T_e + (u_e \cdot u_e)/2 \right) \qquad (6.9d)$$

where i denotes the specific eleven species of the air mixture and h_i^o is the standard heat of formation for all reacting species. The kinetic energy of the electron $(u_e \cdot u_e/2)$ is also included in the last term.

6.7 QUANTUM JUMPS MODELING

The detailed balancing principle plays an important role in the development of quantum modeling for high-enthalpy hypersonic flows, and it has been considered the universal law of nature [19]. The characteristic feature of physical equilibrium is that for any process that disturbs the distribution, there is an inverse process with equal frequency and just offsets each other's effects returning to equilibrium. In other words, under the condition of thermodynamic equilibrium, the differential rates for each micro-process and its corresponding inverse process are equal. This principle has even applied equally well to the macro-process.

From the principle of detailed balance, the classic result by Landau and Teller [23] becomes the backbone for the quantum chemical-physical kinetics modeling. The energy exchange between translational and vibrational modes is approximated by Equation (6.7c).

$$Q_{t,v} = \rho \left[(e_v)_e - e_v \right] / \tau_{vt} \qquad (6.9a)$$

$$(e_v)_e = R\Theta_v / \left[\exp(\Theta_v/T) - 1 \right] \qquad (6.9b)$$

$$\tau_{vt} = k_1 T^{5/6} \exp(k_2/T)^{1/3} / p[1 - \exp(-\Theta_v/T)]$$

where Θ_v is the characteristic temperature of the vibrational excitation, and the notation τ_{vt} is the relaxation time for vibration transition.

The energy transfer among the vibrational modes of different species is also modeled by the ladder climbing relation via a single quantum level. A large amount of experimental measurements [24] are available on the vibration relaxation of a simple system. Especially, the database is collected for the excitation probability of diatomic molecules by the translation-to-vibration transition via collision with an atom, excited or unexcited molecule. An empirical equation correlated from data on relaxation time guided by the Landau-Teller theory is developed by Millikan and White [25].

$$\log p\tau_{vt} = 5\times10^{-4}\mu^{1/2}\Theta_v^{4/3}T^{-1/3} - 0.15\mu^{1/4} - 8.0 \tag{6.10a}$$

or

$$\tau_{vT} = 1.16\times10^{-3}\mu^{1/2}\Theta_v^{4/3}\left[T^{-1/3} - 0.15\mu^{1/4} - 18.42\right] \tag{6.10b}$$

where the symbol $\mu = m_i m_j/(m_i + m_j)$ is the reduced mass of colliding partners. Once the relaxation rate is known, the energy transfer can be obtained by Equation (6.9a).

From flight data of reentering space vehicles, the overall dissociation rate has shown strong dependence on the population density and the probability of the vibration quantum levels. In the earth reentering-temperature range, the simple harmonic vibrator model fails when the relaxation time for vibration becomes comparable to that for dissociation. The excited vibration quantum level nearly determines the overall rate of dissociation, or the dissociation is preferentially from the higher vibration quantum levels. The preferential dissociation model by Teanor-Marrone is therefore often adopted for the energy transfer between electronic and vibrational excitations at this upper limit of vibration excitation states [26]. The basic idea is that the rate of molecular dissociation can occur preferentially from a higher vibrational quantum level and is derived from the probability of dissociation from the excited vibrational states.

$$Q_{e,v} = \sum_i \rho_i \left[(e_t - e_v)/\tau_{vt} - (\bar{e}_{t,v} - e_v)/x_i\,(dn_i/dt)_f + (\bar{e}_{v,t} - e_v)/x_i\,(dn_i/dt)_b\right] \tag{6.11}$$

In the coupled vibration and dissociation (CVD) excitation model Equation (6.11), n_i denotes the number density of the interacting particles. The subscribe f and b indicate the rate of change in species number density for the forward and backward transition processes. The symbols $e_{t,v}$ and $e_{v,t}$ characterize the average energy lost from vibration in dissociation and gained by recombination. The symbol e_v is the energy corresponding to the Boltzmann distribution of the vibrational excitation or under the equilibrium condition.

The energy transfer between electron-ion collisions has been approximated by a collection of formulations via the Coulomb logarithm through the elastic collisions between electron and molecule, and the elastic collision between electron and neutral species collision [27–32]. All these models are derived from some unique insights and are partially supported by experimental observations. As the consequence, they all have a limited and uncertain range of accurately applicable restrictions.

For energy transfer by electrons through elastic collisions with ions, the model equation is a function of the Coulomb logarithm;

$$Q_{e,i} = 1.21\times10^{20}\,x_e x_i \left[(T - T_e)/T_e^{3/2}\right]\ln\Lambda \tag{6.12a}$$

Another important energy exchange between electrons and heavy neutral species is approximated by an empirical equation given as.

$$Q_{et} = 3.378 \times 10^{10} x_e x_t \sqrt{T_e} \, (T - T_e) \left[1 - \left(1 + T_e/T_t^* \right)^{-1} \right] \tag{6.12b}$$

In Equation (6.12b), the average translational temperature of neutral species is designated as T_t^* and the notations x's are the molar fractions of species in electronic and translation degrees of freedom. This important subject of modeling electronic quantum transition in high-enthalpy hypersonic flows remains to be a necessarily sustained research focus in basic research for hypersonic flows.

The more recently model of vibration-electron transition is the Chernyi-Losev-Macheret-Potapkin model [33].

$$Q_{ev} = 2 \times 10^{-16} \sum_v n_e n_i v_{e,v} P_{1,0,v} + \exp{-\left(1.44 v_{e,v}/T_e\right)}/\left[1 - \exp(1.44 v_{e,v}/T_e\right] \tag{6.12c}$$

$$+ \exp{-\left(1.44 v_{e,v}/T_v\right)}/\left[1 - \exp(1.44 v_{e,v}/T_v\right]$$

$$p_{1,10,v} = 0.45 \times 10^{-19} \exp{\left(-10000/T_e\right)}$$

In Equation (6.12c), the symbols n_i's denote the number density of species i and $v_{e,v}$ which characterizes the collision frequency between electron-vibration modes of a specific species. Where the notation $P_{1,0,v}$ is the probability of excitation at the first vibrational level by electron impact with a molecule, and $v_{e,v}$ describes the characteristic frequency of the molecule vibration. In the above formulation, T_e, T_V denote the electronic and vibration temperatures and n_e, n_V represent the concentrations of electrons and molecules. The electronic-vibration kinetics plays a significant role for strong shock waves. There are two reasons for the particular formulation; first, transition is the resonant interaction of electrons with excited vibrational states of diatomic molecules like nitrogen and oxygen. The second reason is the fact that the electronic temperature determines the intensity of several significant radiative processes including the spectral emissivity of diatomic molecules from electronically excited states.

The total energy transfer in the vibrational and electronic modes is just the sum of each possible energy mechanisms, $Q_{v,\Sigma} = Q_{t,v} + Q_{v,v} + Q_{e,v}$ and $Q_{e,\Sigma} = Q_{e,v} + Q_{e,i}$. The internal mode energy exchanges given by Equations (6.8a)–(6.8e) are just a typical implementation of the quantum chemical-physical models. Additional internal modes of energy transfer or energy cascading processes between internal modes can be added to the sum according to the specific circumstance. It is obvious, some of these models are developed from a rigorous theoretic consideration but others are simply devised by empirical means. Therefore, there are ample rooms for improvement as a promising new frontier for basic research.

In fact, the kinetics of the species exchanges among the quantum states of vibration-translation, vibration-vibration, and vibration-dissociation have been studied by solving the master equation for population density distribution [34–37]. From the basic research viewpoint, sustaining efforts by more fundamental experimental investigations coupled with the *ab initio* (from the beginning) computational approaches could be the most fruitful [38].

6.8 VALIDATING BY FLIGHT DATA

The high-enthalpy hypersonic environments are nearly impossible to duplicate by ground-based testing facilities, and any assessment for the quantum transition modeling under hypersonic conditions can only be conducted by verifying by flight test measurement. For this reason, a wide range of data collecting space flight probes have been devised and operated over the past years until today. The earth reentry Stardust sample return probe is the fastest manmade space vehicle and is characterizing the most typical high-enthalpy hypersonic condition [39]. The Stardust configuration considered is a 60° one-half angle spherical cone with a nose radius of 0.2286 m and a base radius of 0.4064 m. The corner radius of this configuration is merely 0.02 m. The nominal freestream conditions for computational simulation can be summarized as the air density of 2.34×10^{-4} kg/m, and the ambient temperature of 238 K at the reentry velocity of 11.137 km/s. For thermal protection, the probe utilized the ablative coating by the most widely adopted silicone-impregnated reusable ceramic (SIRCA) or phenolic impregnated carbon ablator (PICA) reinforced by secondary impregnation of polymethyl methacrylate (PMMA) to maximize the pyrolysis gas generation so to minimize the char surface recession [31]. The chemical composition within the shock layer over the ablating probe increases to 18 chemical species including the molecular species C_2, C_3, CO, CO_2, CN, and HCN; the atom C, H, and positively charged C^+ and H^+.

In spite of these added complexities by the multi-phase ablating phenomena, comparable results are still obtained by chemical kinetic models of the Stardust capsule [30,32]. At the peak heating condition, the chemical components distribution illustrates the essential and unique features along the stagnation streamline within the shock layer. There are visible differences in the calculated composition between the results of the two quantum chemical-physics models. The discrepancy also appeared in predicted standoff distances for the bow shock wave, and yet the calculated heat transfer rates are within a few percent of each other. The disparity among computational simulations is contributed by the accumulated computational error by different numerical resolution criteria plus quantum transition modeling that becomes the main quandary in requirements between engineering and scientific disciplines.

Figure 6.4 displays the chemical compositions of all 18 chemical species along the stagnation streamline within the shock layer. The incoming flow enters from the right side of the graph; the bow shock immediately compresses the air at a standoff distance of around 1.2 cm from the stagnation point. The ablating chemical compositions are concentrated mostly over the solid surface over the stagnation region as expected. In the shock layer, the gas mixture change rapidly but the species concentration in the mole fraction for both molecular and atomic nitrogen is in the order of $2.0 \times 10^{-2} < x_N < 4.0 \times 10^{-1}$ in the major portion of the shock layer. The atomic oxygen persists in the range around $1.1 \times 10^{-1} < x_o < 2.0 \times 10^{-1}$ between the bow shock and the stagnation point. Only the molecular oxygen depletes sharply from

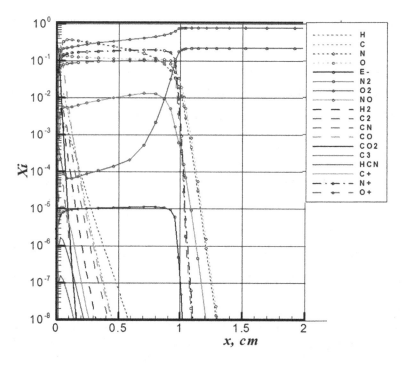

FIGURE 6.4   Shock layer chemical composition in Stardust probe.

the value of $x_{o_2} = 1.8 \times 10^{-1}$ to the value of $x_{o_2} = 2 \times 10^{-4}$ toward the stagnation point. The electron and the equal amount of positively charged ion only attain a maximum concentration of $x_e = 10^{-5}$. In terms of the particle number density per unit volume, it amounts to $n_e = 2.0 \times 10^{12}/cm^3$, in comparison with the atomic nitrogen and oxygen in the value around $n_i = 10^{16}/cm^3$. The specific comparisons indicate that the ionized species only appears only as a tracing amount in the shock layer from typical earth reentering hypersonic flows. However, the presence of ionized air profoundly alters the behavior of electromagnetic or microwave wave propagation in the now electrically conducting air medium.

The electromagnetic wave will be dissipated and particularly when the frequency of the transmitting wave is lower than the plasma frequency of the globally neutral ionized air or plasma, the incident wave is completely reflected [19,40]. The plasma frequency of ionized air around the earth reentering space vehicles like the Stardust probe usually has the value of about 8.7×10^9 Hz, which is within the H and X bands of the microwave frequency. The dissipated or reflected microwave leads to the communication blackout phenomenon [41].

In view of the fact that the electronic quantum modeling is derived mostly from the empirical formulation and the ionized species are presented only at a perturbation level, a close assessment of the relative significance is warranted. The translation-electron, vibration-electron collisions, and the energy cascading from electronic to vibrational excitations are examined to ascertain the dominant mechanisms. The advantage of computational simulations is easily accomplished by just suppressing the models of quantum energy transfer via the quantum chemical-physics process for the simulated condition. Two identical computational simulations are performed side-by-side; one of the computational results includes the complete quantum transition models, and the other just omits the electronic transition model. The translation, vibration, and electronic temperatures along the stagnation streamline of two different models are given in Figure 6.5. For this purpose, the temperatures of different internal degrees of excitations are calculated by assuming the Boltzmann distributions are still applicable.

In this presentation, the numerical results of the full complement of model equations are depicted on the right-hand side of the figure. First of all, the vibrational temperatures of O_2, N_2, and NO between the two different models are essentially identical. The computed translational temperatures are bracketed between the values of 36,400 and 39,000 K to fall within the range of all known numerical results [21,30–32]. The two comparative results are essentially identical; the vibrational temperatures of N_2, O_2, and NO calculated by assuming a Maxwell-Boltzmann distribution are unaffected. The lower translational temperature from the full complement of energy conservation equations correctly reflects the fact that a portion of the converted kinetic energy from the hypersonic freestream is distributed to the electronic excitation. The difference in the predicted translation temperature is <7% and there is no creditable verification by either experimental observations or theoretical analyses. However, the unpredicted population density of the ionized species has a significant impact on the radiative energy exchange. The shortage of a validating experimental database becomes a glaring concern.

A temporal progression is depicted in Figure 6.6 for the sequence of quantum energy transitions at the elapsed time from 48 to 54 seconds measured from the reentry interface reference point. Again, the translational temperature of the gas mixture

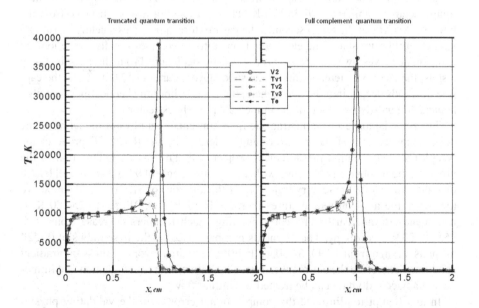

FIGURE 6.5 Thermodynamic state with/without the electronic quantum transition.

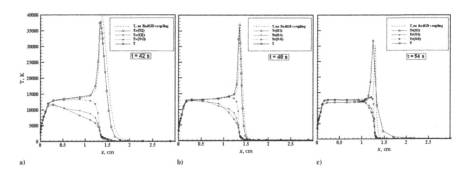

FIGURE 6.6   Progress of internal distributions in earth reentry by the quantum transition.

and vibrational temperatures of molecular nitrogen, oxygen, and nitric oxide in the
shock layer are obtained assuming the Maxwell distribution is still held approximately [32,42]. The translation temperature is also double-checked by adopting two
different chemical kinetic models.

At the initial stage of reentry, most of the kinetic energy from the oncoming stream
is converted mostly into translation mode; immediately downstream to the bow shock,
a maximum translational temperature is 36,400 K. The vibration temperature of nitric
oxide increases faster than both the molecular nitrogen and oxygen. In the immediate
post-shock region, the vibrational temperatures of N_2, O_2, and NO reach a value of
11,000 K. But at a location of 0.02 cm upstream to the stagnation point, all internal
modes equilibrate with the translational temperature. After a time elapse of 6 seconds,
the temperature distribution along the stagnation streamline remains similar. The maximum electron temperature is 14,500 K but only exists for a short distance from the
bow shock. Only at the $t = 54$ seconds, the gas mixture approaches a nearly thermodynamic equilibrium state. The electronic temperature decreases rapidly as the ionized
component moves away from the bow shock by depletion. From the computational
results, the electronic temperature in the shock layer is around 12,000 K. As the capsule descends deeply into the atmosphere, the nonequilibrium chemical-physical phenomena are mostly restricted in the immediate post-shock region.

The best database for validating the quantum chemical-physics modeling is the
direct measurement of the electron number density by the RAM-CII probe of the
NASA radio attenuation measurement project [43]. The basic configuration of the
probe is a hemispheric-nosed cone with a semi-cone angle of 9° and an overall length
of 1.2954 m. Along the reentry trajectory at the altitudes of 81, 71, and 61 km above
the earth, the ambient temperatures increase from 198.64, 219.58, to 254.80 K as
it descends toward earth. The corresponding Mach numbers decrease from 28.4,
25.9, to 23.9 accordingly. The Reynolds number based on the nose radius of 0.1524
m spans a range from 1,590, 6,280, to 19,500 and is in a favorable pressure gradient
condition. Therefore the entire investigated domain within the stagnation and immediately adjacent domain can be treated as laminar flow.

In an attempt to eliminate the computational error from the validating physics
process, a grid refinement study is carried out to eliminate the numerical resolution

concern. Two simulations are generated by the identical initial values and boundary conditions on two axisymmetric body-conformal meshes with clustered grid spacing by the grid-point density of (163 × 93) and (327 × 187). The convergent criterion for numerical solutions is based on the relative L2 Norm of dependent variables in consecutive temporal advancements. The diminishing residual iterative process drives the residual to a minimum, when the residual of the differencing approximation meets a predetermined relative error tolerance, 10^{-5}, thus the formulation will produce a consistent accurate solution. Again, the only noticeable discrepancies between the results of the grid resolution study are confined to the transient state of the shock jump. It reflects itself as a pure numerical artifact induced by the piecewise continuous solution.

The comparison of species concentrations generated on the fine and the coarse mesh system is presented in Figure 6.7. The number densities in cubic centimeter along the stagnation streamline consist of 11 molecular, atomic, and ionized species, N_2, O_2, NO, N, O, N^+, O^+, NO^+, $O2^+$, NO^+, and e^-. The overall quantum chemical-physics transitions are clearly displayed by the bow shock compression until the recombination process to a lower temperature on the stagnation point. The electronic excitations of atomic oxygen O and nitrogen H are actually leading and followed by the nitro oxide NO, the rapid depletion of positively charged oxygen also becomes very clear. The computational results over the species number density range by seven orders of magnitude ranging from 10^{12} to 10^{18}. The computational results agree well with other records in the open literature [38,44]. Two interesting observations can be made in this presentation: First, the affinity among the results is discernible even extending to the tracing elements of molecular oxygen O_2 and ionized nitrogen molecules N_2^+. Second, the sum of the number densities of the positively charged ionized species equals the number density of the free electrons to reflect the intrinsic globally neutral property of plasma.

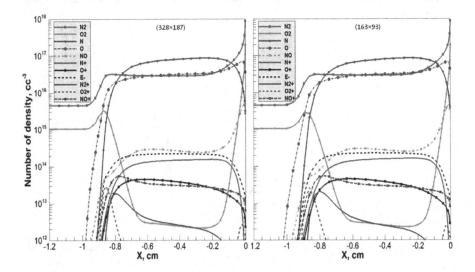

FIGURE 6.7 Comparison of chemical species in number density in RAM-CII shock layer.

The vibrational temperatures of molecular nitrogen, oxygen, and nitro oxide, as well as the electronic temperature within the shock layer, are depicted in Figure 6.8. For the purpose of establishing a common reference, the temperatures of the non-equilibrium gas mixture are calculated by assuming the Maxwellian distributions for all equilibrium/nonequilibrium internal degrees of excitation. The standoff distance defined by the temperature jump is 0.79 nose radius and agrees well with previous simulations. The maximum vibration temperature is achieved by the molecular nitro oxide at 18,900 K, a lower value of 16,400 K is attained by molecular oxygen. The calculated electronic temperature has a maximum value of 12,500 K in the immediately post-shock region and nearly coincided with the vibrational temperature of the molecular nitrogen over the entire domain of the shock layer. The calculated maximum electronic temperature within the shock layer attains a reasonable agreement with all known published results. Another outstanding feature of the nonequilibrium high enthalpy flow in the thin shock layer is that all thermal conditions of the internal degrees of freedom reach an equilibrated state shortly downstream to bow shock, or only 0.2 cm upstream to the stagnation point.

The only noticeable discrepancies between the results of the grid resolution study are the overshoot peak temperatures and the ramping-up portion of the pressure jumps. The maximum disparity in peak temperatures has the values of 18,250 K for the coarse mesh solution versus the value of 18,750 K for the solution generated on the fine mesh and is confined within the transient state of shock jump. It reflects itself as a pure numerical artifact at the shock jump. Again the disparities of computational results by difference numerical resolution are confined at the overshooting peak values as they have been pointed out earlier.

A rare exception to the lack of experimental verification for the prediction of the ionized component is from the flight test data of the RAM-CII. The major portion of electron number density data by the probe is measured by a reflectometer and a checking point by an electrostatic probe at the distance around eight (8) nose

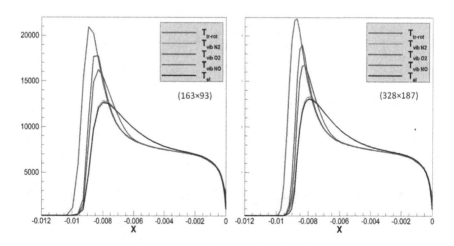

FIGURE 6.8   Multi-temperature profile within the RAM-CII shock layer.

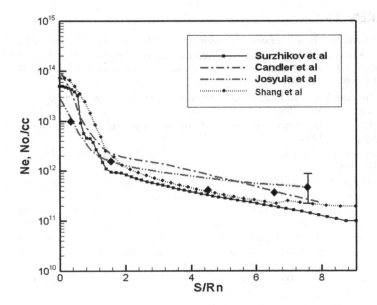

FIGURE 6.9 Validating quantum chemical-physics modeling by RAM-CII probe data.

radii from the stagnation point [43]. The data from the reflectometer represents the averaged peak value at the sample location and the probe data is the time-averaged value among the innermost and the outermost probes. The estimated error bar due to the body motion of the probe data covers a range by the peak-to-peak density fluctuation from $3.0 \times 10^{11}/cm^3$ to $1.2 \times 10^{12}/cm^3$. From this information, the computational error revealed by the grid refinement study is sufficient for the present purpose. The numerical error through the grid refinement study has a magnitude less than the experimental data scattering. By the grid-independent numerical analysis of the quantum chemical-physics modeling, the evidence for a sufficient numerical resolution has prevailed. This assessment ensures the numerical simulations have sufficient accuracy for the sought-after physical phenomena. In essence, the evidence for a sufficient numerical resolution is provided by the fact that the numerical error through the grid refinement study has a magnitude less than the experimental data scattering.

Figure 6.9 depicted the comparison with the electron number density measurements along the RAM-CII probe. In spite of a noticeable difference in the predicted temperature profiles in the shock layer by the computational simulations, the agreement with the flight data is reasonable. In general, all numerical simulations over-predicted the electron number density near the stagnation region where the non-equilibrium state dominated. All computational results are comparable with earlier numerical results by Candler and MacCormack [45], and the effort by Josyula et al. [46]. The latter effort incorporates the state-specific vibration-translation and dissociation rate through the master kinetic equation rather than depending on the relaxation quantum transition models. Indirectly, this comparison reflects the progress or the lack of it over nearly the past 30 years.

REFERENCES

1. Leighton, R.B., *Principles of modern physics*, McGraw-Hill Inc. New York, 1959.
2. Moore, W.J., *Physical chemistry*, Prentice-Hall Inc., Englewood Cliff, NJ, 1963.
3. Clarke, J.F. and McChesney, M., *The dynamics of real gases*, Butterworths, Washington, DC, 1964.
4. Vincenti, W.G. and Kruger, C.H., *Introduction to physical gas dynamics*, John Wiley & Sons, New York, 1965.
5. Herzberg, G., *Atomic spectra and atomic structure*, 2nd Ed., Dover Publisher, New York, 1944.
6. Herzberg G., *Spectra of diatomic molecules*, Van Nostrand Reinhold Co.: Washington, DC, 1950.
7. Zel'dovich, Y.B. and Raizer, Y.P., *Physics of shock waves and high-temperature hydrodynamic phenomena*, Dover publication, Mineola, 2002.
8. Shang, J.S. and Surzhikov, S.T., Simulating stardust earth reentry with radiation heat transfer, *Journal of Space Craft and Rockets*, Vol. 48, No. 3, 2011, pp. 385–396.
9. McHarg, M.G., Stenbaek-Nielsen, H.C., and Kanmae, T., Observations of the stardust sample return capsule entry using a high frame rate slit-less Echelle Spectrograph, AIAA Paper 2008-1210, Reno, NV, 2008.
10. Courant, R. and Herbert, D., *Methods of mathematical physics*, Vol. 1, Interscience Publishers, New York, 1965.
11. Mott, I.N. and Sneddon, N.F., *Wave mechanics and its applications*, Oxford University Press, Oxford, 1948.
12. Chernyi, G.G., *Introduction to hypersonic flow*, Academic Press, New York, 1961.
13. Shang, J.S. and Surzhikov, S.T., *Plasma dynamics for aerospace engineering*, Cambridge University Press, Cambridge, 2018.
14. Shang, J.S., Solving schemes for computational magneto-aerodynamics, *Journal of Scientific Computing*, Vol. 25, 2006, pp. 289–306.
15. Shang, J.S., Kimmel, R.L, Menart, J. and Surzhikov, S.T., Hypersonic flow control using plasma actuators, *Journal of Propulsion and Power*, Vol. 24(5), 2008, pp. 923–934.
16. Krause, J.D., *Electromagnetics*, 1st Ed, McGraw-Hill, New York, 1953.
17. Jahn R.G, *Physics of electric propulsion*, McGraw-Hill, New York, 1968.
18. Raizer, Y.P., *Gas discharge physics*, Springer-Verlag, Berlin, 1991.
19. Mitchner, M. and Kruger, C.H., *Partially ionized gases*, John Wiley & Sons, New York, 1973.
20. Alfven, H., *Cosmical electrodynamics*, Clarendon Press, Oxford, 1950.
21. Shang, J.S. and Surzhikov, S.T., Nonequilibrium radiative hypersonic flow simulation, *Journal of Progress in Aerospace Sciences*, Vol. 53, 2012, pp. 46–65.
22. Shang, J.S., Surzhikov, S.T., and Yan, H., Hypersonic nonequilibrium flow simulation based on kinetics models, *Frontiers in Aerospace Engineering*, Vol. 1, No.1, 2012, pp. 1–12.
23. Landau, L. and Teller, E., Theory of sound dispersion, *Physik, Zeitschrift der Sowjetunion*, Vol. 10, 1936, pp. 34–38.
24. Herzfeld, K.F., *Relaxation phenomena in gas, Vol. 1 of High speed aerodynamics and jet propulsion*, Oxford University Press, London, 1955.
25. Millikan, R.C. and White, D.R., Systematics of vibrational relaxation, *Journal of Chemical Physics*, Vol. 39, no. 12, 1963, pp. 3209–3213.
26. Treanor, C.E. and Marrone, P.V., Effect of dissociation on the rate of vibrational relaxation, *Physics of Fluids*, Vol. 5, No. 9, 1962, pp. 1022–1026.
27. Lee, J.H., Basic governing equations for the flight regime of Aeroassisted orbital transfer vehicles, In: Nelson, H.F. (Ed.), *Progress in aeronautics and astronautics: Thermal design of aeroassisted orbital transfer vehicles*, Vol. 96, AIAA, New York, 1985, pp. 3–53.

28. Park, C., Review of chemical kinetics problems of future NASA missions, I Earth entries, *Journal of Thermophysics and Heat Transfer*, Vol. 7, No. 3, 1993, pp. 385–388.

29. Chernyi, G.G., Losev, S.A., Macheret, S.O., and Potapkin, B.V., Physical and chemical processes in gas dynamics: Cross sections and rate constants. In: Zarchan, P. (Ed.), *Volume 1: Progress in astronautics and aeronautics*, Vol. 196, AIAA, New York, 1995.

30. Olynick, D.R., Chen, Y.K., and Tauber, M.E., Aerodynamics of the stardust sample return capsule, *Journal of Spacecraft and Rockets*, Vol. 36, No. 3, 1999, pp. 442–462.

31. Chen, Y.K. and Milos, F.S., Navier-Stokes solutions with finite rate ablation for planetary mission earth reentries, *Journal of Spacecraft and Rockets*, Vol. 42, No. 6, 2005, pp. 961–970.

32. Shang, J. and Surzhikov, S., Simulating nonequilibrium flow for ablating Earth reentry, *Journal of Spacecraft and Rockets*, Vol. 47, No. 5, 2010, pp. 806–816.

33. Chernyi, G.G., Losev, S.A., Macheret, S.O., and Potapkin, B.V., Physical and chemical processes in gas dynamics: Physical and chemical kinetics and thermodynamics of gases and plasmas. In: Zarchan, P. (Ed.), *Volume 2: Progress in Astronautics and Aeronautics*, Vol. 197, AIAA, New York, 2004.

34. Abe, T., Inelastic collision model for vibrational-translational and vibrational-vibrational energy transfer in the direct simulation Monte Carlo method, *Physics of Fluids*, Vol. 6, No. 9, 1994, pp. 3175–3179.

35. Adamovich, I., Macheret, S., Rich, J., and Treanor, C., Vibrational relaxation and dissociation behind shock waves Part 2: Master Equation Modeling, *AIAA Journal*, Vol. 33, No. 6, 1995, pp. 1070–1075.

36. Adamovich, I.V., Three-dimensional analytic model of vibrational energy transfer in molecule-molecule collisions, *AIAA Journal*, Vol. 39, No. 10, 2001, pp. 1916–1925.

37. Josyula, E., Bailey, W., and Ruffin, S., Reactive and nonreactive vibration energy exchanges in nonequilibrium hypersonic flows, *Physics of Fluids*, Vol. 15, No. 10, 2003, pp. 3223–3235.

38. Shang, J.S. and Yan, H., High-enthalpy hypersonic flows, *Advances in Aerodynamics*, Vol. 2, No. 19, 2020, pp. 1–39.

39. Desai, P., Lyons, D., Tooley, J., and Kangas, J., Entry, descent, and landing operations analysis for the Stardust entry capsule, *Journal of Spacecraft and Rockets*, Vol. 45 No. 6, 2008, pp. 1262–1268.

40. Shang, J.S., *Computational electromagnetic-aerodynamics*, John Wiley & Sons, Hoboken, NJ, 2016.

41. Smoot, L.D. and Underwood, D.L., Prediction of microwave attenuation characteristics of rocket exhausts, *Journal of Space and Rockets*, Vol. 3, No.3, 1966, pp. 302–309.

42. Surzhikov, S.T. and Shang, J.S., Coupled radiation-gasdynamic model for Stardust earth entry simulation, *Journal of Spacecraft and Rockets*, Vol. 49, No. 5, 2012, pp. 875–888.

43. Jones, L.J. and Cross, A.E., Electrostatic probe measurements of plasma parameters for two reentry flight experiments at 25,000 feet per second, NASA TN D 66-17, 1972.

44. Tchuen, G., Cameroon, B., and Zeitoun, D.Z., Effects of chemistry in nonequilibrium hypersonic flow around blunt bodies, *Journal of Thermophysics and Heat Transfer*, Vol. 23, No. 3, 2009, pp. 433–442.

45. Candler, G.V. and MacCormack, R.W., Computation of weakly ionized hypersonic flows in thermochemical nonequilibrium, *Journal of Thermophysics*, Vol. 5, No. 3, 1991, pp. 266–272.

46. Josyula, E., Bailey, W., and Suchyta, C., Dissociation modeling in hypersonic flows using state-to-state kinetics, *Journal of Thermophysics and Heat Transfer*, Vol. 25, No. 1, 2011, pp. 34–47.

7 Statistical Thermodynamics

7.1 MICROSCOPIC STATE OF THE GAS MIXTURE

The individual microscopic behavior of gas for practical applications is really a daunting multiple-body problem. Just remember that under the standard atmospheric condition, the Avogadro number is 10^{23} molecules or mass points per unit volume. The gas mixture in the microscopic scale is best depicted as molecular chaos; the path of the center of mass is carried out on an irregular zigzag pattern and consists of a straying mean free path between collisions. The collisions are occurring in time and space in an extremely irregular manner. The collision rate is dependent on the random particle velocities and the locations of the colliding partners. Nevertheless, the gas particles whether atoms or molecules must still obey the fundamental laws of mechanics. For the multi-body problem, the necessary initial conditions are largely unknown and thus unsolvable. Therefore, the overall macroscopic thermodynamic behavior of a gas system must be linked to details of the myriads of individual particles through the of theoretical probability. In the gaseous states, the particles are separated most of the time by distances that are far away from the characteristic dimension of colliding particles. The momentum and energy transfers between particles are exclusively dependent upon the collision process. In Chapter 2, the equation of motion by the general dynamical theory is provided by the Liouville theorem in the Hamiltonian form [1–3].

$$dq_i/dt = \partial H/\partial p_i; \quad dp_i/dt = \partial H/\partial q_i \tag{7.1}$$

The coordinates p_i and the momenta q_i for three-dimensional particle motion are simply (x, y, z) and (mu_x, mu_y, mu_z). By assuming the collision process to elastic and restricting the particles' interaction only via binary encounters, the microscopic motion can be described by the Boltzmann equation.

$$\partial f/\partial t + c_i \, \partial f/\partial x_i + F_i \, \partial f/\partial c_i = \tag{7.2}$$

$$\iint \left[f(c_i')f(x_i') - f(c_i)f(x_i) \right] g(d\sigma/d\Omega) \sin\theta \, d\theta \, d\varphi \, d^3 c$$

The formulation is a general description of the laws for the conservative property of a dynamic system. The distribution function is defined for the number of particles to have velocities in the range of the specular velocity c_i, and positions in the range of dx_i around x_i, $f(c_i, x_i, t)$.

DOI: 10.1201/9781003212362-9

It may be noticed that the distribution function has no explicit reference of the microscopic structure of the particle dynamics to the macroscopic characteristics of a thermodynamic system in terms of pressure, density, velocities, and internal energy states. There must be a clear link between the macroscopic behaviors of a thermodynamic system to the detailed microscopic constituent parts. The linkage across the states is relying on an effective averaging process or by statistical means. From the statistical viewpoint, a complete knowledge of the dynamic state is not required, but only some features of physical significance is of interest. Again, the quantification by statistic means is by a probabilistic value which is not practical for deterministically scientific and engineering applications. Statistic mechanics is the critical connection between the macroscopic and microscopic behavior of a gas system.

The link between the microscopic behavior of a gas system and the measurable properties of aerodynamic interest such as energy, pressure, temperature, and velocity is known as statistic mechanics. In principle, this scientific discipline attempts to describe the averaged behavior of a point mass with an energy state, e_i by the probability or statistics determination. It is the connection between the overall macroscopic behaviors of a thermodynamic system to the detail of its myriads of interactions in chaotic random motion.

Random and disorder motion is mathematically related to probability, in the sense a number of distributions can be represented at a given state. An excellent example is the relationship interweaving the extensive variable of thermodynamic states, entropy S, and the probability of the distribution function. The total entropy of a combined two separate thermodynamic systems is the sum of the two. The combined probability of the two systems is the product of the two separate probabilities. The only possible functional relation of the entropy is a logarithmic function according to the Boltzmann H theorem $S = k \log \Omega$, and the increasing entropy increases with the number of distinguishable probabilities, the proportional constant k must be positive [1,2]. The total possible arrangements in organizing the particles with allowable states are paramount important in statistical mechanics and it is the essential link between thermodynamics and statistics. The variable Ω is a function of the internal energy, specific volume, and number of the particles.

7.2 THERMODYNAMIC EQUILIBRIUM STATE

The thermodynamic equilibrium condition of a gaseous system resides in the dynamic equilibrium state of gas particle motion in a microscopic scale. Equilibrium is the most likely state in which the collisions between molecules no longer affect the distribution function. This special state of microscopic dynamics can be viewed as the total cancellation of perturbations arriving from opposite directions. From the gas kinetic theory, the dynamic equilibrium state of a gas dynamics system is that the contribution by the collision integral must be vanished [1,3].

$$\left[\partial f(x_i, c_i, t)/\partial t\right]_c = \iint \left[f(x_i, c_i', t)(x_j, c_j', t) - f(x_i, c_i, t)f(x_j, c_j, t)\right]g\,(d\sigma/d\Omega)d^3\,c_i = 0$$

$$(7.3)$$

Therefore $f(x_i,c_i',t)f(x_j,c_j',t) = f(x_i,c_i,t)f(x_j,c_j,t)$ is a sufficient and necessary condition for dynamic equilibrium.

For a closed phase space, the gases are at a uniform temperature T and pressure P throughout, and the gas mixture is chemically inert and distributed uniformly within the control volume. The system is in a complete equilibrium dynamic state. For an open phase space, the chemical reactions are still explicitly excluded from the system and the number of different species is independent of each other. From the foregoing discussions, thermodynamic equilibrium is an axiomatic concept. In essence, under the equilibrium state, there are neither net macroscopic flows of mass nor energy exchange within the system nor with its surroundings.

In the most probable state or the thermodynamic equilibrium condition discussed in Section 7.1, the entropy of the system is known and is traditionally further approximated by Refs. [3,4].

$$S = k \log \Omega \simeq k \log T_{\max} \qquad (7.4a)$$

In a homogeneous thermodynamic system, any change in the system proceeds from a state of higher free energy to a lower state. For this reason, the free energy is considered as the thermodynamic potentials; the Helmholtz potential, F, and the Gibbs free energy, G. These potentials in a gas mixture are a function of P, V, and T. and the different chemical constituents. The chemical composition is chosen by the mole of the molecule to be the chemical measure, with the symbols of N_i; $i = 1,2,3,....$ As the consequence, the complete differential of the free energy becomes.

$$dG = (\partial G/\partial T)_{p,N_i}\, dT + (\partial G/\partial p)_{T,N_i}\, dp + \Sigma(\partial G/\partial n_i)_{T,P,N_j}\, dN_i \qquad (7.4b)$$

Therefore, thermodynamic equilibrium is the unique stable stationary state that has usually reached after a long period of interactions with the surrounding and becomes the most probable state of a system [3,4]. Under the thermodynamic equilibrium state, the definition of the energy of the thermodynamic system is the most important relation of thermodynamic quantities for a reversible process of an open chemically reacting system.

$$dE = TdS - pdV + \Sigma\mu_i dN_i \qquad (7.4c)$$

In Equation (7.4b), the symbol μ_i denotes the chemical potentials of species i per N_i mole, which is a coefficient introduced by Gibbs or the chemical potential $\mu_i = (\partial G/\partial N_i)_{T,p,N_j}$.

The important relationships among thermodynamic quantities with chemical reactions are expanded to include the enthalpy H, Helmholtz potential F, and Gibbs free energy G.

$$H = E + pV$$
$$F = E - TS \qquad (7.4d)$$
$$G = H - TS$$

The corresponding reversible change including equilibrium chemical reactions are.

$$dH = TdS + Vdp + \Sigma\mu_i dN_i$$

$$dF = -SdT - pdV + \Sigma\mu_i dN_i \qquad (7.4e)$$

$$dG = -SdT + Vdp + \Sigma\mu_i dN_i$$

Once in the equilibrium state, the entropy S of the system is maximized, and these thermodynamic potentials are minimized. The conclusion is easily observed because at the uniform pressure and temperature, the work is restricted to the product of P and V; the condition of equilibrium is that the thermodynamic potentials must be the minimum $dF = 0$ and $dG = 0$.

7.3 INTERNAL DEGREES OF FREEDOM

The existence of internal degrees of freedom or the internal excitation of atoms and molecules is clearly displayed by the atomic and molecular spectra. The spectrum of atoms consists of sharp lines and those of molecules appear as bands, which are densely packed line structures. The molecular spectra at the same time give the possible molecular energy levels. Spectra are the results of the emission and absorption of finite quanta of radiation when quantum transitions occur between quantum energy levels. In an atom, the energy levels represent different allowable states of the orbiting electrons around the nucleus. A molecule also emits and absorbs energy in quantum transitions between different electronic energy levels. In addition, a molecule can also change its energy level through the vibrations of the atoms within the molecule and by the change in the rotational energy of the molecule.

In the theory of molecular spectra, the first approximation considers the internal energy of a molecule to be expressed as the sum of rotational, vibrational, and electronic excitations. The separation of internal energy into these categories however is an approximation because atoms of a rapidly rotating molecule can be pushed apart by centrifugal acceleration. This interaction affects the vibrational motion of the atoms. From the experimental data and the energy levels of electronic excitation are usually much greater than the vibrational energy levels in turn are much larger than that between rotational levels. A typical sketch energy level diagram of the rotational and vibrational energy levels is depicted in Figure 7.1. The first level of the molecule corresponding to the absence of vibrational excitation, lies slightly above, because of the zero=point vibrational energy. In each electronic state, there are a large number of vibrational quantum levels and in each vibrational level, there are a large number of rotational levels.

The challenge of expanding hypersonic flow beyond the realm of classic gas kinetic theory requires including the high-enthalpy effects. When the air temperature is elevated above 1,500 K, the properties of air differ considerably from the perfect gas, because the excitations of the internal structure of gas particles become significant. From the spectroscopic data, the quantum state of oxygen is found to transit within the Schumann-Runge wave band, and the nitrogen molecule shifts in the first, second positive, and the Vegard-Kaplan forbidden wave bands. The nitric oxide shows up in

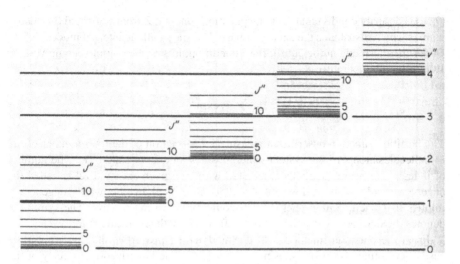

FIGURE 7.1   Quantized energy levels diagram of a molecule.

the spectral data in the beta and gamma bands [5]. The energy content of different quantum states reveals energy levels in electron volts ranging from 5.47 to 11.1 ev (1 ev = 1,6022 × 1,012 erg). The energy in 1 ev per excited molecule has a substantial value of 23.95 kcal/mole and is contributed by the internal degree of freedom of the molecule by the molecular rotational, vibrational plus electronic motions. The internal degrees of freedom of molecules include the translational, rotational, vibrational, and electronic modes [3,5,6].

The electronic mode has also been referred to as nuclear excitation and can be further split into dissociation and ionization. The internal degree of freedom of an atom without looking into atomic sub-particles is much simpler than the molecule because it only constitutes of electron's translational motion and different quantum level energy of electrons. All these internal excitations are quantized, and the bridge between the microstate of individual particles and the macroscopic properties of the gas mixture is linked by statistic mechanics through theories of probability and distribution functions. Only in an equilibrium state, the internal energy distributions of a gas molecule or atom have been made easily understood by quantum mechanics through the partition functions. Under the equilibrium states and from the statistic thermodynamics, the total energy of a molecule is additive. In other words, the energy of a molecule is measured above its zero-energy or ground state and is the sum of the energy of translational, rotational, vibrational, and electronic degrees of freedom. For atoms, the total energy includes only the translational and electronic modes.

$$e_i = e_t + e_r + e_v + e_e \tag{7.5a}$$

$$e_i = e_t + e_e$$

The energy transfer by inelastic collisions is quantum-restricted; the energy is transferred through discrete states and the transfer process is instantaneous because there is no continuity between quantum states. The sum of the internal energy by Equations

(7.5a) is described individually by the partition function Z from statistical thermodynamics which distributes the energy of particles among all the internal modes [3,5,7]. Under the thermodynamic equilibrium condition, it is a function of specific volume V, and static temperature T.

$$Z(V,T) = \sum_i g_i \exp\left(-e_i/\kappa T\right) \tag{7.5b}$$

The partition function describes a measure for a fraction of the total number of the ith molecules/atoms of the system which possess an energy state of e_i. The number of independent wave functions associated with any given energy level is called the degeneracy or the statistical weight of that quantum level. In other words, degeneracy implies that one quantum energy level could arrive in many ways. The symbol g_i denotes the degeneracy which explicitly affirms how the molecule is formed.

From the partition functions, all thermodynamic properties of a system can be easily determined from statistics mechanics [3,6]. The factorization property of the partition functions is only valid so long as each energy mode can be assigned an energy level that is independent of the other energy modes. The schematic quantum spectrum of the internal energy of a gas molecule is depicted in Figure 7.2, a quantum transition may occur within a quantum state and across different states. Both the rotational and vibrational degrees of freedom however have limited discrete quantum levels before the dissociation of gas molecule occurred. The energy states of the translation mode are so tightly packed relative to the datum; they are practically continuous and contribute solely to the partial pressure of the gas mixture. All other internal degrees of freedom are distinctive quantum phenomena that are independent of the specific volume of the system, and thus do not contribute to the partial pressure of the gas mixture [3]. In addition, the rotational degree of freedom requires a very few number of collisions to reach a fully excited state, for this reason, it is always considered to be equilibrated with the translation mode [6,7].

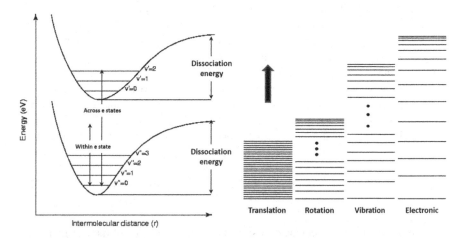

FIGURE 7.2   Schematics quantum spectra of internal degrees of freedom.

Although the actual transition process or the quantum jump is instantaneous, accumulating sufficient energy to reach the next quantum state still requires enough collision to occur. The time lag between jumps is often labeled as the relaxation phenomenon [3,5,8]. When the jump is limited only to the next adjacent quantum state, the process is designated as the ladder-climbing process. Otherwise, the transition over multiple quanta by a single collision is known as the big-bang process [3,6,9].

7.4 PARTITION FUNCTIONS

The Maxwell distribution under collision equilibrium conditions is the sole analytic connection between individual particle dynamics to the thermodynamic variables. In fact, the distribution function establishes a definable relationship between the properties of microscopic and macroscopic states under the equilibrium condition. The sum of the most probable states is the partition function Z, which distributes the energy of all particles among all the particle groups. In essence, it defines the most probable state for the existence and independence between internal modes across the microscopic and macroscopic description of gas. The independent variables of the partition function are the specific volume and temperature of a thermodynamic system.

$$Z(V,T) = \sum_i g_i \left\{ \exp(-e_n/kT) \cdot \exp(-e_{ie}/kT) \cdot \exp(-e_v/kT) \right\} \qquad (7.6)$$

where the notation g_i is the so-called degeneracy, which provides the detailed formation for the internal atom/molecule structure [3,7].

From quantum mechanics, the molecular partition functions of each internal degree of freedom are obtained by integrating quantum solutions over all permissible quantum spectrums of each mode. Individual partition functions in equilibrium conditions for the translational, rotational, and vibrational by different physical models can be obtained, except the electronic excitation due to its multiple ground states cannot be described by a single expression [3,10]. The partition function for electronic excitation remains as the specific quantum solutions of individual species.

For translational motion, the discrete energy levels of translational motion in three-dimensional space according to de Broglie's hypothesis are $mv = h/\lambda$ and let $l = n\lambda/2, n = 1, 2, 3$ and must have the allowed quantum energy levels [3,11]. Along each coordinate, the quantum numbers, n_1, n_2, n_3 have discrete values extended to infinity.

$$e_t = \left(h^2/8m \right) \left[\left(n_1/\lambda_1 \right)^2 + \left(n_2/\lambda_2 \right)^2 + \left(n_3/\lambda_3 \right)^2 \right] \qquad (7.7a)$$

The translational partition function of a particle with permissible translational energy levels is the product aligned with three coordinates.

$$Z_t = \Sigma \exp\left(h^2/8mkTl_1^2 \right) n_1^2 \cdot \Sigma \exp\left(h^2/8mkTl_2^2 \right) n_2^2 \cdot \Sigma \exp\left(h^2/8mkTl_3^2 \right) n_3^2 \qquad (7.7b)$$

The summation procedures of the partition function along each coordinate direction are identical, so the identical derivations are carried out only for a single coordinate. The sum is approximated by integration with respect to the quantum number n. Let $\chi = \sqrt{h^2/8mkTl_i^2}\, n$ to simplify the summing process.

$$\left(1/\sqrt{h^2/8mkTl_i^2}\right)\Sigma \exp\left(-\sqrt{h^2/8mkTl_i^2}\, n\right)^2 \Delta\left(\sqrt{h^2/8mkTl_i^2}\, n\right) \tag{7.7c}$$

$$\approx \left(1/\sqrt{h^2/8mkTl_i^2}\right)\int_0^\infty \exp\left(-\chi^2\right)d\chi$$

The integrand of $\int_0^\infty \exp\left(-\chi^2\right)d\chi$ is just $\sqrt{\pi}/2$, the summing result becomes.

$$\left(1/\sqrt{h^2/8mkTl_i^2}\right)\Sigma \exp\left(-\sqrt{h^2/8mkTl_i^2}\, n\right)^2 \Delta\left(\sqrt{h^2/8mkTl_i^2}\, n\right) \tag{7.7d}$$

$$= \sqrt{\pi}\Big/\left(2\sqrt{h^2/8mkTl_i^2}\right)$$

Substituting Equation (7.7d) into (7.7b), the partition function of the translation mode acquires the following form.

$$Z_t = \left(2\pi mkT/h^2\right)^{3/2} l_1 l_2 l_3 \tag{7.7e}$$

Or simply,

$$Z_t = \left(2\pi mkT/h^2\right)^{3/2} V \tag{7.7f}$$

Since the composition of air consists of most diatomic molecules, the rotational mode is focused on this group of molecules. For diatomic molecules in rotational motion, the simplest model of a rotating model is a rigid dumbbell. The significant moment of inertia I only have about each two of the orthogonal axes which are perpendicular to the inter-nuclear axis of the molecule. Quantum mechanics yields the rotational energy levels as.

$$e_r = J(J+1)h^2/8\pi^2 I, \; J = 0,1,2,\ldots \tag{7.8a}$$

The rotational wave functions are $\varphi(r,t)_r = NP_J^k(\cos\theta)\exp(ik\phi)$ and the N designating normalization factor [7]. For each quantum number J, there is $2J + 1$ wave function characterized by the allowable value of k which run from $-J$ to $+J$. Therefore, the appropriate degeneracy of the rotational degree of freedom of a diatomic molecule is $g_r = 2J+1$. The value corresponds to the possible orientation of angular momentum J, the degeneracy is rising from the nuclear spinning orientation in counting the distinguishable molecular states. The degeneracy of rotational excitation simply expresses the statistical weight in a term of the degree of degeneracy.

The summing processes for a hetero-nuclear molecule such as the nitric oxide NO and homo-nuclear molecules such as oxygen and nitrogen molecules O_2 and N_2 are different. The former are summing overall sequential numbers, $J = 1, 2, 3,...$ and the latter are summing over either odd or even numbers. The partition function of the rigid rotor after the summing process becomes.

$$Z_r = \Sigma(2J+1)\exp\left[-J(J+1)h^2/8\pi^2IkT\right] \tag{7.8b}$$

The summation of all the allowed quantum numbers has simply approximated the summation by integration.

$$\Sigma(2J+1)\exp\left[-J(J+1)h^2/8\pi^2IkT\right] \tag{7.8c}$$

$$\cong \left(8\pi^2I\,kT/h^2\right)\int_0^\infty \exp(-\chi)d\chi = 8\pi^2IkT\,/\,h^2$$

The partition function of rotational mode acquires the following form.

$$Z_r = 8\pi IkT/\sigma\,h^2 \tag{7.8d}$$

The symbol σ is known as the symmetric factor, for the heteronuclear molecules it assigns a value of unity and for the homonuclear molecules designates a value of 2.

The characteristic rotational temperature is defined as $\Theta_r = h^2/8\pi^2Ik$ in assessing the degree of excitation of rotational mode. The values for most components of the air are less than a few degrees in Kelvin [3]. For this reason, the rotational degree of air components is always considered to be equilibrated with the translational mode [6,12,13].

The vibrational excitation and dissociation are closely interrelated; the dissociation of molecules is a limiting state of the vibration degree of freedom when the atoms are separated from a molecule. The simplest simple harmonic oscillator model of diatomic molecules is two masses connected by a spring and oscillating around an equilibrium distance. The discrete energy levels are given as

$$e_v = (n_v + 1/2)h\nu \tag{7.9a}$$

There are no degenerate levels of vibrational energy with a frequency of $\nu = \sqrt{f/2\pi^2m}$ and there are two unique features of the quantum solution. First, the ground state energy level is $h\nu/2$. Quantum mechanics requires the oscillator always has no-vanished residual energy at the referential state. Second, the adjacent quantum energy levels of the simple harmonic model are independent of the values of the quantum number n_v, which implies that the atoms continue to vibrate even when the dissociation occurred. The second feature is inconsistent with the physics of a dissociated molecule.

Regardless of the shortcoming of the simple harmonic model, the partition function of vibrational mode according to the model is obtainable.

$$Z_v = \Sigma\exp\left[-\left(n_v + 1/2\right)h\nu/kT\right] \tag{7.9b}$$

Since all n_v are permitted, the sum is a simple geometric progression, and the result is obtained straightforwardly [3].

$$Z_v = \exp(-h\nu/2kT)\Big/\big[1-\exp(-h\nu/kT)\big] \qquad (7.9c)$$

$$= \big\lceil 2\sinh(h\nu/2kT)\big\rceil^{-1}$$

The detailed and proper potential curve for the diatomic molecule is generally consisting of two types of bonding forces between atoms and a molecule. The ionic binding or the bonding is by electromagnetic force through the attraction or repulsion from the elementary electric charges of opposite polarity. The homopolar binding is a quantum mechanics manifestation of diatomic gases. To duplicate the physics, a widely adopted potential for an anharmonic oscillator is developed by Morse [7].

$$V = d_e\Big\{1-\exp\big[-a(r-r_e)^2\big]\Big\} \qquad (7.9d)$$

In Equation (7.9d), d_e and a are empirical constants related to dissociation energy and vibrational frequency of the lowest state. The equilibrium r_e is the separation distance between atoms where the force between atoms is zero, or the gas is dissociated. If the vibration frequency is synonymous with the vibrational quantum number, the characteristic vibrational and dissociation temperature are definable as $\Theta_d = d_e/k$ and $\Theta_v = h\nu/k$ are identified from the potential function respectively. The characteristic vibrational temperatures of the main air composition are in the range from 2 to 3,000 K ($\Theta_{O_2} = 2,230$ K, $\Theta_{NO} = 2,719$ K, $\Theta_{N_2} = 3,374$ K). For oxygen, $\Theta_d = 59,000$ K, $\Theta_v = 2,230$ K, so there are about 50 quantum states of the vibrational excitation for oxygen molecule.

The remedy for the shortcoming describing the vibrational vibration excitation is the anharmonic oscillator, which has the partition function as

$$Z_v = \Sigma \exp\big[-(v+1/2)\Theta_v/T\big]\cdot\exp\big[(v+1/2)^2(\Theta_v^2/4\Theta_d T)\big] \qquad (7.9e)$$

However, the widely adopted vibrational partition function based on the anharmonic vibrator is

$$Z_v = \exp(-\Theta/2T)\Big\{\big[1-\exp-(v_{max}-1)(\Theta_v/T)\big]\Big/\big[1-\exp(\Theta_v/T)\big]\Big\} \qquad (7.9f)$$

The potential wells of the simple harmonic vibrator and the anharmonic vibrator are presented in Figure 7.3. The inconsistency of the simple harmonic vibrator is displayed by the disassociation physics; therefore, the simpler model is only an approximation for the vibration excitation near the ground state.

The general partition function for electronic excitation must be retained the genetic form because the ground states of air mixture have multiple quantum states because of electron spin. The components of the multiplet can only be considered separately.

$$Z_e = g_0 \exp(-e_0/kT) + g_1 \exp(-e_1/kT) + g_2 \exp(-e_2/kT) + \cdots \qquad (7.10a)$$

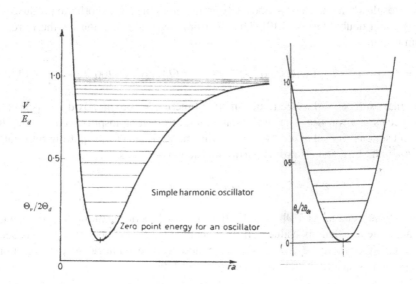

FIGURE 7.3   Vibrational potential wells of the simple harmonic and anharmonic model.

For example, the partition function for molecular oxygen has two coupled electron spins. These two close excited states give the partition function as.

$$Z_{eo_2} = 3\exp\left(-e_o/\kappa T\right)\left[1 + 2/3\exp(-11,300/T) + 1/3\exp(-18,900/T) + ...\right] \qquad (7.10b)$$

The lowest possible electron cloud of atomic oxygen over the ground quantum state is $2p$ and four equivalent electrons are presented to give the terms of $^3P, \, ^1D, \, ^1S$. The lowest and the ground state for atomic oxygen is ^3P. According to the Pauli exclusion principle, the excited electrons must be all parallel when one electron is brought from $2p$ shell into a higher shell. Thus, the triplet terms are formed [10,14].

For the atomic oxygen which has the inverted triplet electronic ground state, the partition function is.

$$Z_{eo} = 5\exp\left(-e_o/\kappa T\right)\left[1 + 3/2\exp(-228/T) + 1/5\exp(-327/T)\right.$$

$$\left.\exp(-22,700/T) + \exp(-48,400/T) + ...\right] \qquad (7.10c)$$

One notes immediately that the characteristic temperatures of 228 K and 327 K are relatively low at the lowest ground states. Therefore, the electronic excitation partition function of atomic oxygen can be approximated as [3].

$$Z_{eo} \cong 9\exp\left(-e_o/\kappa T\right) + ... \qquad (7.10d)$$

The nitrogen atom in its lowest electron configuration has three electrons in the L_2 shell ($^2p_3$). In this state all three $2p$ electrons have parallel spins. Higher excited states result when one electron moves from the $2p$ to the excited orbit to produce the quartet and double terms [3,10]. The electronic partition function for the nitrogen atom is then.

$$Z_{en} = 4\exp\left(-e_o/\kappa T\right)\left[1 + 10/4\exp(-27,500/T) + 6/4\exp(-41,200/T) + ...\right] \quad (7.10e)$$

The final example of the electronic partition function is nitro monoxide. This chemical species is always present in the ionization of air at atmospheric conditions. Its ground state is characterized by $S = 1/2$, thus multiple states are splitting into states of $^2\pi_{1/2}$ and $^2\pi_{3/2}$. The electronic partition function becomes.

$$Z_{eNO} = 2\exp\left(-e_o/\kappa T\right)\left[1 + \exp(-178/T) + ...\right] \quad (7.10f)$$

In the above Equations (7.10b)–(7.10f), the partition function contains any terms with a magnitude less than the value of the exponential value of $\exp(-50,000/T)$ has been truncated. The above electronic partition functions reflect a huge energy level jump across the first quantum state above the ground state.

7.5 THERMODYNAMIC PROPERTIES IN AN EQUILIBRIUM STATE

The statistical approach to thermodynamic state in chemical equilibrium gas mixture is similar to the statistics approach in classic thermodynamics. The only complication is required to account for the total number of different and distinguishable chemical species in the system. The total number of species N_i consists of n_i components and each with an energy level of e_i. The gas mixture is described as

$$N_i = \Sigma n_i, i = 1, 2, 3, ..., n \quad (7.11a)$$

$$E = \Sigma n_i e_i$$

From Equation (7.4b), the partial derivative of the entropy with respect to the internal energy at the constant volume and number of species is

$$(\partial S/\partial E)_{V,N} = -\kappa\beta \quad (7.11b)$$

From the definition of the energy equation under the equilibrium condition Equation (7.4a), it gives $\beta = -(1/kT)$.

Recall the partition function of the total number of the species as $Z = \Sigma g_i \exp(-e_i/kT)$. From Stirling's formula, Gibbs' free energy now can be written as [3].

$$F = -kT\Sigma \log\left(Z_i^N/N_i!\right) \quad (7.11c)$$

From Equation (7.4c), the pressure of the gas mixture system under the equilibrium condition is derived from the partial derivative of the Gibbs free energy with respect to volume at a constant temperature and number species process.

$$p = -(\partial F/\partial V)_{T,N} = kT\left(\partial \log Z/\partial V\right)_{T,N} \quad (7.11d)$$

Since the partition function Z is linearly depended on the volume, then.

$$p = kT\Sigma(N_i/V) = kT\Sigma n_i \tag{7.11e}$$

In short, the above equation is Dalton's law for the partial pressure of a gas mixture.

The internal energy in terms of the partition function is simply differentiated by the partition function of each species of the gas mixture with respect to temperature under a constant volume process.

$$(\partial Z_i/\partial T)_V = (1/kT^2)\Sigma g_i \exp(-e_i/kT) \tag{7.12a}$$

The sum of all internal energies of all species of the gas mixture is the internal energy of the complete system.

$$\Sigma n_i e_i = NkT^2(\partial \log Z_i/\partial T)_V \tag{7.12b}$$

The internal energy of the equilibrium gas mixture is therefore.

$$E = \Sigma N_i kT^2(\partial \log Z_i/\partial T)_V \tag{7.12c}$$

From Equation (7.4c), the entropy of a gas mixture under thermodynamic equilibrium conditions is easily obtainable.

$$S = \Sigma N_i kT(\partial \log Z_i/\partial T)_V + k\Sigma \log(Z_i^N/N_i!) \tag{7.12d}$$

All thermodynamic properties: the internal energy, pressure, and entropy of a system in the equilibrium state of a gas mixture can be easily evaluated from the formulations of the statistics thermodynamics [3,6].

$$E = N\kappa T^2(\partial \ln Z/\partial T)_v$$

$$P = N\kappa T(\partial \ln Z/\partial V)_t \tag{7.13}$$

$$S = N\kappa\{[\ln(Z/N)+1]T(\partial \ln Z/\partial T)\}_V$$

It may be noticed that only the translational mode, Equation (7.7f) has an explicit dependence on the specific volume of a gas mixture system. The particle in translational motion can contribute to the partial pressure of the thermodynamic system that includes the electron of ionized air. All other internal modes of the gas would not be able to alter the pressure of a thermodynamic equilibrium condition [6,14,15].

7.6 FACTORIZATION OF PARTITION FUNCTIONS

In a molecule consisting of multiple atoms, the nuclei of the atoms are separated by some oscillatory distances with electrons gyrating within a cloud of shells and bonding all atoms together. The concentration of mass of atoms with rotation in the molecule stores the energy of rotation. In a diatomic gas molecule, the energy associated with the internuclear axis is negligible for the molecular oxygen and nitrogen as it was discussed previously. Similarly, the vibrational of atoms also generates vibrational energy. The internal electronic energy traditionally includes the dissociation degree of freedom but more recently a finer differentiation of the nuclear internal degree of freedom is no longer emphasized.

All the internal degrees of freedom, rotational, vibrational, and electronic or nuclear modes are quantized. Each of the internal modes possesses a different and whole set of quantum numbers to define a particular energy state of a particular atom or molecule. From Equation (7.5a), the total energy of a molecule is the sum of the translational, rotational, vibrational, and electronic modes $e_i = e_t + e_r + e_v + e_e$. On the other hand, an atom has only the translational and electronic degrees of freedom thus two kinds of energy $e_i = e_t + e_e$. The partition functions of a molecule are resulting from the statistic mechanics which represent a function of probability and thus can be written as

$$Z(V,T) = \sum_i g_i \left\{ \exp(-e_t/kT) \cdot \exp(-e_r/kT) \cdot \exp(-e_v/kT) \cdot \exp(-e_e/kT) \right\} \quad (7.14a)$$

Each internal mode of energy at quantum levels may be degenerate and may also have exceptions, the degeneracy of molecule should be given in general as

$$g_i = g_t g_r g_v g_e \quad (7.14b)$$

For molecules in a gas mixture and under the thermodynamic equilibrium condition, all the energy of the internal degree of freedom depends on the identical temperature and specific Volume. Therefore, should be expressed as the following:

$$Z(V,T) = \sum_i g_t \left\{ \exp(-e_t/kT) \right\} \cdot \sum_i g_r \left\{ \exp(-e_r/kT) \right\} \cdot \quad (7.14c)$$

$$\sum_i g_v \left\{ \exp(-e_v/kT) \right\} \cdot \sum_i g_e \left\{ \exp(-e_e/kT) \right\}$$

The partition function has been factorized by the contribution from all the possible internal degrees of freedom.

$$Z(V,T) = Z_t(V,T) \cdot Z_r(V,T) \cdot Z_v(V,T) \cdot Z_e(V,T) \quad (7.14d)$$

Similarly, the partition function for an atom without the internal degree of freedom of rotation and vibration can be given as

$$Z(V,T) = Z_t(V,T) \cdot Z_e(V,T) \quad (7.14e)$$

Under the thermodynamic equilibrium condition, the factorization of the partition function property holds only as each energy mode with the assigned energy levels which are independent of the others.

7.7 ENERGY DISTRIBUTION OF INTERNAL DEGREES OF FREEDOM

The energy distribution and its relative magnitude from each internal degree of freedom can be easily evaluated from Equation (7.13) with their partition function. The equation is derived for any molecule or atom species in a gaseous mixture at the thermodynamic equilibrium condition. The internal translational energy for a group of the mixed species is given by substituting the translation partition function (7.7f) into Equation (7.13).

$$E_t = NkT^2 \left\{ \partial \log \left[\left(2\pi mkT/h^2 \right)^{3/2} / \partial T \right] \right\}_v \qquad (7.15a)$$

$$= 3/2\, NkT^2 \left(h^2/2\pi mkT \right) \left[\left(2\pi mk/h^2 \right) (\partial T/\partial T)_v \right]$$

$$= 3NkT/2$$

The internal rotational energy of the identical group of the species is obtained by substituting the rotational partition function, Equation (7.8d) into (7.13).

$$E_r = NkT^2 \left\{ \partial \log \left[\left(8\pi IkT/\sigma h^2 \right) / \partial T \right] \right\}_v \qquad (7.15b)$$

$$= NkT^2 \left(\sigma h^2/8\pi IkT \right) \left(8\pi Ik/\sigma h^2 \right) \left(\partial T/\partial T \right)_v$$

$$= NkT$$

The vibrational energy of the identical group of the species based on the harmonic vibrator of Equation (7.9c) becomes.

$$E_v = NkT^2 \left\{ \partial \log [2\sinh(hv/2kT)]^{-1} / \partial T \right\}_V$$

$$= NkT^2 \left\{ \sinh(hv/2kT) \left\{ \partial [(1/2)\csc h(hv/2kT)/\partial T \right\} \right\}_V \qquad (7.15c)$$

$$= NkT^2 \sinh(hv/2kT) \coth(hv/2kT) \csc(hv/2kT)(hv/2kT^2)$$

$$= N(hv/2) \coth(hv/2kT)$$

To be consistent with other energy of internal degree of freedom, the vibrational energy for each species of the gas mixture can be expressed as

$$E_v = NkT(hv/2kT) \coth(hv/2kT) \qquad (7.15d)$$

Recall the harmonic vibrator is not consistent with the dissociation physics of gas; the vibrational energy must be limited to a maximum vibrational frequency as indicated by the anharmonic vibrator [3].

Because of the existence of a multipet ground state of ionization gas species, the electronic energy for gases is facing difficulties for express the model in a common analytic form. From Equation (7.10a), the expression of electronic energy becomes.

$$E_e = NkT^2 \left\{ \partial \log \left[\Sigma g_i \exp(-e_i/kT) \right] / \partial T \right\}_V \tag{7.15e}$$

$$= NkT^2 \left[\Sigma g_i \exp(-e_i/kT) \right]^{-1} \left\{ \partial \left[\Sigma g_i \exp(-e_i/kT) \right] / \partial T \right\}_V$$

At this stage, only a typical term can be obtained like the oxygen atom. It yields the typical expression as the following.

$$E_e = NkT^2 \left\{ \partial \log \left[g_e \exp(-e_e/kT) \right] / \partial T \right\}_V \tag{7.15f}$$

$$= NkT^2 \left[g_e \exp(e_e/kT) \right]^{-1} \left\{ \partial \left[g_e \exp(e_e/kT) \right] / \partial T \right\}_V$$

$$= Ne_e$$

Or simply

$$E_e = NkT \left(e_e/kT \right) \tag{7.15g}$$

The internal energy distribution by a single component of a gas mixture from the translational, rotational, vibrational, and electronic modes is

$$E = (3/2)NkT + NkT + NkT(hv/2kT)\coth(hv/2kT) + NkT \left(e_e/kT \right) \tag{7.16a}$$

To express the energy for a system of the gas mixture, the unit of energy is first converted to the energy per molecular mole. The universal gas constant $R = 8.3145$ J/mole·K is the product of the Avogadro number at the standard ATM condition ($N_A = 6.0226 \times 10^{23}$/mole) and the Boltzmann constant, which is the molecular equivalent to the Boltzmann constant. In all practical applications, the energy unit is traditionally provided in terms of the unit mass of the gas and the specific energy which is measured by the product of a number of a mole of gas mass for the individual molecule $e = E/Nm$. The relationship between the partition function and the specific energy per unit mass becomes

$$e = (Nk/mN)T^2(\partial \ln Z/\partial T)_v = RT^2(\partial \ln Z/\partial T)_v \tag{7.16b}$$

And it follows:

$$e = RT \left[5/2 + (hv/2kT)\coth(hv/2kT) + \left(e_e/kT \right) \right] \tag{7.16c}$$

From Equation (7.16), an external energy source will distribute the internal energy of the gas mixture by the two new additional internal degrees of freedom in form of the vibrational and electronic excitations. After the thermodynamic system reaches a new equilibrium condition, the internal energy has to redistribute to additional internal excitations; the static temperature will be lower than the perfect gas.

The enthalpy of the gas mixture by definition appears as

$$h = e + pV = e + RT$$

$$= RT\left[7/2 + (hv/2kT)\coth(hv/2kT) + \left(e_e/kT\right)\right] \tag{7.16d}$$

The internal energy of an atom has only translational and electronic excitations, the internal energy becomes

$$e = RT\left[3/2 + \left(e_e/kT\right)\right] \tag{7.16e}$$

And the enthalpy also adds the contribution from the electronic degrees of freedom to appear as

$$h = RT\left[5/2 + \left(e_e/kT\right)\right] \tag{7.16f}$$

An interesting observation from the effects of realistic thermodynamics including the internal degrees of freedom to gas dynamics is made easily by the specific heat ratios. The specific heat at the constant volume and constant pressure processes are

$$c_v = (\partial e/\partial T)_v \tag{7.17a}$$

$$c_p = (\partial h/\partial T)_p$$

The two specific heats with additional internal excitations become

$$c_v = 5/2 + (hv/2kT)^2 \csc h^2(kv/2kT) + (\partial e_e/\partial T)_v \tag{7.17b}$$

$$c_p = 7/2 + (hv/2kT)^2 \csc h^2(kv/2kT) + (\partial e_e/\partial T)_p$$

The specific heat ratio includes all internal degrees of freedom is then

$$\gamma = \left[7/2 + (hv/2kT)^2 \csc h^2(kv/2kT) + (\partial e_e/\partial T)_p\right]/\big| \tag{7.17c}$$

$$\left[5/2 + (hv/2kT)^2 \csc h^2(kv/2kT) + (\partial e_e/\partial T)_v\right]$$

For the perfect gas by neglecting the vibrational and electronic degrees of freedom, the specific heat ratio of the air molecule is constant of 7/5 or 1.4. At excited internal degrees of freedom, the specific heat ratio will be less than the perfect gas value in general by the added positive contributions via the vibrational and electronic internal excitations. At extremely high-enthalpy hypersonic flows, the specific heat ratio may even approach the value of unity to substantially alter the results of the Rankine-Hugonoit condition even under the thermodynamic equilibrium condition. Meanwhile, the decreased molecular specific heat ratio reduces proportionally the standoff distance of the bow shock envelope over a hypersonic blunt body. The altered thermodynamic properties of the internal structure of molecules and atoms have shown significant impacts on hypersonic flows.

REFERENCE

1. Chapman, S. and Cowling, T.G., *The mathematical theory of non-uniform gases*, Cambridge University Press, New York, 1964.
2. Kennard, E.H., *Kinetic theory of gases, with an introduction to statistical mechanics*, McGraw-Hill, New York, 1938.
3. Clarke, J.F. and McChesney, M., *The dynamics of real gases*, Butterworths & Co, Washington, DC, 1964.
4. Adkins, C.J., *Equilibrium thermodynamics*, 3rd Ed., McGraw-Hill, London, 1983.
5. Zel'dovich, Y.B. and Raizer, Y.P., *Physics of shock waves and high-temperature hydrodynamic phenomena*, Dover Publication, Mineola, New York, 2002.
6. Shang, J.S., *Computational electromagnetic-aerodynamics*, John Wiley & Sons, Hoboken, NJ, 2016.
7. Moore, W.J., *Physical chemistry*, Prentice-Hall Inc., Englewood Cliff, NJ, 1963.
8. Millikan, R.C., and White, D.R., Systematic of vibrational relaxation, *Journal of Chemical Physics*, Vol. 39, No. 12, 1963, pp. 3209–3213.
9. Shang, J.S., and Surzhikov, S.T., *Plasma dynamics for aerospace engineering*, Cambridge University Press, Cambridge, UK, 2018.
10. Herzberg, G., *Atomic spectra and atomic structure*, Dover Publication, New York, 1944.
11. Leighton, R.B., *Principles of modern physics*, McGraw-Hill, New York, 1959.
12. Park, C., Review of chemical kinetics problems of future NASA missions, I Earth entries, *Journal of Thermophysics and Heat Transfer*, Vol. 7, No. 3, 1993, pp. 385–388.
13. Park, C., *Nonequilibrium hypersonic aerodynamics*, John-Wiley & Sons, New York, 1989.
14. Herzberg, G., *Molecular spectra and molecular structures*, Vol. I, Van Nostrand, New York, 1950.
15. Shang, J.S., Landmarks and new frontiers of computational fluid dynamics, *Journal of Advances in Aerodynamics*, Vol. 1, No. 5, 2019, pp. 1–36.

8 Nonequilibrium Chemical Reactions

8.1 LAW OF MASS ACTION

One of the most outstanding properties of high-enthalpy hypersonic flow lies in the chemical physics of the air encountered in the high-temperature environment. The unique condition is the direct consequence of the energy conversion process brought about by an extremely strong shock wave. The air composition downstream to a normal shock at the altitude of 6,000 m is depicted in Figure 8.1 consists of mainly the molecular nitrogen, oxygen, nitric oxide, atomic oxygen, and nitrogen. At incoming airspeed over 1,500 m/s, the molecule oxygen and nitrogen begin to dissociate and the free electron and positively charged oxygen and nitrogen ions will be presented. All the aforementioned chemical reactions, including dissociation, are endothermic. The concentration of all chemical species in mole fraction changes according to a specific generation and depletion rate according to the quantum chemical-physics process. In the case where the surface ablating phenomenon occurred, the chemical composition around a hypersonic vehicle will be much more complex and introduces the multi-phase issues.

The typical environment of hypersonic earth reentry is characterized by a speed of around 10 km/s or a Mach number of about 30, and the air temperature within the bow shock envelope exceeds 20,000 K, which leads to an intensive chemical reaction, rapid thermal excitation including air ionization. All the chemical reactions

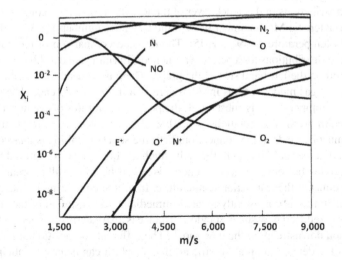

FIGURE 8.1 Air composition downstream to a normal shock at high altitude.

DOI: 10.1201/9781003212362-10

initiated by a strong compressing shock must be accomplished in the thin shock layer, which is a fraction of the characteristic length of the leading-edge radius, thus the process is far from thermodynamic and chemical reaction equilibrium. Departure from thermodynamic and chemical equilibrium is known to have substantial effects on the flow field structure and the propagation of disturbance [1,2]. It impacts on the thermodynamic properties of the gas medium are drastic, and the nonequilibrium chemical reactions become an increasingly important aspect of hypersonic flows.

Our understanding of chemical kinetics is based on the inter-atomic and inter-molecular collision process, therefore the radiation cannot be excluded but will be treated separately because of the huge difference in time scales. For nonequilibrium chemical reactions, the concept of local equilibrium must be used by the analytic procedure. The relaxation time scale in a nonequilibrium situation is thus not unique. For this reason, the designation of a chemical reacting flow field as equilibrium, frozen, and finite-rate by comparing the relative time scales between molecular reaction and organized gas motion lacks rigor. The relaxation time scale may be adopted only as a convenient way to simplify analysis.

A chemical reaction by collision processes always contains reactants, activated complexes, and products. A transitory compound is formed by accumulating a sufficient amount of activation energy for a reaction to take place. This amount of energy, required to move the reactant over the energy barrier in order for the reaction to begin is usually greater than the standard heat of formation [3]. Conditions of the chemical reacting system controlling the rate of chemical reactions in macroscopic states are numerous, such as concentrations of the chemical composition, temperature, pressure, and the presence of catalyst or inhibitor. However the complexity of chemical reactions has not been completely resolved, because, on microscopic scales, the energy of colliding particles, and orientation of particles at collision, and the so-called steric factor are insoluble issues in quantum chemistry [4].

The impacts of nonequilibrium chemical reactions and quantum transitions to the high-enthalpy hypersonic flow are presented in Figure 8.2. The static temperature and pressure distributions in the shock layer of the RAM-CII spherical cone are computed at the characteristic Mach number of 27.9, the Reynolds number of 1.38×10^4, and the freestream temperature of 198.6 K [5]. The detailed comparison of the translational temperature distributions for a perfect gas, nonequilibrium, and equilibrium conditions are depicted on the left-hand side of the graph. There is no need to describe the result of the perfect gas model but it is inserted here as a frame of reference. The difference in the static temperatures is substantial; the peak temperatures of the nonequilibrium and the equilibrium computations are 20,787.46 and 5,855.44 K respectively. Two remarks must be made at the connection; first, the shock definition is unacceptable by the adopted numerical procedure through which the bow shock is captured by 14 grid points. Second, the converted kinetic energy is redistributed into all internal degrees of freedom reducing the translation temperature. In all the comparing results, the maximum temperature is consistently located immediately downstream of the shock. The significantly reduced bow shock standoff distance is also clearly observed by the chemical reaction downstream to the bow shock. The different composition of air composition alters the density jump across the normal shock. It can be easily anticipated that the conductive heat transfer rates at the stagnation region will be substantially different.

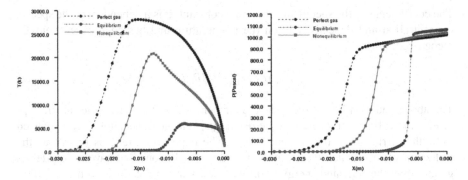

FIGURE 8.2 Impacts of nonequilibrium chemical reactions and quantum transition to high-enthalpy hypersonic flow over RAM-CII probe.

The right-hand side of Figure 8.2 presents the comparison of the static pressure along the stagnation streamline for the same three cases studied. There is only a slight difference between the results of the perfect gas and nonequilibrium model from the equilibrium condition on the body surface. This small difference is due to the existence of a larger fraction of free electrons at the equilibrium condition. Electronic excitation is the only internal mode of the high-enthalpy gas that can contribute to the static pressure of the mixture. All other internal degrees of excitations are quantum phenomena that will not contribute to any net exchange of momentum with the ambient. In fact, the equilibrium computation generates the highest mass fraction of the free electrons which attains a value of 7.16×10^{-8} in comparison with the nonequilibrium limit.

A nonequilibrium chemical reaction is an irreversible process. The chemical composition change in a system takes place toward its equilibrium value. Any chemical reaction can be expressed as a group of simple one-step reactions according to permissible chemical kinetics.

$$\Sigma v_i' x_i \rightleftarrows \Sigma v_i'' x_i \qquad (8.1a)$$

where the symbols v' and v'' are referred to as the stoichiometric coefficients for reactants and products. These coefficients are integers denoting the number of molecules that partake in the reaction. x_i is the species concentration in mole per unit volume of the reacting species. Equation (8.1a) is a statement of conservation of the total number of atoms of species in any one-step reaction process.

The law of mass action is an empirical relationship that has been confirmed by numerical experimental observations. The law states that the rate of generation or depletion of a reacting species is proportional to the products of the species concentration to a power equal to the corresponding stoichiometric coefficient.

$$dc_i/dt = \kappa \prod_{i=1}^{n} x_i^{v_i} \qquad (8.1b)$$

The coefficient κ is the specific reaction rate constant. It is independent of the species concentration and depends only on the temperature and has been experimentally recognized.

$$\kappa = aT^b \exp\left(-\varepsilon_a / R_u T\right) \tag{8.1c}$$

The above equation is referred to as the Arrhenius law. The leading term, aT^b represents the collision frequency of the reacting species and the symbol ε_a is referred to as the Arrhenius activation energy [6]. The exponential term is known as the Boltzmann factor which indicates the fraction of collisions that have energy levels greater than the activated energy to initiate the chemical reaction. All these parameters are unique for a specific chemical reaction.

In order to define the equilibrium condition of a chemical reaction, the chemical reaction is assumed to proceed simultaneously by the forward and reverse or backward reactions. From Equation (8.1b), the forward and backward reaction rates are.

$$R_{i,f} = k_{i,f}(T) \prod_{i=1}^{n} x_i^{v_i'} \tag{8.1d}$$

$$R_{i,b} = k_{i,b}(T) \prod_{i=1}^{n} x_i^{v_i''}$$

The net reaction rate of change in concentration of any species is the combined result of all permissible individual elementary single simple-step reactions. In a complex overall chemical reaction consisting of a number of elementary reactions, the rate of reaction is determined by the slower rate of an elementary process and the stoichiometric coefficients may not govern the overall reaction rate of the system. The chemical reaction rate is dependent upon the product of the forward and backward reaction rate constants and the species concentration.

$$\left(dx_i / dt\right)_j = \left(v_{i,j}'' - v_{i,j}'\right) k_{j,f} \prod_{j=1}^{n} x_j^{v_{i,j}'} + \left(v_{i,j}' - v_{i,j}''\right) k_{j,b} \prod_{j=1}^{n} x_j^{v_{i,j}''} \tag{8.2a}$$

In Equation (8.2a), the subscribe j denotes the jth forward or backward reaction. In order to apply the law of mass action, the rate constants are assumed to be independent of whether the system is in equilibrium or not. The commonly accepted assumption is consistent with the location equilibrium approach. The reaction rate constant is defined as

$$k_j = k_{j,f} / k_{j,b} = \prod_{j=1}^{n} x_j^{v_{i,j}''} \Big/ \prod_{j=1}^{n} x_j^{v_{i,j}'} \tag{8.2b}$$

The rate constant k_j is considered only to be the function of temperature but not the nonequilibrium composition. The relationship is equally applicable to the electronic degree of freedom such as disassociation and ionization. Only when the chemically reacting gas mixture is in equilibrium, the forward and backward reactions are in dynamic balance, and the net of change in composition ceases.

$$\left(dx_i/dt\right)_j = \left(v_{i,j}'' - v_{i,j}'\right)k_j\left[\prod_{i=1}^{n} x_i^{v_{i,j}'} - \left(1/k_{f,j}\right)\prod_{i=1}^{n} x_i^{v_{i,j}''}\right] \tag{8.2c}$$

In practical applications, the net of change for the i species concentration in mole fraction is expressed in the species density and molecular weight M_i. The net rate of composition change is the balance between all individual elementary one-step forward and backward reactions.

$$dw_i/dt = k_{f,i}(T)\prod_{i=1}^{n}\left(\rho_i/M_i\right)^{v_{i,j}'} - k_{b,i}\prod_{i=1}^{n}\left(\rho_i/M_i\right)^{v_{i,j}''} \tag{8.2d}$$

Therefore, the finite rate chemical reaction or the nonequilibrium reaction is the sum of all permissible simple elementary chemical reactions.

$$dw_i/dt = M_i\sum_{j=1}^{J}\left(v_{i,j}'' - v_{i,j}'\right)\left\{k_{f,j}\prod_{k=1}^{N_{kf}}\left(\rho_k/M_k\right)^{v_{k,j}'} - k_{b,j}\prod_{k=1}^{N_{kb}}\left(\rho_k/M_k\right)^{v_{k,j}''}\right\} \tag{8.2e}$$

8.2 CONDITION OF EQUILIBRIUM CHEMICAL REACTION

In a chemical reacting system, it is an essential process toward a new equilibrium state, which infers the entropy of the system to be a maximum. The chemical potential μ_i is identified as the partial Gibbs potential; $\mu_i = h_i - Ts_i$; $\mu_i = g_i$ by Equation (7.4c) $G_i = H - TS$ [3]. The direction of chemical reaction therefore always proceeds from a state of higher to a lower state of the chemical potential

$$\sum_{i=1}^{n} v_{i,j}'\mu_i > \sum_{i=1}^{n} v_{i,j}''\mu_i \tag{8.3a}$$

In Equation (8.3a), the chemical potential of species i per molecule is expressed as μ_i. The first law of thermodynamics including chemical reaction becomes [3,7];

$$dE = TdS - PdV + \sum_{i=1}^{n} \mu_i dN_i \tag{8.3b}$$

The chemical reactions are still explicitly excluded from the system by the condition that N_i shall be varied independently of each other. For an irreversible change occurring in a constant mass mixture, the following generalization of thermodynamic quantities, the Gibbs thermodynamic potential with chemical reactions can be written as

$$dG = -SdT + Vdp + \sum_{i=1}^{n} \mu_i dN_i \tag{8.3c}$$

For a perfect gas, the internal energy is a function of temperature only, therefore

$$Tds_i = dh_i - \left(1/\rho_i\right)dp_i \tag{8.3d}$$

Recall that the partial pressure of the species i is $p_i = \rho_i (R/w_i) T$. It follows that

$$T(\partial s_i / \partial p_i)_T = -1/\rho_i \quad \text{and} \quad T(\partial s_i / \partial p_i)_T = -R/p_i w_i \qquad (8.3e)$$

Integrate the above equations at a constant temperature to get a useful result involving chemical reactions.

$$s_i(p_i, T) - s_i(p^*, T) = -\left(\frac{R}{w_i}\right) \log\left(\frac{p_i}{p^*}\right) \qquad (8.3f)$$

In Equation (8.3f) the notation $P*$ is just a reference pressure. The above equation is derived for a gas in local thermodynamic equilibrium. For a thermally perfect gas is the same as assuming the individual molecules of the gas exert no long-range force on each other. It means that the potential energy of interaction is zero, or the internal energy of the gas is not a function of the density. The entropy S of the system will be a maximum once equilibrium has been reached and simultaneously the Gibbs potential has attained the minimum value for a fixed thermodynamic state.

At the chemical equilibrium state, the chemical potentials of the forward and backward reaction are equal as indicated by Equation (8.3a). Under this condition, it yields.

$$\sum_{i=1}^{n} (v_i'' - v_i') \mu_{e,i} + RT\left[\prod_{i=1}^{n} (p_i/p^*)^{(v_i'' - v_i')}\right] = 0 \qquad (8.4a)$$

Since the partial pressure of each chemical species is uniquely determined by the species concentration, the system is in chemical equilibrium if the partial pressures of the reactants satisfy the following condition:

$$\prod_{i=1}^{n} (p_i/p^*)^{(v_i' - v_i'')} = \exp(-\Delta\mu/RT) \qquad (8.4b)$$

In Equation (8.4b), the increment of the chemical potential of the system is simplified and expressed as $\Sigma(v_i'' - v_i') \mu_{e,i} = \Delta\mu$. For convenience, an arbitrarily chosen value of the reference pressure can be set as one atmosphere. The law of mass action leads to the conclusion that when partial pressures of all reacting species satisfy Equation (8.4b), then the chemical reacting system is in the equilibrium condition. Under the chemical equilibrium condition, an equation for the equilibrium constant dependent on temperature can be then given as

$$\kappa(T) = \prod_{i=1}^{n} p_i^{(v_i' - v_i'')} \qquad (8.4c)$$

The rate of chemical reaction is usually studied as a function of the species concentration of the reactants and the temperature. The rates of forward and backward reactions are determined from the Arrhenius equations

$$\kappa(T)_{f/b,i} = A_{f/b,i}T^n \exp\left(i\varepsilon_{f/b,i}/kT\right) \tag{8.4d}$$

And the net rate of progression of the nth reaction is simply the difference of $k_f - k_b$. Arrhenius recognized that one of the most striking features of the chemical reaction rate is the strong temperature dependence. His observation led to the Arrhenius hypothesis. In fact, the Arrhenius equation has been given by Equation (8.1c).

For a nonequilibrium process, the total differential of the Gibbs free energy $(G = H - TS)$ is

$$dG = \left(\frac{\partial G}{\partial T}\right) + \left(\frac{\partial G}{\partial p}\right)dp + Tds_{\text{irrev}} \tag{8.5a}$$

The relationship between the species concentration and the chemical reaction rate can be found in the first law of thermodynamics $dG = Vdp - SdT$. By a term-by-term comparison, it is shown that

$$Tds_{\text{irrev}} = \sum_i \left(\partial G/\partial N_i\right)dN_i \tag{8.5b}$$

At equilibrium, $dS_{\text{irrev}} = 0$, or the summation of the Gibbs free energy for all reacting species vanishes which provides a mean for determining equilibrium chemical composition. The condition for equilibrium has an alternate, namely

$$\sum_{i=1}^{n}(v_i'' - v_i')G_i = 0 \tag{8.5c}$$

Thus the law of mass action becomes formally a relationship between the partial pressure of the reactants and the temperature of the reacting system

$$\kappa_p(T) = \prod_{i=1}^{n}p_i^{v_i} = \exp\left(-\frac{\Delta G}{\kappa T}\right) \tag{8.5d}$$

The values of the chemical reaction rate $\kappa_p(T)$ can be obtained from either the experimental observation or can always be calculated from statistical thermodynamics utilizing the Arrhenius equations Equation (8.4b). These processes are complex and our understanding of the physical-chemical interaction at the molecular and atomic scales is extremely limited. Nevertheless, there are continuous developments of chemical kinetics data starting from the early 1960s until 2000s [8,9].

8.3 COUPLING CHEMICAL KINETIC WITH AERODYNAMICS

In order to complete the complex energy transfer process including all the internal degrees of freedom for high enthalpy hypersonic flows; the chemical kinetics and aerodynamic interactions are coupled by the source term for all chemical species. The rate of a chemical reaction is also dependent on the energy cascading and redistributing among different quantum states of the same species, as well as, between internal degrees of freedom. Although quantum jumps are instantaneous, the relaxation phenomenon between jumps frequently may occur, the phenomenon is determined by the collision frequency and energy levels between quantum states.

Integration of aerodynamics with chemical kinetics has been initiated earlier for propulsive systems and combustion [10] and refined with the significant progress in numerical algorithm and computing technology advancement. Combustion in rocket jet plumes and fuel injectors was successfully simulated [11,12]. These research efforts exemplify a rapidly growing interdisciplinary thrust by computational aerodynamics in conjunction with chemical kinetics.

For computational simulation of nonequilibrium chemical reacting hypersonic flow, all chemical reacting components including the same species at different quantum states are treated as independent reactants and products. The reaction rate, Equation (8.2e) is appended to the mass conversation formulation as a source or a sink for a specific chemical species in terms of mass flux per unit volume. The electrostatic and the Lorentz force also include in the conservation momentum equation as externally applied forces. The governing equations that integrate aerodynamics with nonequilibrium chemical reactions and quantum transition can be summarized as the following; The three global conservation laws of continuity, momentum, and internal energy become [13].

$$\partial \rho_i / \partial t + \nabla \cdot \left[\rho_i (u + u_i) \right] = dw_i / dt \tag{8.6a}$$

$$\partial \rho u / \partial t + \nabla \cdot (\rho u u + pI - \tau) = \rho_e E + (J \times B)$$

$$\partial \rho e / \partial t + \nabla \cdot \left[\rho e (u + u_i) \right] - \kappa \nabla T + \sum \rho_i u_i h_i + q_{rad} + u \cdot p\overline{\overline{I}} + u \cdot \overline{\overline{\tau}}$$

$$= E \cdot J + \left[\rho_e E + (J \times B) \right] \cdot (u + u_i) + Q_{vt} - Q_{et}$$

The conservation of vibrational energy and electronic energy that have been derived previously; Equations (6.9a) and (6.9b) are included to close the integrated governing equations system.

$$\partial \rho_i e_{iv} / \partial t + \nabla \cdot \left[\rho_i (u + u_i) e_{iv} + q_{iv} \right] = Q_{V,\Sigma} \tag{8.6b}$$

$$\partial \rho_e e_e / \partial t + \nabla \cdot \left[\rho_e (u + u_i) e_e + u \cdot p_e I + q_e \right]$$

$$= E \cdot J + \left[\rho_e E + (J \times B) \right] \cdot (u + u_i) + Q_{e,\Sigma}$$

The associated initial values and boundary conditions are identical as solving the Navier-Stokes equations, except the catalytic and non-catalytic conditions on the medium interface are required to determine a unique solution. In essence, these boundary conditions describe whether the chemical reactions are permitted or not to continue on the interface. Complications will be encountered on the ablating surface, which will be delineated later.

The reaction rates for all nonequilibrium chemical reactions become the controlling parameters for physical fidelity concerns and are adopted exclusively from the chemical kinetics models developed for hypersonic earth reentry. The early chemical kinetics model for high-temperature air by Dunn and Kang [14] composes of 12 species and 64 elementary chemical reactions and later simplifies to 11 species and 26 reactions. All internal degrees of excitation are assumed to retain thermal equilibrium condition and the reactions of the neutral species are known from the accumulated research results. The rates of coefficients for the recombination of charged species NO^+, O^+, N^+, O_2^+, and N_2^+ are measured from the application conditions.

The most widely adopted Olynick-Chen-Tauber chemical kinetics model, which is continuously refined from Park's original effort and extends its application to include ablation phenomena [15,16]. The carbonaceous heat shields have been implemented on space vehicles entering the atmospherics of Mars and Venus, and reentry to earth; the chemical compositions increased at least with C, C_3, CN, CO, and CO_2. The Park's chemical kinetic model has also been applied to the Mars reentry by a systematic modification with increasing numbers of elementary chemical reactions. The reaction rates also frequently apply to the simulations involving chemical reactions with empirically determined different temperatures for the nonequilibrium internal degrees of freedom [17]. A series of computational validated have been conducted against available shock tube data by the European Space Agency sponsored program [18]. Still, the reaction rate constants in the hypersonic application are known to possess a large and varying degree of uncertainty.

The flight data of the Fire-II experimental program have been used for the validation of physical-chemical models and computer codes for the aerophysics of reentry vehicles for more than 40 years [19,20]. It has been shown that the chemical kinetic models for the nonequilibrium Fire-II simulation lead to different predicted temperatures by a wide range, 3,000–5,000 K within the shock layer [5]. Such disparities have also been shown in the early 1990s [21]. Several recent publications

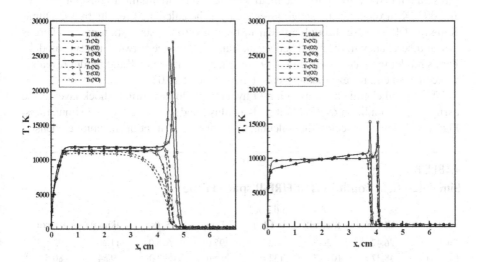

FIGURE 8.3 Results by two chemical kinetic models.

also demonstrate the significant influence of chemical and physical kinetics on the radiative heating of a new generation of space vehicles. These calculations are performed with models of chemical kinetics of Dunn and Kung [14] and Park [15]. Two different models of dissociation are used; the first one takes into account differences between translational and vibrational temperatures for N_2, O_2, and NO in the framework of the CVD model [18].

Two computations performed to correspond to the flight data of the Fire-II experimental probe are presented illustrating the status of the chemical kinetic models. In Figure 8.3, the translation and vibration temperatures of nitro oxide, nitrogen, and oxygen molecules are included only for the selected two stages from its earth reentry flight record. The characteristics of the fight environment are given by the following Table 8.1 (76.42, 48.37 km).

At the earlier stage of reentry, the predicted translation or static temperature by the Park model [15] is persistently higher than that of the Dunn-Kang model. The peak values immediately downstream to the bow shock are 27,000.0 versus 26,300.0 K. After a short distance of 2 cm in the shock layer, an isothermal condition is observed at the values of 12,000.0 and 11,400.0 respectively. However, the Dunn-Kang chemical kinetic model predicts a higher nitrogen molecular vibration temperature than that of the Park model, which adopts an empirical dissociation temperature assumption. A consistent and anticipated trend is also noted for a more rapidly increasing vibration temperature of nitric oxide molecules than molecular oxygen and nitrogen. For aerodynamic structure, the Dunn-Kang model produces a slightly greater shock standoff distance than the comparing Park model by a distance of 5.0 to 4.8 cm. Nevertheless, the maximum disparities from all computational results are confined to 5.2%.

At the later FIRE-II probe reentry stage, $t = 1,645$, both comparing computational simulations indicate that the hypersonic flow reaches an equilibrated thermodynamic state immediately downstream to the shock. The peak static temperature within the shock layer now reduces to 15,000.0 K by both models. The equilibrium temperatures calculated by the Park chemical kinetic model maintain a constant level of 10,000.0 K throughout the shock layer, and yet the calculated results by the Dunn-Kang model decrease through a linear fashion toward the stagnation point. There is a clear discrepancy in the predicted shock standoff distance between the two models. Park's model indicates a value of 4.19 cm, whereas the Dunn-Kang model reveals a value of 3.80 cm, to reveals the greatest difference of 10.03%.

The typical chemical composition of hypersonic flows within a shock layer at the earth reentry condition over a blunt body is illustrated by computational simulations. Figure 8.4 depicts species the calculated species concentration in number density

TABLE 8.1

Simulated fight conditions of FIRE-II space Probe

Stage	Alt (km)	P (Pascal)	ρ (kg/m³) $\times 10^{-5}$)	T_∞ (K)	T_w (K)	U_∞ (km/s)	R_n (cm)
$T = 1,634$	76.42	20.8	3.72	195.0	615.0	11.36	93.5
$T = 1,645$	48.37	1079.7	132.4	285.0	1520.0	9.54	80.5

utilizing Park's kinetic model along the stagnation streamline of the RAM-CII probe. At the freestream conditions of $M_\infty = 23.9, T_\infty = 244.0$ K, and the surface temperature of 1,500 K, the peak translational temperature still has a value of 14,490 K, and the calculated vibrational temperature for molecular nitrogen and oxygen are 11,411 and 11,396 K respectively. For computational simulations using two different surface temperatures (300 and 1,500 K), the main feature of the temperature profile is unaltered in that the equilibrated temperature for vibrational, rotational, and translational modes has been reached at the distance of one-third of the shock-layer thickness downstream of the bow shock. This feature is shared by all numerical simulations to reveal vibrational excitation requires a sufficient number of collisions to attain the same thermodynamic state as the translational and rotational modes [22–24].

To ensure the computational results are grid density-independent, most computational simulations have routinely conducted a grid-point refinement study. Drawn from the research efforts; two significant differences stand out in a numerical refinement study. First, the molecular components including the dissociated components N and O are nearly identical to the early simulation by an unrefined grid resolution. The second major difference is the concentrations of the charge carried atomic and molecular species O^+, N^+, O_2^+, and N_2^+ have now fallen below a value $<10^{12}/\text{cm}^3$. Therefore the number density of electrons is mostly balanced by the species NO^+ as it is expected to behave as the globally neutral partially ionized plasma. The molecular oxygen still exhibits the most drastic change from the immediate post-shock location to the stagnation point; its value drops from $5.8 \times 10^{15}/\text{cm}^3$ below the value of $10^{12}/\text{cm}^3$ and rises to the value of $5.3 \times 10^{12}/\text{cm}^3$ at the stagnation point. Finally, the recombination of atomic nitrogen and oxygen as well as molecular nitrogen are noted on the stagnation point. All these observations from the numerical simulations are similar among computational simulations adopting the different chemical kinetic models, but these findings are not still able to be verified by available experimental data.

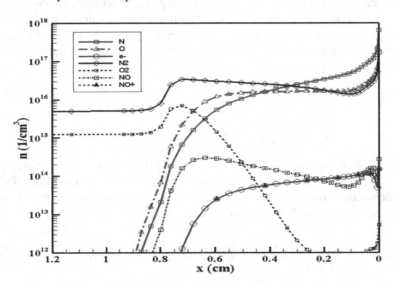

FIGURE 8.4 Air composition in the shock layer, $M_\infty = 23.9, T_\infty = 244.0$ K, $T_w = 1,500$ K.

8.4 MASTER EQUATION FOR PROBABILITY FORMULATION

Drawn from physical fidelity concerns, the vibrational energy quantum transitions in nonequilibrium reactive and nonreactive hypersonic flows attract a lot of analytic research efforts. The generation and depletion of dissociation by the state-to-state dissociating nitrogen have been investigated by different approaches, first was analyzed by the first order perturbation to quantum chemical kinetics by Schwartz-Slawsky-Hertzfeld (SSH) theory [25] and analytical methods [26–28]. The calculated relaxation time by the forced harmonic oscillator dissociation (FHO) models agrees well with experimental data [27]. The vibrational relaxation and the non-equilibrium dissociation at hypersonic temperatures may be satisfactorily by either the SSH theory or the FHO model, but the SSH theory is clearly beyond its valid theoretical domain [27,29].

The FHO model approach is a non-perturbation analytic method to calculate quantum transition probabilities for translation-vibration quantum transition based on the forced harmonic oscillator model. It is well known that the master kinetics equation is a set of first-order phenomenological differential equations describing the time evolution for the probability of discrete states, and is applicable to the high-temperature nonequilibrium gas phenomena with quantum transitions. The simplicity of the master equation makes the forced harmonic oscillator rate model attractive by a direct Monte Carlo simulation. Otherwise, a more rigorous approach through three-dimensional calculations by conducting a very large number of collisions with different orientations can be very cumbersome and time-consuming.

The approximate energy transfer among translation and vibration internal degrees of freedom has often assumed that the quantum jumps follow the ladder-climbing process [24]. The process can also be systematically determined from the population density distribution in a quantum state through the generation and depletion process. For an anharmonic oscillator, the quantum jumps are describable by the changing vibrational quantum level through inelastic collisions between diatomic molecules of the i, to i', quantum state as well as the dissociation process i, j to i', j' quanta. The model of the transition process also includes molecular nitrogen dissociation [24,29].

$$N_2(i) + N_2 \rightleftarrows N_2(i') + N_2; \, i' = i \pm 1 \tag{8.7a}$$

$$N_2(i) + N_2(j) \rightleftarrows N_2(i') + N_2(j'); \, i' = i \pm 1, j' = j \pm 1$$

$$N_2(i) + N_2 \rightleftarrows 2N + N_2$$

The quantum jump is determined numerically by solving the master equation for the population density distribution. The kinetic rate of the quantum transition for the vibration-translation from i to i' state and vibration-vibration transition from i, j to i', j' is given the population distribution function.

$$dw_i/dt = \left(1/M_{N_2}\right)\left\{\sum_{i'}\left[(k_{vt}(i';i)\rho_{i'}\rho - k_{vt}(i;i')\rho_i\rho\right]\right\} \tag{8.7b}$$

$$+ \sum_{j,i',j'}\left[k_{vv}\left(i',j';i,j\right)\rho_{i'}\rho_{j'} - k_{vv}\left(i,j;i',j'\right)\rho_i\rho_j\right]$$

The vibrational energy of a nitrogen molecule is the sum of energy in different quanta.

$$e_v = \sum_i \left(\rho_i / \rho_{N_2} \right) \varepsilon_i \tag{8.7c}$$

The quantum state is determined by the solution of the master equation, and the ratio $\left(\rho_i / \rho_{N_2} \right)$ defines the fraction population of the ith vibration quantum level. Therefore, once the population density distribution is known, the vibration quantum energy level i can be determined by a third-order approximation.

$$\varepsilon_i / hc = \omega_e (i - 1/2) - \omega_e x_e (i - 1/2)^2 + \omega_e y_e (i - 1/2)^3 \tag{8.7d}$$

In Equation (8.7c), the notation c designates the speed of light, and the spectroscopic constants are $\omega_e = 2358.57 \text{ cm}^{-1}$, $\omega_e x_e = 14.324 \text{ cm}^{-1}$, and $\omega_e y_e = -0.00226 \text{ cm}^{-1}$.

A comparative investigation of the quantum chemical physics is critical by the traditional chemical kinetics molding and solution by the master equation for a better understanding and improving the physical-based modeling. Figure 8.5 presents the computational simulations of the nonequilibrium vibrational temperatures predicted by the CVD model and master equation at the peak heating condition of Stardust at an altitude of 59.8 km, a velocity of 11.13 km/s, and an ambient temperature of 238.5 K.

The quantum transition models merging into nonequilibrium chemical kinetic formulation with the quantum chemical-physics models have been successfully applied to earth reentry simulations from the RAMC-II probe to the fastest man-made sample return capsule Stardust.

The maximum difference in predicted vibrational temperature of molecular nitrogen and oxygen is around 12.5% versus the solutions by the master equation [29]. However, a greater discrepancy is detected for the predicted translational temperatures and the bow shock wave standoff distances. A major difference is that the solution of the master equation is focused on the vibration-translation and vibration-vibration

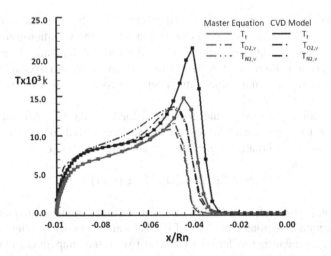

FIGURE 8.5   Comparison of temperatures generated by the kinetic model and master equation.

quantum transitions, the ionization phenomenon is not explicitly included. The other causes of discrepancies can be easily traced back to the computational accuracy in numerical resolution and the adopted numerical algorithm. The general agreement between published simulation results of the species concentration reflects the current status of nonequilibrium chemical kinetics models in practical applications, but with very limited verification from experimental observations.

Another translation-vibration transition has been developed for possibly more detailed energy transfer processes including the dissociation. The forced harmonic oscillator (FMO) model is adopted to evaluate both the single-quantum (V-T) and double-quantum vibration-vibration-translation (V-V-T). The state-resolved vibrational kinetic are solved by the following master equation [27].

Specifically, the forced harmonic oscillator of multiple jump rate models has been adopted for simulating vibrational relaxation. These approaches have generated the most reliable theoretical data available. A good agreement is exhibited for both the temperature and the quantum number of a single-quantum and the double-quantum vibration-vibration-translation jump in the temperature range of 200 K < T < 8,000 K, and for a vibrational quantum number up to the molecule dissociation [27].

$$df/dt = \sum_i \left[k_a(i;m) f_i - k_a(m;i) f_m \right] \left(x_a/\mu_a \right) \rho \tag{8.8a}$$

$$+ \sum_{i,j,n} \left[k_m(i,j;m,n) f_i f_j - k_m(m,n;i,j) f_m f_n \right] \left(x_m/\mu_m \right) \rho$$

$$- f_m \left(1/x_m \right) \left(dx_m/dt \right) \left[k_a(m \rightarrow) f_m - \rho k_a(\rightarrow m) \right]$$

$$\left[\left(x_a/\mu_a \right)^2 \left(\mu_m/x_m \right) \right] \left(x_a/\mu_a \right) \rho$$

$$- \left[k_m(m \rightarrow) f_m - \rho k_m(\rightarrow m) \right] \left[\left(x_a/\mu_a \right)^2 \left(\mu_m/x_m \right) \right] \left(x_m/\mu_m \right) \rho$$

In Equation (8.6c), the mth vibration quantum level of a diatomic molecule is designated as f_m. The notations $k_a(i;m)$ and $k_m(i, j;m, n)$ are the vibration-translation and vibration-vibration-translation rates. The detailed rates of dissociation and recombination are given as $k_a(m \rightarrow)$, $k_a(\rightarrow m)$, $k_m(m \rightarrow)$, and $k_m(\rightarrow m)$. Again in the equation, the mass fraction and reduced mass between collision partners are designated as x and μ respectively.

The relaxation phenomena are considered for both the simultaneous vibration transition of the vibration-vibration-translation process and the nearly isothermal relaxation of a diluted diatomic gas. The vibrational energy is given as

$$E_v = \Sigma E_v f_v = \Sigma \omega_e v \left[1 + x_e(v+1) \right] f_v \tag{8.8b}$$

From a comparison of the vibration-translation time for nitrogen molecule, the Forced Harmonic oscillator model and the Schwartz-Slawsky-Herzfeld theory are in general agreement but differ by a factor of two in the temperature range around 50,000 K [27]. At the lower temperature range, the Forced harmonic oscillator model closely agrees with the empirical result by Millikan and White.

8.5 AB INITIO, THE FIRST-PRINCIPAL APPROACH

The rapid and remarkable progress in high-performance computation technology promulgates a basic research paradigm switching from analytic to computational methodology. There is no doubt that the analytic results are far superior for describing and understanding science than the computational simulation, which provides a focused and isolated solution for an investigated problem that only on rare occasions can be extrapolated to similar or related phenomena. Therefore, computational simulations belong to the same category as experimental observations as a database development tool. However, for any interdisciplinary science that is severely limited by our analytic and experimental ability in studying complex physics such as quantum chemical physics, computational simulations are the only option. The idea of direct computational simulations for quantum chemical physics has been advocated nearly half a century ago, but unrealizable by the limited computational capability [30,31].

A scientific breakthrough opportunity presents itself by employing the unprecedented high-performance supercomputing capability to answer the formidable challenge through the *ab initio* (first principle) approach to quantum chemical physics [31,32]. In other words, the nonequilibrium chemical reactions and energy transfer by the quantum transitions among internal excitations can be calculated directly from chemical kinetics and quantum physics. Born and Oppenheimer have observed that the nucleus in a molecule is nearly stationary with respect to the oscillating electronic clouds. The *ab initio* approach is based on the Born-Oppenheimer approximation by separating the wave function of the nuclei from the electron, the approximation is also applicable to the molecule.

At present, the high-performance supercomputing capability can be used in quantum chemical-physics research through the *ab initio* approach. In essence, a direct calculation via the governing equations of quantum chemical physics is feasible for nonequilibrium chemical reactions, the energy transfer by the quantum transitions among internal excitations, and the optic properties of electromagnetic waves for radiation. In the *ab initio* computational process, the potential energy surface (PES) of the atom/molecule structure becomes the center of computational quantum chemistry.

The approach of *ab initio* is based on the Born-Oppenheimer approximation by separating the wave function of the nuclei from the electrons, and the approximation is also applicable to the molecule. In 1927, Born and Oppenheimer showed a very good approximation that the nuclei in a molecule are stationary with respect to electrons [33]. In this regard, the Schrödinger equation for a molecule may be separated into an electronic and molecular equation. The Schrödinger equation for a single electron is first solved for energy by the wave function of an atom and then adds the electronic energy to the internuclear repulsion to get the total energy. A profound consequence of the Born-Oppenheimer approximation is that a molecule has a finite dimension. The nuclei surround by a cloud of electrons that binds them in fixed relative positions and define the surface of the molecule and define the boundary of the molecule. Therefore, the nuclei that have permanent geometric parameters of a molecule are the nuclear coordinates. The energy of a molecule is a function of the electron coordinates. The approximation makes the molecular energy state versus the nuclei coordinates possible. However, the required computational resource by processing a huge number of nuclei coordinates is still prohibitive achieving a statistically stationary solution.

Wave mechanics originated from the work of de Broglie in 1923 which defined a relationship between the wavelength of a particle and its momentum, and led to the wave-particle duality [34]. The wave motion of a particle is governed by the wave equation, its amplitude varies in time and distance is described by $df(x)/dx^2 + \left(4\pi^2 m^2 u^2/h^2\right)f(x) = 0$. Since the kinetic energy is defined as the difference between the total and potential energy of a dynamics system, $mu^2/2 = E - V$, the one-dimensional wave equation governing the wave amplitude becomes $df(x)/dx^2 + \left(4\pi^2 m/h^2\right)(E-V)f(x) = 0$.

The Schrödinger wave equation for a completely dynamic system is the basis for nearly all quantum computational chemistry. The wave function φ_i is the description of an electron as a wave and is a probabilistic description of electronic behavior, which is the classic Eigen-value problem and written as.

$$H\varphi_i = E\varphi_i \qquad (8.9a)$$

where the Hamiltonian operator for electron dynamics in an electromagnetic field is.

$$H = \sum_i \nabla_i^2/2m_i + \sum_i q_i q_j/r_{ij} \qquad (8.9b)$$

The Schrödinger wave equation is the historical union of classical wave mechanics with quantum mechanics [31].

$$-\left[\sum_i \nabla_i^2/2m_i + \sum_{i<j} q_i q_j/r_{ij}\right]\varphi_i = E\varphi_i \qquad (8.9c)$$

In Equation (8.9c) the wave function is denoted as φ_i and the second-order spatial derivative ∇_i^2 is the Laplace operator, and the symbol E designates the energy of the system. The Schrödinger wave equation is an eigenvalue problem. Thus, this wave function is often known as the eigenfunction, and accordingly, E is the associated eigenvalue, which is the energy or the electric potential of a system. For reference, the important ground-level energy is defined as the energy value that corresponds to having all nuclei and electrons placed at an infinite distance from one another.

A system of particles of n degrees of freedom is determined completely by the values at a given time t, coordinates $q_1, q_2, ..., q_n$, and the conjugate momenta $p_1, p_2, ..., p_n$. The state of the system is described by the values of a wave function φ at a given time and the coordinates. It describes the probability of electronics bringing in certain locations but cannot predict exactly where the electrons are located. The wave function is also called probability amplitude because it is the square of the wave function that yields probabilities. The total probability of the system must be unity.

$$\int \varphi^*(q_1, q_2, ..., q_n, t)\varphi(q_1, q_2, ..., q_n, t)\partial r = 1 \qquad (8.9d)$$

The most common *ab initio* calculation is based on the Hartree-Fock approximation, in which the primary premise is the central field approximation, namely only the coulombic force by charged particles repulsion is taken into consideration [31,33]. The Hamiltonian operator of the Schrödinger equation for computational chemistry has frequently been simplified by the Hartree-Fock approximation by writing the Hamiltonian in the atomic units as

$$H = -\Sigma \nabla_i^2/2 - \Sigma Z_i/r_{ij} + \Sigma 1/r_{ij} \tag{8.9e}$$

In Equation (8.9e), the first term is associated with the kinetic energy of electrons, the second term describes the attraction between the electrons and nuclei, and the third term represents the repulsion among all electrons. The variation of energy from the Hamiltonian operator is simply the eigenvalue, and the total energy is determined by the integral over the entire phase space $E = \int \varphi^* H \varphi dr$.

One of the advantages of the Hartree-Fock approximation is that it breaks the many-electron Schrödinger equation into many simpler one-electron equations. Each one-electron equation is solved and to be identified as an orbital and the associated energy is called orbital energy. The orbital described the behavior of an electron in the net field of all the other electrons [33]. Still, the Schrodinger equation for a single electron is then solved for the energy and wave function of an atom or a molecule, the required computational resource is truly prohibitive.

8.6 POTENTIAL ENERGY SURFACE AND CRITICAL POINT OF TRANSITION

From the Heisenberg's uncertainty principle; a molecule has a defined momentum and position, therefore it away possesses kinetic and potential energy. In fact, a molecule is never stationary with zero kinetic energy. From the femtosecond spectroscopy experiments that the molecules actually vibrate in many directions until an energetically accessible reaction occurred. For this reason, the rational description of chemical reactions should be a statistical average of all possible energy paths than just the minimum path. The potential energy (PES) gives the energy at any location of the nuclei of a chemical reaction system. The Potential energy surface is critical in understanding the relationship between potential energy and molecular geometry. Computational chemistry determines the structure and energy of molecules in transition states involving chemical reactions.

The basic approach by the first principle is rigorous and straightforward, and is building on the fundamental equation of modern physics that describes how the electrons in a molecule behave. In addition, the solution is focused on the energy distribution by quantum mechanics. In fact, the most useful concept in science is the energy of a system, which can predict what molecular processes are likely to occur, or able to occur. In order to examine these phenomena; the entire potential energy surfaces must be computed. In fact, the PES is a plot of the collection nuclei and electronic energy versus the molecule geometric coordinates. In other words, the PES provides a visualization and an understanding of the relationship between potential energy and molecular geometry. Therefore, the *ab initio* approach is to determine the structure and energy of molecules and the transition states involving chemical reactions.

Following the idea of the Eyring's transition-state theory, the initiation of a chemical reaction or the transition of a quantum jump can be easily demonstrated by a simple schematic presentation. In Figure 8.6, the partition of atoms in a molecule for either heteronuclear or homonuclear is provided by a zero-flux surface S that passes through the critical point bond path with a saddle point C. On the zero-flux potential energy surface, the electromagnetic repulsive force is the minimum.

In most practical applications of *Ab initio* approach, the first step is the generation of the potential energy based on first principles by solving the Schrödinger equation for the electron motion on a fixed nuclear geometry. Then, the computation is proceeding to compute the reaction cross section and rate coefficient for applications. These cross sections are determined by classic trajectories for individual collision partners. A large number of collisions are computed with randomly selected initial orientations and phases, and typically millions of trajectories are carried out for a set of cross sections. Finally, the cross sections are converted to state-to-state reaction rate coefficients by Boltzmann weighting for determining the specific energy levels.

From the most recent progress in *Ab initio* research; computations have been simplified by semi-empirical methods by solving the Schrödinger equation for the forces between nuclei from the electron wave motion at a grid of fixed nuclear geometries. Then the analytic representation of the quantum chemical energies and forces on the grid of geometries is constructed for the potential energy surface (PES). The potential energy surface of the atom/molecule structure becomes the center of the *ab initio* approach because a stationary point (saddle point) on the PES is closely linked to the transition state.

In gist, the *ab initio* computational quantum chemistry is the study of PES stationary points and the pathway connecting them. Stationary points correspond to actual molecules with a finite lifetime and are energy minima. The stationary point of the PES is one at which the first derivative of the potential energy with respect to each geometry parameter vanishes.

$$\partial E/\partial q_1 = \partial E/\partial q_2 = \ldots = 0 \qquad\qquad (8.10a)$$

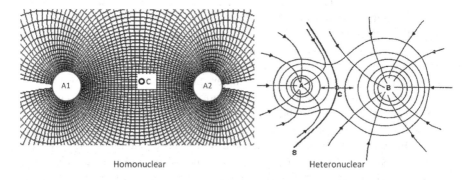

Homonuclear   Heteronuclear

FIGURE 8.6   Schematic of the zero-flux surface of the critical point bond path.

In Equation (8.10a), the partial derivative emphasizes that each derivative of E is just one of the geometric degree of freedom q.

For a transition state

$$\partial^2 E/\partial q^2 > 0 \tag{8.10b}$$

Equation (8.10b) holds for all q, except along the reaction coordinate, and along the reaction coordinate.

$$\partial^2 E/\partial q^2 < 0 \tag{8.10c}$$

A transition state is a thermodynamic concept and the distinction is made between the transition state and transition structure. According to Eyring's transition-state theory, the chemical species are in a kind of equilibrium with the reactant. Since equilibrium constants are determined by the Gibbs free energy difference, the transition state is a free energy maximum along the reaction coordinate. The species is often called an activated complex. The transition structure is thus a saddle point on an enthalpy surface.

These stationary points correspond to the actual molecules with a finite lifetime are energy minima. The transition state connecting the two minima represents a maximum along the direction of the intrinsic reaction coordinate, but along all other directions is a minimum. This is a characteristic of a saddle-shape surface, thus the transition state is a saddle point. A more detailed description can be found in Eyring's transition state theory, but for our purpose, the saddle point is a transition structure on the calculated PES [30]. The required topology knowledge has been developed by CFD to study surface shear patterns for three-dimensional separated

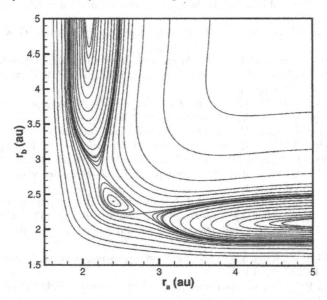

FIGURE 8.7 Potential energy surface of N and N_2 exchange reaction.

flows. The complex topological formations of noddle, saddle, and focus singulari-ties were established more than 30 years ago by the work of Lighthill, Tobak, and Peaks [35,36].

Figure 8.7 presents the contours of the PES in a two-dimensional plane that is produced from 3,326 *ab initio* points of the N–N bond of a nitrogen molecule [37]. The computational simulation shows the double barriers of two transition states that are connected by a shallow energy well. The two barriers are symmetric with respect to the interchange of two nitrogen atoms, and the bond distance is less than 3 au. The calculated bond angle is fixed at 119°.

The required research for interdisciplinary CFD involved quantum chemical physics has clearly revealed that the pathfinding effort is unmistakably pointing to the need by returning to basic scientific research. These undertakings, no doubt are forbidden and arduous, but they also offer undisputable scientific and technological leadership worldwide.

REFERENCE

1. Hayes, W.D. and Probstein, R.F., *Hypersonic flow theory*, Academic Press, New York, 1959.
2. Cox, R.N. and Crabtree, L.F., *Elements of hypersonic aerodynamics*, Academic Press, New York, 1965.
3. Kuo, K.K., *Principles of combustion*, 2nd Ed., John Wiley & Sons, Hoboken, NJ, 2005.
4. Clarke, J.F. and McChesney, M., *The dynamics of real gases*, Butterworths, Washington, DC, 1964.
5. Shang, J.S., Surzhikov, S.T., and Yan, H., Hypersonic nonequilibrium flow simulation based on kinetics models, *Frontiers in Aerospace Engineering*, Vol. 1, No.1, 2012, pp. 1–12.
6. Vincenti, W.G. and Kruger, C.H., *Introduction to physical gas dynamics*, John Wiley & Sons, New York, 1965.
7. Anderson, J.D., *Hypersonic and high temperature gas dynamics*, McGraw-Hill, New York, 1989.
8. Shang, J.S., Numerical simulation of hypersonic flows, Chapter 3. In: Murthy, T. (Ed.), *Computational methods in hypersonic aerodynamics*, Kluwar Academic Publishers, Boston, MA, 1991, pp. 81–114.
9. Shang, J.S. and Surzhikov, S.T., *Plasma dynamics for aerospace engineering*, Cambridge University Press, Cambridge, UK, 2018.
10. Patankar, S.V., Pratap, V.S., and Spalding, D.B., Prediction of laminar flow and heat transfer in helically wield pipes, *Journal of Fluid Mechanics*, Vol. 62, 1974, pp. 539–551.
11. Markatos, N.C., Spalding, D.B., Tatchell, D.G., and Mace, A.C.H., Flow and combustion in the base-wall region of a rocket exhaust plume, *Combustion Science and Technology*, Vol. 28, 1982, pp. 15–29.
12. Weidner, E.H. and Drummond, J.P., Numerical study of a staged fuel injection for supersonic combustion, *AIAA Journal*, Vol. 20, No. 10, 1982, pp. 1426–1427.
13. Shang, J.S., *Computational electromagnetic-aerodynamics*, John Wiley & Sons, Hoboken, NJ, 2016.
14. Dunn, M.G. and Kang, S.W., Theoretical and experimental studies of reentry plasmas, NASA CR 2232, April 1973.
15. Park, C., Review of chemical kinetics problems of future NASA missions, I earth entries, *Journal of Thermophysics and Heat Transfer*, Vol. 7, No. 3, 1993, pp. 385–395.

16. Olynick, D.R., Chen, Y.K., and Tauber, M.E., Aerodynamics of the stardust sample return capsule, *Journal of Spacecraft and Rockets*, Vol. 36, No. 3, 1999, pp. 442–462.
17. Park, C., Jaffe, R., and Partridge, H., Chemical-kinetic parameters of hyperbolic Earth entry, *Journal of Thermophysics and Heat Transfer*, Vol. 15, No. 1, 2001, pp. 76–90.
18. Surzhikov, S.T., Convective and radiation heating of MSRO, predicted by different kinetic models, *Radiation of High Temperature Gases in Atmospheric Reentry*, Rome, Italy, September 2006.
19. Cauchon, D.L, Radiative heating results from the fire II flight experiment at a reentry velocity of 11.4 km/s, NASA TM-X-1402, 1967.
20. Johnston, C.O., Hollis, B.R., and Sutton, K., Nonequilibrium stagnation-line radiative heating for fire-II, *JSR*, Vol. 45, No. 6, 2008, pp. 1185–1195.
21. Park, C., *Nonequilibrium hypersonic aerothermodynamics*, Interscience Publication, New York, 1990.
22. Graham, G.V., and MacCormack, R.W., Computation of weakly ionized hypersonic flows in thermochemical nonequilibrium, *Journal of Thermophysics*, Vol. 5, No. 3, 1991, pp. 266–272.
23. Scalabrin, L.C. and Boyd, I.D., Numerical simulation of weakly ionized hypersonic flow of reentry configurations, AIAA 2006-3773, San Francisco, June 2006.
24. Josyula, E., Bailey, W.F., and Ruffin, S.M., Reactive and nonreactive vibrational energy exchanges in nonequilibrium hypersonic flows, *Physics of Fluid*, Vol. 15, 2003, pp. 3223–3235.
25. Schwartz, Z.I., Slawsky, Z.J., and Hertzfeld, K. F., Calculation of vibrational relaxation times in gases, *The Journal of Chemical Physics*, Vol. 20, 1952, pp. 1591–1599.
26. Billing, B.D. and Fisher, E.R., V-V and V-T rate coefficients in diatomic nitrogen by a quantum classical model, *The Journal of Chemical Physics*, Vol. 43, 1979, pp. 395–401.
27. Adamovich, I., Macheret, S., Rich, J., and Treanor, C., Vibrational relaxation and dissociation behind shock waves, Part 2: Master equation modeling, *AIAA Journal*, Vol. 33, 1995, pp. 1070–1075.
28. Captelli, M. et al., *Non Equilibrium vibrational kinetics*, Springer Verlag, Berlin, Germany, 1980.
29. Josyula, E., Bailey, W., and Suchyta, C., Dissociation modeling in hypersonic flows using state-to-state kinetics, *Journal of Thermophysics and Heat Transfer*, Vol. 25, No. 1, 2011, pp. 34–47.
30. Eyring, H., Walter, J., and Kimball, G.E., *Quantum chemistry,* John Wiley & Sons, New York, 1944.
31. Lewars, E.G., *Computational chemistry: Introduction to the theory and applications of molecule and quantum mechanics*, Kluwer Academic Publishers, New York, 2004.
32. Levine, L.N., *Quantum chemistry*, 5th Ed., Prentice Hall, Upper Saddle River, NJ, 2000.
33. Young, D.C., *Computational chemistry*, John Wiley & Sons, New York, 2001.
34. Leighton, R.B., *Principles of modern physics*, McGraw-Hill, New York, 1959.
35. Lighthill, M.J., Attachment and separation in three-dimensional flow. In: Rosenhead, L. (Ed.), *Laminar boundary layers*, Vol. II, 2.6, Oxford University Press, Oxford, England, 1963, pp. 72–82.
36. Tobak, M. and Peaks D.J., Topology of three-dimensional separated flows, *Annual Review of Fluid Mechanics*, 1982, Vol. 14, pp. 61–85.
37. Wang, D., Stallcop, J.R., Huo, W.M., Dateo, C.E., Schwenke, D.W., and Partridge, H., Quantal study of the exchange reaction for $N + N_2$ using an ab initio potential energy surface, *The Journal of Chemical Physics,* Vol. 118, 2003, pp. 2186–2189.

9 Transport Property of Multi-Species Gas

9.1 COEFFICIENTS OF TRANSPORT PROPERTIES

The transport properties of a gas mixture are the consequence of a thermodynamic system departing from the collision equilibrium state. Under the collision nonequilibrium condition, all thermodynamic cycles now become non-isentropic. Defining and derivation of viscosity, thermal conduction, and diffusion coefficients are landmark achievements of the kinetic theory of diluted gas mixtures. The transport property of any individual gaseous species is governed by the inter-molecular forces during the collision process [1,2].

To be consistent with the kinetic model of the internal structure of a gas; transport properties of the gas mixture for thermal diffusion, molecular viscosity, and thermal conductivity need to be calculated from the Boltzmann equation by the Chapman-Enskog expansion. The infinity series solution has been given by Equation (2.33), after substituting the Enskog series into the collision integral of the Boltzmann equation, the non-vanished integrals approximation beyond the Maxwell distribution or the zero-order term is the net results of nonequilibrium collisions. The higher order terms of the equation give to arise the transport properties of the air mixture, Equation (2.37). The approximated collision integral should contain both the inelastic and elastic contribution, under the assumption that the number of inelastic collisions is small in comparison with the elastic counterpart. The order of magnitude analysis can be written as $f^{(1,el)}(x_i, c_i, t) = f^{(0)}(x_i, c_i, t)\varphi_{el}$.

The results from the kinetic theory for the molecular viscosity and thermal conductivity of individual species are used to generate a global property for a gas mixture. The binary diffusion coefficient of gas species is also obtained from kinetic theory. However, the additional forced diffusion mechanisms induced by the ionized air in an electromagnetic field may be dominated over the ordinary diffusion and need special attention. The treatment of ionized air for the force diffusion by the long-range Coulomb forces is beyond the gas kinetic theory. In addition, when the de Broglie wavelength, $\lambda = h/mu$ is of the order of magnitude of the particle dimension or greater, about one angstrom. Quantum mechanics effects begin to become important. For the hypersonic flows, the situations are frequently encountered.

The classic gas kinetic theory is applicable in determining a single gas species. Thus the Chapman-Enskog theory gives only the viscosity and thermal conductivity for each species of a gas mixture. Furthermore, these transport coefficients are originally derived only for the monatomic gas [3,4].

DOI: 10.1201/9781003212362-11

The molecular viscosity of a single species is given by the gas kinetic theory as

$$\mu = 2.67 \times 10^{-5} \sqrt{M_i T} \, / \, \sigma_i^2 \Omega^{(2,2)} \tag{9.1a}$$

The polyatomic molecules in the air, chiefly the nitrogen and oxygen molecules possess rotational and vibration internal degrees of freedom in addition to the translation mode. The existence of these internal modes can lead to non-adiabatic or inelastic collisions that the total kinetic energy may not be preserved. Therefore, the conductive heat transfer to polyatomic molecules is only a reasonable approximation. The thermal conductivities for monatomic and polyatomic molecules are

$$\kappa_{i,m} = 1.989 \times 10^{-4} \sqrt{T/M_i} \, / \, \sigma_i^2 \Omega^{(2,2)} \tag{9.1b}$$

$$\kappa_{i,p} = 2.519 \times 10^{-4} \sqrt{T/M_i} \, / \, \sigma_i^2 \Omega^{(2,2)} \tag{9.1c}$$

For most aerodynamic problems, the air mixture components of oxygen and nitrogen are similar in low-temperature environments. At the high-temperature domain, the dissociated or ionized air mixture is essentially binary, namely, the colliding particles can be divided into light and heavy particles. The expression for the binary diffusion coefficient in the practical unit is.

$$d_{i,j} = 1.858 \times 10^{-3} \sqrt{T^3 \left(M_i + M_j \right) / M_i M_j} \Big/ \sigma_{i,j}^2 \Omega^{(1,1)} \tag{9.1d}$$

In Equations (9.1a)–(9.1c), the molecular weight of species i, collision cross section, and collision integrals are designated as $M_i, \sigma_i,$ and $\Omega^{(i,j)}$ respectively. The required collision integrals and cross sections can be obtained by either a non-polarized or a polarized model intermolecular potential.

Forced diffusion is one of the complications in the analysis of ionized gas in hypersonic flows. The diffusive mass flux is generated by external forces exerted on the electrically charged species which is excluded by the kinetic theory of dilute gases. In an electromagnetic field, additional random motions by the electrostatic force and by Lorentz acceleration are recognized as new mechanisms for the diffusion process. In diffusing ionized air, the velocities of individual electrically charged species can be significantly different from each but are constrained by the Debye length of plasma as ambipolar diffusion [5]. The shielding distance between a pair of electrons and an ion is the intrinsic property of plasma maintaining electric neutrality [6]. This particular state requires an enormously large electrostatic force between an electron and an ion. The definition of the Debye shielding length in thermodynamic equilibrium conditions is $\lambda_d = \sqrt{\varepsilon k T_e / e^2 n} = 69 \sqrt{T_e/n}$. It is the smallest characteristic dimension of plasma. In a typical magneto-hydrodynamic generator, it has the value of 1.14×10^{-7} m.

In an ionized gas, the random or thermal motion of a charged particle is the result of mutual collisions plus forces acting on the particle by an electromagnetic field. The acceleration of an electron has a magnitude of eE/m in the negative direction of E. After numerous collisions, the average kinetic energy reaches a constant value and the force diffusion has an average velocity. This motion has a well-known drift velocity. A similar drift velocity also is attained by the positively charged ion but in the positive E direction, due to the greater mass of the ions its drift velocity is much slower than that of the electron.

The averaged energy gain between collisions is dependent on the ratio of E/p, which is often referred to as the reduced electrical field [7]. The proportionality constant between the drift velocity and the electric field $u_e = -\mu_e E$ is called mobility.

$$\mu_e = u_e/E = e/m_e \nu \qquad (9.2a)$$

In Equation (9.2a), ν is the averaged collision frequency for momentum transfer, and the mobility of the drift velocity of ion can also be given as.

$$\mu_i = e/m_i \nu \qquad (9.2b)$$

The self-diffusion or ordinary diffusion is proportional to the concentration gradient of the number of particle density ∇n. The diffusion coefficient d determined by the elementary kinetic theory of gases is proportional to the mean random velocity and the mean free path between collisions. The rate of flow particles per unit area is

$$\Gamma = -d\nabla n; \quad d \sim \lambda u \qquad (9.3a)$$

From the viewpoint of the conservation of number particle density, the mass flux density is Fick's second law of diffusion [3]. This relationship is valid for both electrons and ions. Since the mean free path of electrons is greater than that of ions, the electron's random velocity is much greater; it follows that $d_e > d_i$. The diffusion of charged particles is related to mobility; both arise from random motion and unbalanced collision force. In the one-dimensional motion of an iron, $u_i = \Gamma_i/n_i = (d_i/n_i)\partial n_i/\partial x = (d_i/p_i)\partial p_i/\partial x$. The gradient of the partial pressure $\partial p_i/\partial x$ must be balanced by the total force acting on the ion. The force is exerted by the electric field onto charged particles, thus $eEn_i = \partial p_i/\partial x$. By the definition of mobility, we have $d_i/\mu_i = p_i/en_i = kT/e$. This relationship between mobility and diffusion or $\mu_i = ed_i/kT$ is known as the Einstein relation [7,8, 9].

The often-encountered charge separation is the result of the disparity in the diffusion of electrons and ions in a strong electric field. The electromagnetic field augments the drift velocity of the ions and retards the electrons. From the charge conservation equation the net charged number density of electrons and ions must be identical. The plasma will establish the required electric field in the system to slow the more mobile electrons so that the electron moving rate is the same as that of the slower ions. When this process reaches a state of local equilibrium for the drift velocity, this resultant process is called ambipolar diffusion.

The simple charge conservation principle required the charged number flex density of both ions and electrons must be equal. From this requirement, the limiting and reasonable estimated value of the ambipolar diffusion coefficient can be found. Since the flux of diffusing electrons and ions must be balanced at all times, the following equality holds; $-d_e\nabla n_e - n_e\mu_e E = -d_i\nabla n_i + n_i\mu_i E$.

For the globally neutral plasma and singly charged ions, the number of electrons and ions must be equal; $n_i = n_e = n$. The electric field intensity that satisfies the equality condition is then, $E = (d_i - d_e)/(\mu_i + \mu_e)[\nabla n/n]$. The Ambipolar diffusion is given as.

$$d_a \nabla n = -d_e \nabla n - n\mu_e E = -d_i \nabla n + n\mu_i E \qquad (9.3b)$$

Eliminating the identical electric intensity E from Equation (9.3b) to yield the ambipolar diffusion coefficient in the absence of any externally applies magnetic field.

$$d_a = \frac{d_e \mu_i + d_i \mu_e}{\mu_e + \mu_i} \qquad (9.3c)$$

The unique force diffusion phenomenon appears only in high-enthalpy hypersonic flows in ionized air. In general, the influence on momentum and energy transfer is discernible and the effect can be significant in the presence of an externally applied magnetic field. The ambipolar diffusion formulation including the externally applied magnetic field has also been developed [10].

9.2 INTERMOLECULAR FORCES

The cornerstone of the gas kinetic theory is the binary collision processes between particles. The laws of classic mechanics describe the trajectories in the collision only between rigid elastic molecules, but the molecular structure does affect the collision process. The physics of intermolecular forces are generated by the binding electric charges within the electromagnetic fields. The bond constitutes an electric dipole, which is by definition an equal positive and negative charge, separated by a distance. A dipole is characterized by its dipole moment and is a vector having the magnitude and the direction connecting the negative to the positive charge. From the electrostatic, induction, and dispersion the forces take the form of dipoles and quadruples [11]. In general, the intermolecular forces during the collision process are attractive at the long range and repulsive at the short range and are derivable from the intermolecular potential.

$$\varphi(r) = \int_r^\infty F(r)\, dr; \; F(r) = -\partial\varphi(r)/\partial r \qquad (9.4a)$$

The intermolecular potential for the binary collision process on a fixed mass point coordinate origin is a spherically symmetrical function.

A wide range of inter-force molecular is possible, depending on the structure of the colliding partners. For electrically neutral molecules, the force field that bonds the molecular structure is basically the electromagnetic force generates by the nuclear and evolving electrons. The long-range interparticle forces are contributed by the electrostatic charges (Coulomb), electrostatic charge-dipole, charge-quadrupole, dipoles, and induced dipoles [4]. The long-range attractive forces usually decrease in magnitude proportional to r^{-7}, and the short-range repulsion generally decreases faster than the attraction. The long-range attraction between molecules is attributed by the induced dipole by the electron cloud which is unsymmetrical and dispersed over the nucleus. The short-range force can only be described by quantum mechanics.

At high-enthalpy hypersonic flow conditions, the dissociation of air molecules leads to a significant concentration of free-moving electrically charged which are composed of the dispersion force components. From the prior discussions of quantum transition in Chapter 6, the nitrogen atom has four spin states as $^1\Sigma$, $^3\Sigma$, $^5\Sigma$, and $^7\Sigma$. The forces contribute by these different quantum states are exclusively attractive and repulsive. Similarly, oxygen has the collision complex potentials of the singlet $^1\Sigma$, and triplet $^3\Sigma$, and the spectroscopically undetermined $^5\Sigma$ [4]. Therefore, the strong diatomic molecules in the exited electronic states possess large cross sections by the large number of the displaced electrons in respect to their ground state. The inter-molecular potential in principle can be determined by quantum mechanics, but some uncertainties still remind unresolved even to date.

The practical intermolecular potentials are usually adopted in a simple functional form which preserves the essential physical details for cross sections and collision integrals calculations. Therefore, there are always limitations and shortcomings in all the approximated potentials. The most widely known intermolecular potential functions are listed in the following.

1. Rigid-sphere potential only gives the short-range stiff repulsive force and ignores completely the long-range attraction. It presents the simplest rigid elastic sphere collision model.

$$\varphi(r) = \infty, r \le b$$
$$\varphi(r) = 0, r > b \tag{9.4b}$$

In Equation (9.4b), the symbol b usually is designated as the impact parameter in collision mechanics, for the rigid-sphere model is simply the minimum separation distance between the two colliding rigid spheres.

2. Point center repulsion potential follows the conclusion of the gas kinetic theory that the mean kinetic theory of all gas molecules must be defined by the temperature in equilibrium. The potential energy will carry little bearing upon the inter-molecule trajectory; the repulsive region of the potential curve in the immediately adjacent distance between colliding particles will determine the collision dynamics. This behavior gives the analytic base for the repulsive potential model, and it is appropriate when the gas mixture temperature $T \gg \varepsilon/k$.

$$\varphi(r) = d/r^\sigma \tag{9.4c}$$

The notations d and σ are free parameters and are to be determined by matching the measured transport coefficient [12].

3. Morse potential is the combination of a short-range repulsion and long-range attraction inter-molecule interaction, analogous to the Lennard-Jones potential [13]. From kinetic theory research, the small attractive force at large distances is found to affect the free path more than the large repulsions at close encounters. The resultant free path is quite different from that appropriate to viscosity and the other transport phenomena [14]. Thus, the short-range attraction term is appended to gain a higher degree of physical fidelity to the intermolecular potential model.

$$\varphi(r) = D_e \left[e^{2\alpha(r_e - r)} - 2e^{-\alpha(r - r_e)} \right] \qquad (9.4d)$$

In Equation (9.4d), r_e denotes the equilibrium bond distance and D_e designated the potential well depth. The parameter α controls the width of the potential well, $\alpha = \sqrt{k_e/2D_e}$ and k_e is the force constant at the potential function $\varphi(r)$ attaining the minimum value. Since the zero of potential energy is arbitrary, the Morse potential has also been given as $F(r) = D_e \left[1 - e^{\alpha(r - r_e)} \right]^2$.

4. Lennar-Jones potential is a realistic and simple intermolecular potential and becomes an archetype model that describes an electrically neutral atom or molecule. The model is usually limited in its application to air temperature <3,000 K for transport properties evaluations. The potential function was known as early as 1924 by Jones's two early publications [15].

$$\varphi(r) = 4\varepsilon \left[(\sigma/r)^{12} - (\sigma/r)^{6} \right] \qquad (9.4e)$$

In Equation (9.4e), ε is the depth of the potential well and σ is the value of separation distance r at $\varphi(r) = 0$ or $r = r_{min} - 2^{1/6}\sigma$ where $\varphi(r) = -\varepsilon$, Figure 9.1.

The potential function describes those two colliding particles repel each other at a moderate distance and do not interact at an infinite distance.

In the high-enthalpy hypersonic domain, the inter-molecule and inter-atom interactions are dominated by the short-range repulsive force that is generated from the remote-acting electrostatic field in the electrically conducting ionized air. The potential functions thus employ different models for dissociating and the ionizing phenomena of air [16].

5. Exponential repulsive model is mainly focused on the complicated N_2–N_2, N_2–N, and N–N interchange interactions for dissociation [16].

$$\varphi(r) = \varphi_o e^{-r/\sigma} \qquad (9.4f)$$

In Equation (9.4f), the parameters φ_o have a value range <10^{-4} electron voltage associated with the dissociation potential. The electron voltage is an energy unit (1 ev $= 1.6022 \times 10^{-12}$ erg) and the collision cross sections also have a chosen range of $3.16 < \sigma < 5.12$ Å.

6. Screened Coulomb potential is specially applied for the ion-ion and electron-electron interaction.

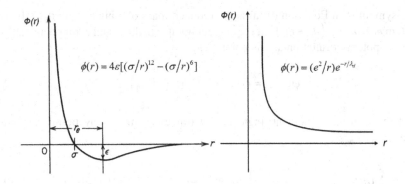

FIGURE 9.1 Schematic of typical intermolecular potential functions.

$$\varphi(r) = \left(e^2/r\right)e^{-r/\lambda_d} \qquad (9.4g)$$

In Equation (9.4g), the symbol λ_d denotes the Debye shielding length.

The most widely adopted intermolecular potential function for high-enthalpy hypersonic flows are the Lennar-Jones and the screened coulomb models. A schematic of these models is presented in Figure 9.1 to illustrate the short-range repulsive and long-range attraction. The notation e of Equation (9.4g) designates the elementary electric charge that has the physics unit of 1.6022×10^{-19} C.

9.3 COLLISION CROSS SECTION

The formal definition for the total collision cross section between collision partners is the ratio of the number of collisions per unit time, $1/\tau$ and the coming particle flux density, which is defined as the product of the particle number density and their relative velocities $n\left(u_i - u_j\right)$ [17]. Therefore, the collision cross section has the physical unit of area; $\sigma^{(t)} = \tau^{-1}/n\left(u_i - u_j\right)$ and may interpret as the effective geometrical blocking area a particle presents to colliding partners. The total cross section includes both the elastic and inelastic collision cross sections. For example, the collision cross section between atomic hydrogen and an incident electron with the energy of 100 ev has an elastic collision probability of 12.2%, and for ionization becomes 43.5%. The total cross section is 2.16×10^{-20} m^2. The incident electron at a higher energy level will reduce the total collision cross section.

In a binary collision, the coordinate origin of the two-body problem (i, j) is fixed on the mass center of particle j, and the symbol b represents the closest distance between the colliding partners. From equations of motion including the conservation of angular momentum and energy; the trajectory of the incoming particle i is obtainable [4].

$$\bar{m}\left(u_i - u_j\right)^2/2 = \bar{m}(dr/dt)^2/2 + \bar{m}\left(u_i - u_j\right)^2\left(b/r\right)^2/2 + \varphi(r) \qquad (9.5a)$$

The symbol $\bar{m}$ in Equation (9.5a) is the reduced mass of a binary colliding dynamic system $\bar{m} = m_i m_j / (m_i + m_j)$. At the contact point, r_{min} the radial component vanishes, and the potential function at this point yields.

$$\varphi(r_{min}) = \bar{m}(u_i - u_j)^2 \left[1 - (b/r_{min})^2\right] / 2 \tag{9.5b}$$

And the minimum incident angle or the trajectory turn away point can also be determined.

$$\theta_{min} = \int_{\infty}^{r_{min}} (d\theta/dr) dr = -b \int_{\infty}^{r_{min}} dr \bigg/ \int_{r_{inin}}^{\infty} dr/r^2 \left[1 - (b/r)^2 - (2\varphi(r)/\bar{m}(u_i - u_j)^2\right]^{1/2} \tag{9.5c}$$

The term $dr/d\theta$ actual has two roots, the minus sign is chosen because r decreases as θ increases on the incoming particle trajectory. The location of θ_{min} defines the range of possible collisions that can occur and is often referred to as the turning point.

From the Geometry, the scattering angle and the minimum incident angle have a fixed relationship $\chi + 2\theta_{min} = \pi$. The angle of deflection becomes.

$$\chi\left[b,(u_i - u_j)\right] = \pi - 2b \int_{r_{min}}^{\infty} dr/r^2 \left\{\left[1 - (b/r)^2 - 2\varphi(r)\right]/\bar{m}(u_i - u_j)^2\right\}^{1/2} \tag{9.5d}$$

The collision cross section is often best defined by the differential cross section $d\sigma^{(l)}/d\Omega$ for scattering into the solid angle $d\Omega = \sin \chi d\chi d\varphi$ that has been defined by Equation (2.8) as $d\sigma^{(l)} = bdbd\varphi$. The differential cross section is the function of a particular collision process and the relative velocity of the colliding particles. The symbols (χ, φ) are the spherical polar angles, as depicted by the sketched trajectory of colliding particles in Figure 9.2, where φ is the azimuthal and χ is identified as the shattering angle. The total collision cross section is obtained from the integrated result.

$$\sigma^{(l)} = \iint (d\sigma^{(l)}/d\Omega) d\Omega = \iint (d\sigma^{(i)}/d\Omega) \sin \chi \, d\chi \, d\varphi \tag{9.6a}$$

For nearly all circumstance, the scattering is azimuthally symmetric, the integration over the polar angle φ yields.

$$\sigma^{(l)} = 2\pi \int_0^{\pi} (d\sigma^{(l)}/d\Omega) \sin \chi d\chi \tag{9.6b}$$

In fact, it can be shown that the cross section for momentum transfer by elastic collision is [17].

$$\sigma^{(l)} = \int_{4\pi} (1 - \cos \chi) \sigma^{(l)}(\chi, \varphi) d\Omega \tag{9.6c}$$

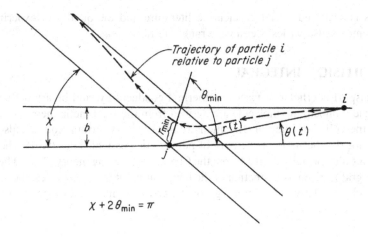

X + 2θ_{min} = π

FIGURE 9.2   Trajectory of colliding particles.

It is interesting to note that for elastic electron-atom collision at sufficiently low energies, the total and elastic collision cross section is approximately identical. Only at the high energy state, the total collision cross section can be less by a factor of one-half to the elastic collision [17]. It shall be also pointed out that the total elastic cross section never includes in the rigorous gas kinetic theory for determining the transport coefficients of gases. For calculating the transport coefficients, the generalized momentum transfer cross sections are given by the modified relation.

$$\sigma^{(l)} = \int_{4\pi} \left(1 - \cos^{i} \chi\right)\left(d\sigma^{(l)}/d\Omega\right)d\Omega \qquad (9.6d)$$

In Equation (9.6d), the notation l is a positive integer either one or two.

For an elastic collision, the intermolecular potential function does have the long-range attraction and the potential function as a step function. The corresponding deflection angle is also has only two distinctive value for $b \le d$; $\chi[b,(u_i - u_j)] = 2\cos^{-1}(b/d)$, and $b \ge d$; $\chi[b,(u_i - u_j)] = 0$. Substitute the discrete value of the deflection angle into Equation (9.6d) to get.

$$\sigma^{(l)} = \left\{1 - \left[(1+(-1)^l\right]/2(1+l)\right\}\pi d^2 \qquad (9.6e)$$

Therefore, the collision cross sections for an elastic rigid sphere are.

$$\sigma^{(1)} = \pi d^2; l = 1$$
$$\sigma^{(2)} = (2/3)\pi d^2; l = 2 \qquad (9.6f)$$

Except for the point centers of the repulsive intermolecular potential model, the angle of deflection, thus the collision cross section and collision integral can only be obtained by numerical integration. The experimental database on collision cross sections is scattered over the scientific literature and the quality of experimental measurements also varies over a wide range [17,18].

9.4 COLLISION INTEGRAL

The transport coefficients: viscosity, thermal conductivity, and binary diffusion of any single monoatomic molecule are derived from the gas kinetic theory as a function of the collision integral, Equations (9.1a)–(9.1c). A formulation has also been devised for the thermal conductivity of polyatomic gas molecules (9.1d). The collision integral is formally derived from the Chapman-Cowling theory [3,14]. The collision integral is given as a function of the temperature of the gas species, the reduced mass, and the relative kinetic energy of the colliding molecules together with the collision cross section.

$$\Omega^{(l,s)} = \sqrt{kT/2\pi\overline{m}} \int_0^\infty e^{-\gamma^2} \gamma^{2s+3} \sigma^{(l)} d\gamma \tag{9.7a}$$

$$\gamma = \overline{m}\left(u_i - u_j\right)^2 / 2kT \tag{9.7b}$$

$$\sigma^{(l)} = 2\pi \int_0^\infty \left(1 - \cos^l \chi\right) b\, db \tag{9.7c}$$

The collision cross section is the integrand over the complete range of impact parameter b for each initial relative velocity $(u_i - u_j)$ between colliding particles. The initial relative kinetic energy γ of the formulation also becomes a source of uncertainty for calculating the collision integral.

An extensive number of efforts have been devoted to describing the transport properties of high-temperature gases and simplified them for practical applications. An outstanding and exhaustive review has been performed by Capitelli and his colleagues [16,19]. The historical development and benchmark progress over the past 30 years are articulated and lay out a strategy for future improvement. The most recent approaches in basic research are directly applied to the formulation of kinetic gas theory through collision integrals and cross sections, Equations (9.6d), (9.7a), and (9.7c). Most collision integrals and collision cross sections for practical applications are obtained through numerical analyses by performing three consecutive integration steps: The first integral determines the classic deflection angle including the impact parameter b.

$$\chi(b,\varepsilon) = \pi - 2b \int_0^\infty \frac{dr}{r^2 \left[1 - (b/r)^2 - \varphi(r)/\varepsilon\right]^{1/2}} \tag{9.8a}$$

In Equation (9.8a), the symbol ε is the relative kinetic energy of the colliding partner as indicated earlier by Equation (9.7b). The second integral is an averaged process for the impact parameter to get the relevant cross section.

$$\sigma^{(l)}(\varepsilon) = \frac{2\pi}{1 - \left[1 + (-1)^l\right]/2/(l+1)} \int_0^\infty \left[b(1 - \cos l\chi)\right] db \tag{9.8b}$$

Finally, an averaging process is conducted over the energy to produce the collision cross section as a function of temperature. A unique feature of this multiple integration processes is the diminished dependence of the collision integrals on the potential functions. The final collision cross section is given as

$$\Omega^{(l,s)}(T) = \frac{4(l+1)}{(s+1)!\left[2l+1-1(-1)^l\right]} \left(\frac{1}{2kT}\right) \int_0^\infty \left(\frac{\varepsilon}{kT}\right)^{s+1} \sigma^{(l)}(\varepsilon) \exp\left(-\frac{\varepsilon}{kT}\right) d\varepsilon \tag{9.8c}$$

The collision integrals of the Morse, exponential repulsive, and Lennar-Jones intermolecular potential function have been tabulated, but other models are needed to be calculated [19,20,21].

The classic collision integral by the numerical method is displayed in Figure 9.3 by Hirschfelder et al [3]. The collision integrals $\Omega^{(1,1)}$ and $\Omega^{(2,2)}$ are normalized by the rigid sphere model, Equation (9.6f) viruses the reduced temperature $T^* = kT/\varepsilon$. In the past, the transport properties for low-temperature gas and a mixture of gases were calculated using the tabulated collision integral based on the Lennar-Jones model. The empirically determined force constant and the initial relative kinetic energy ε must be an input. The most important insight is illustrated by the comparison of the rigid sphere elastic collision model with the more realistic Lennard-Jones intermolecular potential model. The influence of the intermolecular force between colliding molecules is significant and the electromagnetic force and quantum mechanics effect in the high-enthalpy environment are not ignorable.

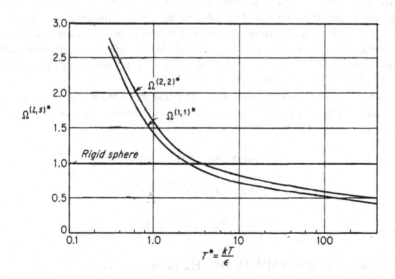

FIGURE 9.3   Comparison of the rigid sphere and Lennar-jones models.

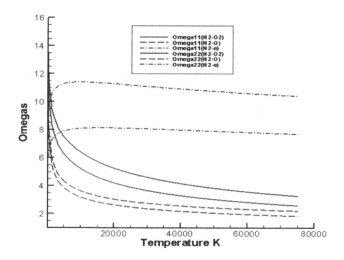

FIGURE 9.4 Collision integrals for N_2-O_2, N_2-O, and N_2-electron interactions.

Some of the distinct characteristics of collision integrals are displayed in Figure 9.4. Especially the magnitudes of collision integrals for the nitrogen molecule and oxygen molecule (N_2-O_2), nitrogen molecule and oxygen atom (N_2-O), as well as nitrogen molecule and electron (N_2-e) collisions are assembled over the temperature range from 300 to 75,000 K. The products of collision integrals and associated cross sections are scaled by a factor of π, and covering the whole range of inter-particle potential by a combination of the exponential repulsive, Lenard-Jones, and polarizability models with weight functions. It may be easily recognized that the collision integrals $\Omega^{(1,1)}$ and $\Omega^{(2,2)}$ for molecular nitrogen-electron interaction stand out and have greater values than the rest. The collision integrals of the molecular nitrogen-oxygen atom have a smaller value than that of the molecular nitrogen and oxygen. The numerical results are in general agreement with the classic observation in that the ratios between $\Omega^{(2,2)}/\Omega^{(1,1)}$ yield a value around 1.1 to have a range from 1.3519 (N_2-e), 1.2611 (N_2-O_2), to 1.196 (N_2-O).

Once the collision integrals are known, the transport coefficients of single gas species; viscosity, thermal conductivity, and binary diffusion are easily calculable by the following formulas [22].

$$\mu_{ij} = 2.6693 \times 10^{-5} \sqrt{2T\bar{m}}/\Omega_{i,j}^{(2,2)}$$

$$k_{ij} = 1.9891 \times 10^{-4} \sqrt{T/\bar{m}}/\Omega_{i,j}^{(2,2)} \qquad (9.8d)$$

$$D_{ij} = 2.268 \times 10^{-3} \sqrt{T^3/\bar{m}}/p\Omega_{i,j}^{(2,2)}$$

9.5 TRANSPORT PROPERTIES OF THE GAS MIXTURE

The classic Chapman-cowling theory gives only the viscosity, thermal conductivity, and binary diffusion for a single gas species. For practical applications, the transport properties of a gas mixture containing n components are approximated by a function of their mole fraction of components and their respective species transport coefficients. These formulas are also known as Wilke's mixing rule which is a logical approximation to the more complex expression by the Chapman-Cowling theory [23]. The mixing rule for a binary gas mixture also reaches an excellent agreement with the gas kinetic theory for a temperature range as high as 15,000 K. The mixing rule gives the viscosity coefficient of a gas mixture as.

$$\mu_{i,j} = \sum \frac{x_i \mu_i}{\sum x_i \phi_{i,j}}$$

$$\phi_{i,j} = \frac{1}{\sqrt{8}} \left(1 + \frac{M_i}{M_j}\right)^{-1/2} \left[1 + \left(\frac{\mu_i}{\mu_j}\right)^{1/2} \left(\frac{M_j}{M_i}\right)^{1/4}\right]^2,$$

$$(9.9a)$$

In Equation (9.9a) the symbol M_i characterized the molecular weight of the species, and the mole fraction x_i is related to the mass fraction $\alpha_i = \rho_i/\rho$; $x_i = (\alpha_i/M_i)\left[\Sigma(\alpha_i/M_i)\right]^{-1}$. The thermal conductivity of a gas mixture by the mixing rule has a similar expression.

$$\kappa_{i,j} = \sum \frac{x_i \kappa_i}{\sum x_i \phi_i}$$

$$\phi_{i,j} = \left[1 + \left(\frac{\kappa_i}{\kappa_j}\right)\left(\frac{M_j}{M_i}\right)^{1/4}\right]^2 \Big/ \left[8\left(1 + \frac{M_i}{M_j}\right)\right]^{1/2}$$

$$(9.9b)$$

The diffusion coefficient of a gas mixture is treated by the idea of a binary mixture, especially for dissociated and ionized air. The air mixture in the environment is naturally divided into heavy and light components. Since the momentum and energy fluxes contribute to energy transport through mass diffusion, the approximation by the diffusion coefficient shall be less significant via a binary mixture. The binary diffusion coefficient is approximated as.

$$d_{ij} = d_{ji} = 2.628 \times 10^{-3} \sqrt{mT^3} \big/ p\Omega^{(1,1)}$$

$$(9.9c)$$

All transport coefficients of a very wide range of temperatures from 200 to 28,000 K with the eleven species of air mixture have been calculated for a better understanding of computational results. According to the study of Capitelli et al., the calculated collision integrals are not dramatically dependent on the specific form of the potential models [16]. The dimensionless collision integral has been generated by a triple-integral process whose value depends on the dynamics of binary collisions, thus is still controlled by the intermolecular force law. The variation of these products of collision integral and cross-section is less pronounced in the high-temperature region to reflect a linear dependence on the local temperature and molecular weight of the species.

Comparative studies on high-temperature transport properties have been conducted by Palmer et al. [24] and Fertig et al. [25]. All the calculated results are based on the Chapman-Cowling theory using the most recently generated collision integrals. The verifications with experimental data and other analytic developments are also included to highlight the progress in transport properties over the years. The transport properties of air plasma are also generated over a wide pressure and temperature to give a glance of state-of-the-art achievements in this research area [26].

Figure 9.5 presents the calculated viscosity coefficients by Ferig et al. [25] and Palmer et al. [24] based on the Chapman-Cowling formulation but different collision integral calculations. these comparing results exhibit excellent agreement with respect to each other. Both calculations show the viscosity coefficient reaching the maximum value of around 10,000 K, beyond this temperature the interaction with ionized species begins to dominate the gas mixture leading to a decreasing and rising trend as the gas mixture temperature is above 20,000 K. The viscosity in the even higher temperature domain indicates a mild elevating trend until the air mixture temperature attained the value of 30,000 K. However, the calculation by Palmer et al. [24] terminates at the temperature at 20,000 K (Figure 9.6).

The more complex behavior of the thermal conductivities in the high-temperature range is observed at different thermal and chemical nonequilibrium conditions. From the result by Fertig et al. [25], the thermal conductivity of the high-temperature air clearly reveals suddenly increasing effects by the oxygen dissociation of about 3,800 K and the nitrogen dissociation of around 70,000 K. As the air mixture temperature is further elevated, the presence of partially ionized air species continuously pushes the magnitude of the conductivities upward. This trend persists until the radiative heat transfer takes place diminishing the contribution by thermal conduction, then increases again after attaining a stable balance among the conduction, convection, and radiation heat transfer mechanisms. The comparison with experimental data

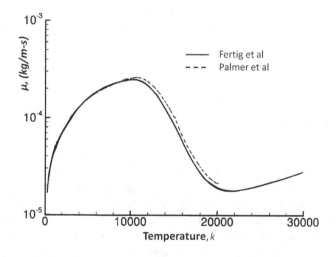

FIGURE 9.5   Viscosity coefficient of high-temperature air mixture.

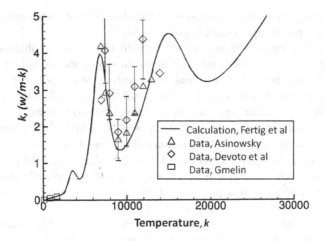

FIGURE 9.6 Comparison of high-temperature thermal conductivity gas mixture with data.

with Gmelin [27], Asinovsky et al. [28], and Devoto et al. [29] reaches only a general agreement. The data scattering band from experimental observation is substantial, except for the data by Gmelin at the lowest temperature range, therefore there is no conclusive validation that can be drawn from the specific comparison.

Although there is substantial progress in calculating the transport properties of high-temperature air mixture, the specific comparison of the calculated transport properties still yields a difference of 16.3%–24.2% in viscosity and thermal conductivity for high-temperature air using different sets of collision integrals in the temperature range from 300 to 30,000 K [24,30]. A continuous basic research for transport property of high-enthalpy hypersonic flows is still warranted.

9.6 ABLATION

The kinetic energy of any reentry vehicle at hypersonic speed must convert and dissipate in the form of thermal energy, a part of the energy is deposited on the surface of the reentry vehicle. The heat transfer rate varies widely depending on the reentry speed; at a low earth orbital (LEO) reentry speed of 8 km/s and Galileo reentry speed up to 47.7 km/s, the rate spanning a range from 2.3×10^2 to 1.7×10^5 kw/m². The heat transfer rate is strongly dependent on the surface material of the reentry vehicle whether it is a non-catalytic or a full-catalytic material. The ablative surface for thermal protection is widely used for dissipating phase-changing latent heat and absorbing the converted thermal energy and ultimately minimizes the conductive heat transfer to the substrate. The transport phenomena become very complex because it involves multiple phase exchange in the process. Intriguing ablation pattern has been found on the recovered reentry vehicle and meteorites and it has been studied for decades [31–33]. In all, the fine details of the ablation surface pattern are most recognizable as the crosshatched and regmaglypt. The cross-hatching pattern angle follows closely with the Mach angle according to the external hypersonic flow.

This behavior suggests the propagation of disturbance waves as the cause of a higher local ablating rate along the pattern grooves [34].

To formulate the numerical simulation for thermo-chemical ablation four basic mechanisms must be considered; heat absorption by conduction into the ablative material, vaporization of the material and pyrolysis of the polymer matrix, mass exchange between ablative surface and external flow, as well as the radiation heat transfer. The most challenging issue for modeling the thermo-chemical interaction is the pyrolysis reaction that generates decomposition gas and solid carbonaceous char residue. Meanwhile, the thermal expansion and the depletion of solid material lead to an increase in the porosity and permeability of the ablating material [35]. Chemical-physical interaction of ablation is sensitive to surface catalysis and species diffusion rates, therefore accurate transport property models and surface catalysis models are required to predict the ablative phenomenon [36,37].

For nonequilibrium hypersonic flows, the initial values and boundary conditions for numerical simulations have been established. The implicit assumption for the species conservation equations at the solid surface is either non-catalytic or fully catalytic. However, the mostly widely used condition is to simply let the chemical reaction control by the local thermodynamic condition identical to the interior domain. For which the species concentration in mass fraction, $\alpha_i = \rho_i/\rho$, at the surface is assumed to be the same as the oncoming stream $\alpha_i(0) = \alpha_i$. Such an idealized boundary condition for species conservation is referred to as super-catalytic; technically, it is assumed that the solid surface is fully catalytic. The other limiting condition is the non-catalytic condition in which the gradient of the species concentration is set to

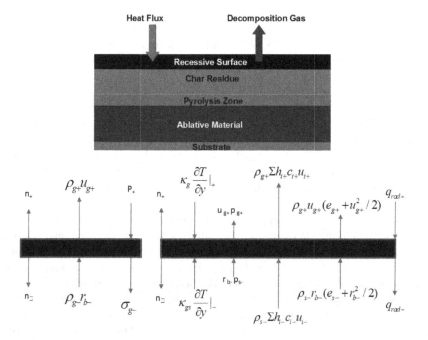

FIGURE 9.7 Schematic of ablative layer structure and exchanging fluxes.

zero, $n \cdot \nabla \alpha_i = 0$. On an ablating surface, the interface boundary conditions are much more complex than the limiting conditions (Figure 9.7).

Most research efforts for describing the ablating surface have been concentrated on the interface boundary conditions that are formulated either from physical observations or empirical formulations from unique insights. The basic formulation describes the decomposition rate, mass balance, and energy balance equations with an assumed dimensionless mass blowing rate B' [36].

$$\partial \rho_i / \partial t = -B_i e^{-E_i/T} \rho_{vi} \left[(\rho_i - \rho_{ci})/\rho_{vi} \right]^{\psi_i}$$

$$\partial \rho / \partial t = \partial \dot{m}_g / \partial y\, \rho_e u_e C_H (H_r - h_{ew}) + \dot{m}_c h_c + \dot{m}_g h_g + \rho_e u_e C_M \left[\sum \left(Z_{je} - Z_{jw}^* \right) h_j^{T_w} - B' h_w \right]$$

$$+ \alpha_w q_{rw} - F \sigma \varepsilon_w T_w^4 - q_{cw} = 0$$

$$B' = \dot{m} / \rho_e u_e C_m$$

$$(9.10)$$

By a rigorous approach, the interface boundary conditions can be derived from the governing equations on the surface of the control volume providing appropriate boundary conditions for the conservation laws [38,39]. These interface conditions are derived from the Reynolds transport theorem [40].

The boundary condition for species conservation equations become.

$$\left[\rho_{i+} (u_+ + u_{i+} - u_{b+}) - \rho_{i-} (u_- + u_{i-} - u_{b-}) \right] \cdot n_+ A = \frac{\lim}{v \to 0} \left[\iiint w_i\, dV \right.$$

$$\left. - \frac{d}{dt} \iiint \rho_i\, dV \right] \to 0 \qquad (9.11a)$$

For momentum conservation equations, the outward normal component of force and momentum exchange is balanced by the integral over the control volume.

$$\left[\rho_+ u_+ (u_+ - u_{b+}) - \rho_- u_- (u_- - u_{b-}) - (\tilde{\sigma}_+ - \tilde{\sigma}_-) \right] \cdot n_+ A = \frac{\lim}{v \to 0} \left[\iiint \sum_{i=1}^{n} \rho_i f_i\, dV \right.$$

$$\left. - \frac{d}{dt} \iiint \rho u\, dV \right] \to 0 \qquad (9.11b)$$

For the internal energy conservation equation, the convective, conductive, and radiative energy transfer and work done by the normal stress over the ablating surface are balanced by the volume integral over the control space enclosed in the ablating interface.

$$\left[\rho_+ e_+ (u_+ - u_{b+}) - \rho_- e_- (u_- - u_{b-}) + q_+ - q_- + u_+ \cdot \tilde{\sigma}_+ - u_- \cdot \tilde{\sigma}_- \right] \cdot n_+ A$$

$$= \frac{\lim}{v \to 0} \left[\iiint q\, dV + \iiint \sum_{i=1}^{n} \rho_i f_i \cdot (u + u_i)\, dV - \frac{d}{dt} \iiint \rho e\, dV \right] \to 0 \qquad (9.11c)$$

In Equations (9.11a)–(9.11c), the subscript symbols + and − denote the variables evaluated above or beneath the ablating interface, thus indicate the required descriptions for the non-equilibrium gas emitting from the ablator. The velocity of the recessing surface is indicated as u_b which is permitted to vary from point to point on the control surface. In this formulation, the ejection velocity of pyrolysis gas and the release vapor rate of the sublimated material on the ablating surface are required as input. Therefore, the details of gaseous motion through the porous ablating material are relegated to the research results of the ablative material. The heat transfer term q accommodates the heat by conduction, convection, as well as radiation transfer according to the energy conservation equation. Finally, the normal stress component is designated as $\tilde{\sigma}$ and is the only stress tensor component that can contribute to the work done by the gas media: the non-equilibrium gas, sublimating vapor, and pyrolysis gas of the ablating material.

The formulation of the interface boundary conditions is derived from the eigenvector structure of the governing equations. The present result intends to show that these interface boundary conditions for ablating surfaces are much more complex than that in the existing literature. Additional efforts are still required on how these conditions can be best implemented and maintained computational stability. More importantly, the formulation can also identify how spectral optical properties of the boundary layer are changed by the ablative gases. At the present time, the implementations have not fully verified that the interface boundary condition can satisfy the well-posed criterion for numerical simulation. This issue remains open as a basic research objective of hypersonic ablation simulation.

REFERENCES

1. Bird, R.B., Stewart, W.E., and Lightfoot, E.N., *Transport phenomena*, John Wiley & Sons, New York, 1960.
2. Clarke, J.F., and McChesney, M., *The dynamics of real gases*, Butterworths, Washington, DC, 1964.
3. Hirschfelder, J.O., Curtiss, C.F., and Bird, R.B., *Molecular theory of gases and liquids*, 2ne Ed., John Wiley & Sons, New York, 1954.
4. Dorrance, W.H., *Viscous hypersonic flow*, McGraw-Hill, New York, 1962.
5. Howatson, A.M., *An introduction to gas discharge*, Pergamon Press, Oxford, England, 1975.
6. Langmuir, L., The interaction of electron and positive ion space charge in cathode sheath, *Physical Review*, Vol. 33, No. 6, 1929, p. 954.
7. Raizer, Y.P., *Gas discharge physics*, Springer-Verlag, Berlin, 1991.
8. Surzhikov, S.T. and Shang, J.S., Two-component plasma model for two-dimensional glow discharge in magnetic field, *Journal of Computational Physics*, Vol. 199, No. 2, 2004, pp. 437–464.
9. Surzhikov, S.T., *Computational physics of electric discharges in gas flows*, De Gruyter, Berlin, 2013.
10. Shang, J.S. and Surzhikov, S.T., *Plasma dynamics for aerospace engineering*, Cambridge University Press, Cambridge, UK, 2018.
11. Slater, J.C., *Introduction to chemical physics*, McGraw-Hill, New York, 1939.
12. Chapman, S. and Cowling, T. G., *The mathematical theory of non-uniform gases.* Cambridge University Press, Cambridge, 1958.

13. Morse, P.M., Diatomic molecules according to the wave mechanics II: Vibrational levels, *Physical Review*, Vol. 34, 1929, pp. 57–64.
14. Chapman, S. and Cowling, T.G., *The mathematical theory of nonuniform gases.* Cambridge University Press, London, 1964.
15. Lennard-Jones, J.E., Cohension, *Proceedings of the Physical Society,* Vol. 43, No. 5, 1931, pp. 461–482.
16. Capitelli, M., Gorse, C., Longo, S., and Giordano, D., Collision integrals of high-temperature air species, *Journal of Thermophysics and Heat Transfer*, Vol. 14, No. 2, 2000, pp. 259–268.
17. Mitchner, M., and Kruger, C.H., *Partially ionized gases*, John Wiley & Sons, New York, 1973.
18. Kieffer, L.J., Low-energy electron-collision cross-section data Part III: Total scattering: Differential elastic scattering, *Atomic Data*, Vol. 2 No. 4, 1971, p. 293.
19. Capitelli, M., Celiberto, R., Gorse, C., and Giordano, D., Transport properties of high temperature air components: A review, *Plasma Chemistry and Plasma Processing*, Vol. 16, 1996, pp. 267S–302S.
20. Yos, J.M., Revised transport properties for high temperature and its components, Technical release, Space System Division, AVCO Corp, Wilington, MA, 1967.
21. Gupta, R.N., Yos, J.M., Thompson, R.A., and Lee, K., A review of reaction rates and thermodynamic properties for an 11 species air model for chemical and thermal nonequilibrium calculations to 30,000k, NASA RP-1232, 1990.
22. Hirschfelder, J.O., and Silbey, R., New type of molecular perturbation treatment, *The Journal of Chemical Physics*, Vol. 45, 1966, p. 2188.
23. Wilke, C.R., Viscosity equation for gas mixture, *Journal of Chemical Physics*, Vol. 18, 1950, pp. 517–522.
24. Palmer, G.E. and Wright, M.J., Comparison of methods to compute high-temperature gas viscosity, *Journal of Thermophysics and Heat Transfer*, Vol. 17, 2003, pp. 232–239.
25. Fertig, M., Dohr, A., and Fruhaul, H.H., Transport coefficients for high-temperature nonequilibrium air flows, *Journal of Thermophysics and Heat Transfer*, Vol. 15, 2001, pp. 148–156.
26. D'angola, A., Colonna, G., Gorse, C., and Capitelli, M., Thermodynamics and transport properties in equilibrium air plasma in a wide pressure and temperature range, *The European Physical Journal D*, Vol. 46, 2008, pp. 129–150.
27. Gmelin, L., Gmelin-Handbuch der Anorganishen Chemie, System Nr., 3, Sauerstoff, Velag Chemie Gmbh, Weinheim/Bergstrafie, Germay, 1950.
28. Asinovsky, E., Kirillin, A., Pakhonov, E., and Shabashov, V., Experimental investigation of transport properties of low-temperature plasma by means of electric arc, *Proceeding IEEE*, Vol. 59, 1971, p. 592.
29. Devoto, R.S., Bauder, U.H., Cailletteau, J., and Shires, E., Air transport coefficients from electric arc measurements, *Physics of Fluids*, Vol. 21, 1978, pp. 552–558.
30. Levin, E. and Wright, M.J., Collision integrals for ion-neutral interactions of nitrogen and oxygen, *Journal of Thermophysics and Heat Transfer*, Vol. 18, No. 1, 2004, pp. 143–147.
31. Sutton, G., The initial development of ablation heat protection, an historical perspective, *Journal of Spacecraft & Rocket*, Vol. 19, No. 1, 1982, pp. 3–11.
32. Laganelli, A.L. and Nestler, D.E., Surface ablation patterns: A phenomenology study, *AIAA Journal,* Vol. 7, No. 7, 1969, pp. 1319–1325.
33. Nachtsheim, P.R. and Larson, H.K., Crosshatched ablation patterns in Teflon, *AIAA Journal*, Vol. 9, No. 8., 1971, pp. 1608–1614.
34. Tobak, M., Hypothesis for the origin of cross-hatching, *AIAA Journal*, Vol. 8, No. 2, 1970, pp. 330–334.

35. Suzuki, T., Sawada, K., Yamada, T., and Inatani, Y., Experimental and numerical study of pyrolysis gas pressure in ablating test piece, *Journal of Thermophyiscs and Heat Transfer*, Vol. 19, No. 3, 2005, pp. 266–272.
36. Torre, L., Kenny, J.M., and Maffezzoli, A.M., Degradation behavior of a composite material for thermal protection system part 1: Experimental characterization, *Journal of Material Science*, Vol. 33, No. 12, 1998, pp. 3137–3143.
37. Chen, Y.K. and Milos, F.S., Ablation and thermal response program for spacecraft heat-shield analysis, *Journal of Spacecraft and Rockets*, Vol. 36, 1999, pp. 475–483.
38. Shang, J.S. and Surzhikov, S.T., Nonequilibrium radiative hypersonic flow simulation, *Journal of Progress in Aerospace Sciences*, Vol. 53, 2012, pp. 46–65.
39. Shang, J.S. and Surzhikov, S.T., Simulating stardust earth reentry with radiation heat transfer, *Journal of Spacecraft and Rockets*, Vol. 48, No. 3, 2011, pp. 385–396.
40. Leal, L.S., *Advanced transport phenomena: Fluid mechanics and convective transport processes*, Cambridge University Press, Cambridge, UK, 2007.

10 Dissociation and Ionized Gas Components

10.1 DISSOCIATION AND IONIZATION PROCESSES

The energy transfer by internal degrees of freedom of molecules or atoms follows the ladder climbing process from the translation, rotation, and vibration internal modes. When the molecules gain a sufficient amount of vibration energy to overcome the mutual bonding within the molecule then dissociation occurs. The temperature of a diatomic atomic molecule usually dissociates into atoms in the order of several thousand degrees. The bonding potential is lower for a polyatomic molecule which rarely appears in the air and begin to dissociate at even lower temperature.

If additional energy is continued to be transferred into the gas mixture the ionization of gas will take place. The pulsating and gyrating motion of electrons must maintain a sufficient amount of energy to keep their trajectories in different shells around the nucleus, and can be traced to the binding force of electronic energy which is frequently dominated by the ionization potential. For multi-electron atoms, the interactions between electrostatic repulsion and the interaction with their magnetic moments are quite complex and affect the ionization potential. Once the energy gained by electrons exceeds the ionization potential they become free charges in space. The dissociation and ionization potential of the major species in the air is depicted in Table 10.1 [1].

The composition of air mixture in high-temperature environment is displayed in Figure 10.1 in terms of temperature and pressure conditions [2,3]. At a temperature <2,000 K, the air is essentially describable by the perfect gas law. Species of rare gases such as argon and helium in the air are not included, because they are present only in minute quantities and do not exert discernible behavior in air even at high temperatures. The oxygen molecule molecules begin to dissociate at the temperature of around 2,000 K at the atmospheric pressure condition. The nitrogen follows to dissociate at the temperature of about 4,800 K and the onset condition is dependent on the pressure or rather the density which ties to the complicated combination of a number of possible collisions and rate of energy transfer. At the high-pressure condition, dissociation will happen at a higher temperature. The same behavior

TABLE 10.1

Dissociation and ionization potential of major air components

Species	N_2	O_2	NO	N	O
Dissociation potential (ev)	9.74	5.08	6.50		
Ionization potential (ev)	15.58	16.20	9.25	13.62	14.55

DOI: 10.1201/9781003212362-12

is also displayed by the ionization with a stronger dependence on the pressure or density condition. However, the nitric oxide, atomic oxygen, and nitrogen are presented behind normal shock over 5,000 K [4,5]. From the flight test data from hypersonic reentry, the ionized gas is clearly recorded at a temperature of around 10,000 K [6]. In short, at the high-enthalpy hypersonic flows, the air mixture consists of 11 species including the positively charge nitrogen and oxygen ions and the electrons to form an air plasma.

The formation of nitric oxide $N_2 + O_2 \rightleftarrows 2NO$ in the air does not strongly influence either the dissociation of oxygen and nitrogen molecules or the thermodynamic properties of air. The oxidation of nitrogen required high activation energy, thus the reaction will not proceed at a temperature below 1,500 K. However, at a temperature over 3,000 K, the chemical reaction of nitric oxide establishes very rapidly.

For a diatomic molecule, the chemical physics reaction is simply $A_2 \rightleftarrows 2A$. The number of the dissociated atom is measured by the degree of dissociation α, which is the mass fraction of the dissociated component $\alpha = \rho_d / \rho$. For a binary system, the mass fraction of the diatomic molecule is $1 - \alpha$ and when the gas is complete dissociated $\alpha = 1.0$. The partial pressure of the dissociated component is simply $p = N(1 + \alpha)\rho kT$. At a low degree of dissociation, the change of pressure is small, but the change in internal energy and specific heat of the gas may be appreciable. The characteristic dissociation temperatures or the dissociation energy divided by the Boltzmann constant for oxygen, nitrogen, and nitric oxide are 59,400, 113,000, and 75,500 K respectively. Dissociation always begins at a temperature much lower than the characteristic temperature due to the high statistical weight of the state for dissociation [1]. Dissociation takes place mostly in nonequilibrium conditions, thus the temperature is really corresponding to the internal degree of freedom of the molecule in thermodynamic equilibrium.

The dissociation can be analyzed as a chemical reaction [1,7,8]. From the chemical kinetics or the law of mass action, the degree of dissociations of oxygen or nitrogen

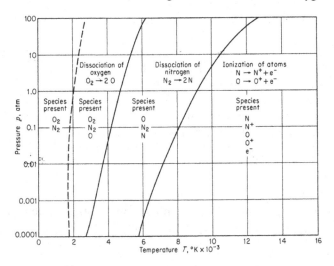

FIGURE 10.1   Composition of air in high-temperature domain.

can be determined by the ratio of the dissociating components in partial pressure rising to the power of each respective stoichiometric coefficient. Recall Equation (8.5d) shows $\kappa_p(T) = \prod_{i=1}^{n} p_i^{v_i} = \exp\left(-\dfrac{\Delta G}{\kappa T}\right)$, by substituting the definition of partial pressure to get.

$$\alpha^2/(1-\alpha) = \left(M_i/4\rho kT\right) K_p(T) \tag{10.1a}$$

In Equation (10.1a), M_m is the molecular weight of the dissociating species. According to the law of mass action, the quantity $K_p(T)$ is the dissociative equilibrium constant.

Ionization of atoms or molecules starts at a much lower characteristic temperature indicated by the ionization potential like that of the dissociation of molecules. The internal energy of ionized gas is now made up of not only the internal energy of the gas but also the potential energy, including the work by removing the electrons and the excitation energy of the electrons. When there are N numbers of electrons in the molecule successive ionization potentials are required to remove all electrons. The ionization begins sooner for the lower ionization potential and lower density condition as observed in Figure 10.1.

The ionization process is equally suitable to be treated as a chemical reaction by removing the m numbers of the electrons by the elementary single-step reaction. In the chemical reacting process, the same species are differentiated into neutral and electrically charged reactants.

$$A_i \rightleftarrows A_i^+ + e \tag{10.1b}$$

The degree of ionization can be derived from the law of mass action from the general expression of the Gibbs free energy and the partition function of ion and electron $Z = \Sigma^{-\varepsilon/kT}$. By factoring the zero-point energy at the ground state, the partition function becomes $Z = ue^{(-\varepsilon_0/kT)}$. The partition function for a free electron consists of the product of the translational partition function and the statistical weight of the free electron to achieve the following expression [1].

$$N_{i^+} N_e/N_i = 2\left(u_{i^+}/u_i\right)\left(2\pi m_e kT/h^2\right)^{3/2} e^{\varepsilon_i/kT} \tag{10.1c}$$

In Equation (10.1c), the symbols N_{i+}, N, and N_e represent the number densities of the ionized gas. For a singly ionized gas, the number densities are reduced to α and $(1-\alpha)$.

$$\alpha^2/(1-\alpha) = k_p(T)/\rho N \tag{10.1d}$$

The relationship is similar to that of dissociation. Ordinarily, the ionization temperature range from 8,000 to 30,000 K for air. At this condition, the gas is singly ionized and the second ionization will begin at twice as higher the ionization potential. The expression for the chemical rate constant is known as the Saha equation to have the following form [1,9].

$$K_p(T) = 2\left(u_1/u_0\right)/\rho N\left(2\pi m_e kT/h^2\right)e^{-\varepsilon_i/kT} \tag{10.1e}$$

The degree of ionization at the very lower value, $\alpha \ll 1$ can be approximated as $\alpha \sim \rho^{-1/2}e^{-I/kT}$. Therefore the degree of ionization will increase very rapidly with temperature but slower with decreasing density. At the equilibrium state, the electron excitation energy can be approximately as $E_{ex} = -kT\,\partial\ln u/\partial T$.

10.2 LIGHTHILL AND SAHA EQUATIONS

The relationship between vibration excitation and dissociation in a molecule is closely linked because energy is stored in vibration and the rate of dissociation is dependent on the vibration state of the molecule. The Lighthill ideal gas nonequilibrium dissociation model has been adopted to describe the dissociation-recombination reactions [10,11]. Research data have shown that the difference in enthalpy between the partially excited vibration degree of freedom in the Lighthill model and the more rigorous gas kinetic model is negligible [12]. In most applications, the vibration excitation is assumed to be in thermodynamic equilibrium with the rest of the internal degrees of freedom, but not always necessarily in a fully excited state. The associated partition function is approximated by the simple harmonic oscillator. This simplification eliminates the ambiguity of having to define the vibration temperature for the non-Maxwellian vibrational species.

The Lighthill ideal dissociation model is restricted to analyzing the hypersonic relaxation phenomenon at a moderated state deviating from chemical equilibrium. The simple classic approach describing the chemical nonequilibrium condition will permit the study of the interaction of heterogeneous gas mixture to be conducted by a step-by-step buildup procedure to gain insight and to provide a better understanding.

In practical application, the air temperature is limited to be lower than 9,000, the ionization will be negligible, and the generation and depletion of nitric oxide in a shock layer will also be omitted. The chemical reactions are limited to dissociation and involve only four species; O_2, O, N_2, and N. The possible elementary chemical reaction between species is then, $A_2 \rightleftarrows 2A$. The source term in the species conservation equation of the Lighthill ideal dissociation model is introduced in Equation (8.6a) $\partial\rho_i/\partial t + \nabla\cdot[\rho_i(u+u_i)] = dw_i/dt$, and is given simply as

$$dw_i\big/dt = \alpha T^s \rho\left[(1-\alpha)e^{-\Theta_d/T} - \alpha^2\left(\rho/\rho_d\right)\right] \qquad (10.2)$$

The Key approximation of the Lighthill model is the constant normalizing dissociation density, ρ_d from his observation. The notation in Equation (10.2) Θ_d designates the characteristic dissociation temperature of species i. The parameters and constants of the dissociation rate equations are summarized for oxygen and nitrogen dissociations in Table 10.2.

Demonstration of the dissociation air computation using the Lighthill model for a hemispherical forebody is depicted in Figure 10.2. The computational simulations are carried out over a hemispherical cylinder with a radius of 3.175 cm, at a Mach number of 12, the stagnation pressure of 5.52×10^3 kPa and temperature of 964 K [13]. The calculated normalized heat transfer versus the body angle Θ is compared with the benchmark result by Fay and Riddell [14], shown by the dashed

TABLE 10.2

Parameters of Lighthill ideal dissociation model

Reaction	C	Seconds	Θ_d (K)	ρ_d (g/cm³)
$O_2 \rightleftarrows 2O$	6.4×10^{23}	-2.0	$59,000$	150.0
$N_2 \rightleftarrows 2N$	4.1×10^{22}	-1.5	$113,000$	130.0

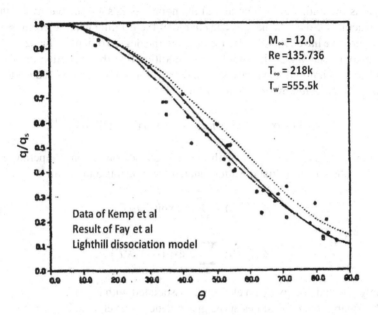

FIGURE 10.2 Verification of Lighthill ideal dissociation model.

line. The computational result is also shown to lie within the data scatter range of the experimental observation of Kemp et al. [15]. The calculation also exhibits a small difference with a verifying calculation, shown by the dashed-dotted line. This discrepancy is attributable to the method of coupling between aerodynamics and chemical reaction calculations; the former uses a cascading coupling and the latter is accomplished by a loosely coupled approach which yields a more accurate result.

The analytic solution for equilibrium ionization is obtained by using the partition functions of a diatomic molecule for the idealized gas mixture of electrons and neutral atoms or molecules.

Thus the ionization or dissociation of atoms and molecules has been treated as a chemical reaction. The Gibbs free energy of a unit mass of ionized gas can be written as

$$G(V,T) = -\sum_i N_i \kappa T \ln \frac{Z_i e}{N_i} - N_e \kappa T \ln \frac{Z_e e}{N_e} \tag{10.3a}$$

where Z_i and Z_e are the partition functions of an ion and an electron. In thermodynamic equilibrium and at the constant temperature and volume (T and V), the Gibbs energy is the minimum with respect to the number of particles.

The simple ionization in an equilibrium state can be analyzed by the law of mass action using the electronic partition function.

$$N_i/N_{i^+}N_e = \left(Z_i/Z_{i^+}Z_e\right)e^{(-\varepsilon_i/kT)} \tag{10.3b}$$

where ε_i is the ionization potential and the notations Z_i's are the partition functions of the reactants. Since the net electric charge of the original neutral gas must always remain zero, we have $N_{i^+} = N_e$. By introducing the degree of ionization, $\alpha = N_{i^+}/N_i$. For a diatomic gas under the simplifying condition that the excitation far above the ground state is ignored, the analytic formulation for ionization will reduce to a simpler form as indicated by Equation (10.1c).

$$\alpha^2/(1-\alpha) = (1/p)\left(2\pi m_e/h^2\right)^{3/2}\left(2\Pi Z_{\alpha^+}/\Pi Z_a\right)e^{(\Theta_i/T)} \tag{10.3c}$$

From the ionization potential ε_i, the characteristic ionization temperature Θ_i is $\Theta_i = \varepsilon_i/k$. The partition functions for ionized components and atomic species are:

$$Z_{i^+}(V,T) = \sum_i g_i \exp\left(-\Theta_i/T\right) \tag{10.3d}$$

$$Z_i(V,T) = \sum_i g_i \exp\left(-\Theta_i/\kappa T\right)$$

The only internal energy by an electron is associated with its spin, and only two permissible quantum energy states at the ground energy level, thus $\Pi Z_{\alpha^+} = 2$.

For diatomic molecules and by recognizing that the masses of ion and neutral molecules are nearly identical, the translational partition functions of these two species Z_{a^+} and Z_a then are equal, and thus can be canceled. The law of mass action becomes [2,16,17].

$$\alpha^2/(1-\alpha) = cT^{3/2}e^{(-\Theta_i/T)}/p \tag{10.3e}$$

Equation (6.8b) is actually the Saha equation first derived from classical thermodynamics. This simple equation provides invaluable insights; at low degrees of ionization for a weakly ionized gas, the degree of ionization is proportional to the three-quarter power of the system temperature and inversely proportional to the square root value of the pressure or density, $p = \rho(1+\alpha)(k/m_e)T$. The degree of ionization increases more rapidly at a higher temperature than lowering the ambient pressure.

10.3 IONIZATION MECHANISMS

Ionization is the upper limit of electronic excitation when a bounded electron leaves the atom and its emitting frequency passes into the continuous spectrum. The lowest ionization potential of the majority of gas atoms and molecules varies between 7 and

15 ev (Θ_i ~80,000–170,000 K). Therefore the statistical weight of a freed electron is very high. The degree of ionization is increasing with temperature and the state of gas rarefaction. More importantly, the equilibrium ion or electron concentration satisfies a relationship identically to the law of mass action for chemical reaction and dissociation.

Ionization processes for atoms and molecules are rather similar although there are some additional complexities for the molecular process because of the increased internal degrees of freedom. In general, the required energy for highly excited states of atoms and molecules may be provided by thermal, chemical, electrical, and radiative resources or by a combination of some even by all mechanisms [1,18,19]. However, the predominant ionizing mechanism is through the collision process. The production and depletion of charged components in a partially ionized gas include additional mechanisms due to the recombination and charge exchange of charges by vastly different collision cross sections and polarity. At the same time, a typical life span of spontaneous decay of an excited state is about 10 ns or 10^{-8} s. Therefore the ionization of gas always exists in a transient state in time and space.

There are eight basic ionization mechanisms:

1. Inelastic collision which is relatively high-kinetic energy particles through thermal excitation. According to the elementary single-step chemical reaction, this particular ionization process can be given as $A + B \rightleftarrows A^+ + e + B$. According to the selection rules for electronic transition configurations $^{2S+1}L_j$; the atomic energy levels are designated by L, S, and J quantum numbers. The oxygen atom therefore has inverted triplet ground states ($^3P_2, ^3P_1, ^3P_0$). The next energy levels above the ground state are two metastable levels of $^1D_2$ and $^1S_0$. The nitrogen ground state is $^4S^0_{3/2}$ and the next energy level is $^2D^0$ (Table 10.3).

   In Practical application to the high-enthalpy hypersonic flows, the inelastic collision is implemented as the nonequilibrium chemical reaction that has been discussed in Chapter 8.

2. Electron impact ionization mechanism occurred when an electron is forcefully extracted from an atom by applied electric field intensity. The chemical reaction follows the single-step chemical reactions:

TABLE 10.3

First excited state and characteristic ionization temperature

species	Quantum state	Ionization potential (ev)	Energy level (V)	Θ_i (K)
O	$^1D_2$	13.60	1.96	22,800
O_2	$^1\Delta_g$	12.20	0.98	11,300
N	$^2D^0$	14.55	2.37	27,500
N_2	$^3\Sigma^+_u$	15.58	6.10	71,000
NO	$^2\Sigma^{-1}$	6.25	5.29	61,400

$E(A) \Rightarrow A^+ + e^-; e^- + A \Rightarrow 2e^- + A^+$. When the electric field is supplied by a direct current, it is referred to as the direct current discharge (DCD). If an alternating electric field is used, it is called the dielectric barrier discharge (DBD). The breakdown voltage can be as low as a few hundred volts in a near vacuum and thousand volts in atmospheric conditions which are governed by *Pachan's law* determined by the gap distance between electrodes [1,20]. Meanwhile, frequencies of the alternating electric field can be slow varying and up to the microwave range. However, the charged particles are most effectively produced by subsequent collisions of electrons with atoms or molecules.

3. Radiative interaction mechanism deriving from radiation can be split into the focused laser beam leading to a discharge and photoionization. The chemical physics process can be presented as $hv + A \Rightarrow A^+ + e^-$. In the microwave range, the excited wave medium is characterized by a small amplitude of electron vibration in comparison with the size of the discharge volume. As the consequence, the electron avalanche is localized and not affected by its surrounding.

The focused coherent microwave achieves sufficient radiation flux density at a confined region where the gas absorbs the laser energy leading to laser-induced optical breakdown [21–23].

The radiative interaction mechanism is the basic approach of the remote energy deposition technique applied to hypersonic flow drag reduction [24]. In application, pulsation and the thin laser beam are focused a diameter upper of a blunt body. The microwave only pulses for a duration of 10 ns at a rate of 10 Hz and the maximum laser energy is 283 mJ per pulse. The laser-induced optical breakdown in air leads to oncoming flow modification. The momentary reductions in surface pressure of 40% and 30% are achieved for the two different configurations in a flow at a Mach number of 3.45.

4. Charge exchange is an atomic ionization by charge exchange is another possible process for some vulnerable atoms. An electron may be lost to a strong valence interaction with another atom, or by resonant charge exchange with an ion. The process can be represented by $A + B \Rightarrow A^+ + B^-$ and $A + B^+ \Rightarrow A^+ + B$ (resonant charge exchange). A molecule in a metastable state collides with another molecule is known as the *Penning process* and is essential for a gas discharge to generate free electrons $A^* + B \Rightarrow A + B^+ + e^-$.

5. Dissociative recombination is the fastest mechanism of bulk recombination in weakly ionized gas, but the recombination process can occur for ions and electrons through a different mechanism. In weakly ionized plasma with sufficient low temperature like plasma generated by electron impact with a high number density of molecular ions, the released energy is mostly transferred into the excitation energy of the atom. The neutralization of charges can be described as,

$$AB^+ + e^- \Rightarrow A + B.$$

6. Three-body recombination mechanism results from a collision with a third particle or two free electrons in a sufficiently high-temperature circumstance that is free of molecular ions. The function of the third particle is to carry away the excessive energy of recombination, the recombination process can be given as; $A^+ + e^- + B \Rightarrow A + B^*$ or $A^+ + e^- + e^- \Rightarrow A^* + e^-$.

7. Dielectric recombination mechanism is mostly by a two-body electronic excitation and through an atomic auto-ionization state; the products may be left in a variety of excited states as follows:

   $A^+ + e^- \Rightarrow A^{**}$, $A^{**} + e^- \Rightarrow A^* + e^-$, and $A^{**} \Rightarrow A + h\nu$. The metastable state of the atom is considered to be in thermodynamic equilibrium, thus the number density of atoms can be determined by the Saha equation.

8. Electron attachment mechanism is frequently the main mechanism of removing electrons from plasma and usually followed by negatively charged ion recombination. The attachment impedes the breakdown process to sustain the ionization or to maintain the electric current. When an electron attaches to a polyatomic molecule, the binding energy tends to distribute into vibrational mode immediately. Therefore the process is complex and has many different paths, some is a single-step process and other requires a multiple-step procedure; $A + e^- \Rightarrow A^-$ and

$$A^- + A^+ \Rightarrow A + A^*, AB + e^- \Rightarrow (AB^-)^{**} \Rightarrow A + B^-.$$

The electron attachment with a molecule is similar to the excitation to the vibrational mode. The interaction produces a stable negatively charged by collision. The physics of the mechanism requires a relationship between cross section and energy that consists of individual resonance. The attachment of electrons to a complex molecule has a relatively long lifetime (10^{-5} to 10^{-4} seconds).

10.4 DYNAMIC MOTION OF CHARGED PARTICLES

Once the air is ionized, the air becomes plasma which is an electrically conductive medium. Two fundamental parameters that associate with the electric properties of plasma are the Debye length and plasma frequency both characterize the macroscopic behavior of a collection of charged particles. The basic property of plasma is its tendency in maintaining electric neutrality [25]. This particular state requires an enormously large electrostatic force between an electron and an ion. The Debye shielding length is given as

$$\lambda_d = \sqrt{\varepsilon k T_e / e^2 n} \tag{10.4a}$$

In Equation (10.4a), the electric permittivity of free space $\varepsilon = 8.8542 \times 10^{-12}$ farad/m. The Debye length is the characteristic dimension of plasma and is equivalent to the mean free path of aerodynamics. In plasma, generated by the electronic impact, the Debye length is 1.14×10^{-7} m at standard atmospheric conditions.

The plasma sheath is another unique feature of plasma at the interface of different media, which is directly connected to the vastly different unit mass between the electron and ion. For a hydrogen atom, the ratio of an ion to electron is 1,836 to 1. Away from the media interface the bonded electron and ion must move at the same speed to maintain global neutrality. Near the boundary, charge separation occurs and the charged particles tend to move at different random motions. Near the interface, boundary-charged particles separate, and the higher collision rate of an electron with a surface is much greater than ion to make the surface acquire a net negative potential [26]. The ions will recombine at the surface and return the plasma to neutral particles; the electrons either recombine or enter the conduction band for an electrically conducting material. Most importantly, the plasma losses the globally neutral property in this region, and the electric potential increases monotonically toward a negative value from the unperturbed neutral state. When the collision process reaches an equilibrium state, the net electric current at the interface surface vanishes. A plasma boundary layer is formed over the interface known as the plasma sheath. It may be anticipated that the plasma sheath often has the same order of magnitude as the Debye shielding length.

The plasma frequency is also the consequence of plasma that has a tendency to be macroscopically neutral, the relative position of paired charges will always return to its neural equilibrium state after a perturbation. The inertia of electrons will unavoidably lead to overshooting and oscillating around the equilibrium condition by a characteristic rate known as the plasma frequency which is determined by the mass and momentum conservation equation of electron.

$$f_e = \sqrt{ne^2/m\varepsilon} \tag{10.4b}$$

The mass of an electron is $m = 9.1095 \times 10^{-25}$ g. At the electron number density around $10^{12}/\text{cm}^3$, the plasma frequency yields a value of 8.7×10^6 kHz in *SI* unit, which is within the microwave H and X bands. The frequency is the controlling parameter to permit microwave communication through plasma. Any radio wave frequency that blows this limit will be reflected from and dissipated within plasma and known as a communication blackout.

The electrostatic force between two singly charged particles q_i(Coulomb) is described by Coulomb's law which is a collinear force between a pair of particles. In an electrically conducting medium, the free charge particle movement will produce electric current, $J = \Sigma\rho_i u_i$ (Amp or coulomb/seconds) and exert additional force on each other. In an isolator, the molecules of a dielectric material will polarize to reduce the net local field intensity E (Newton/Coulomb). The different effects on two kinds of media are described by *a* dielectric constant κ ($k = 1.0005$ in air) and Coulomb's law becomes.

$$F_e = q_i q_j/4\pi\varepsilon k\, r_{i,j}^2 \tag{10.4c}$$

The electric current generates a magnetic field and the orientation of the induced magnetic field is defined by the right-hand rule. The inducing differential magnetic field intensity is governed by the Biot-Savart law for magneto-statics which is the counterpart of Coulomb's law for electrostatics [27].

$$dH = J \times dl/4\pi r^2 \tag{10.4d}$$

In Equation (10.5d) H is the magnetic field strength and is related to the magnetic flux density $B = \mu_m H$ by the magnetic permeability ($\mu_n = 4\pi \times 10^{-1}$ Henry/m) which has the SI unit of Amp/m^2. The differential magnetic force on a current is

$$dF_m = Jdl \times dB \qquad (10.4e)$$

In essence, the interaction of an electric current with an externally applied magnetic field generates a force within the plasma is $F_m = J \times B$, known as the Lorentz force together with the electrostatic force $F_e = qE$. Therefore, the resultant electromagnetic force in ionized air is

$$F = q(E + u \times B) \qquad (10.4f)$$

As a consequence of the Lorentz force, the motion of charged particles either electrons or ions along the orientation of the magnetic field is unaltered, but the normal to the magnetic field is restricted by it. The normal velocity of the charged particle will gyrate around to have a cyclotron frequency or the Larmor frequency for electrons and ions with an elementary electric charge.

$$\omega_b = eB/m \qquad (10.4g)$$

The radius of the circular motion in the plane perpendicular to the magnetic field is balanced by the centrifugal force to have a cyclotron or Larmor radius

$$r_b = mu_\perp/eB \qquad (10.4h)$$

The notation $u_\perp$ in Equation (10.5h) denotes the velocity component of the charge particle perpendicular to the magnetic field. The Larmor frequencies and radiuses are significantly different between the electron and ions due to the substantial disparity between the mass of the two plasma components.

When the electrical field is also present in addition to the magnetic field, the charged particles follow a helical trajectory in a three-dimensional electromagnetic field. In a varying strength of the magnetic field; the trajectory becomes a prolate cycloid or curtate cycloid, or helix with loop [18].

10.5 PLASMA ACTUATORS

The plasma actuators operate on the mechanism of electron impact ionization for ion trust engines, accentuated hypersonic leading edges, and inlet modification with inviscid-viscous interactions, as well as possible laminar-turbulent transition flow control [18,28–31].

The classic formulation for electronic impact ionization is based on a similar law by Townsend which is an empirical formula [20]. The basic process is a complex chain of events, it involves charge accumulation on the cathode, penning penetration-induced secondary emission, and electron cascading. In the formulation, the ionization coefficient α, which measures the number of ionization by electron impact per unit distance, is a function of the reduced electric field $E|/p$. This quotient is also a measure of the energy gained by a charged particle between collisions from the principle of similarity. It is remarkable that Townsend's similarity law holds extremely well in comparison with a large group of experimental data, both in the ionization frequency and degree of ionization [32].

$$(\alpha/p) = 15 \exp\left[-365/(|E|/p)\right] \tag{10.5a}$$

The plasma generation and depletion processes through the complex chemical-physics involved in quantum mechanics are modeled by the inelastic collision for the multi-fluid model [32,33]. The multi-fluid formulation is the traditional magnetohydrodynamics approach by considering plasma consisting only of four constitutions of the neutral electron, and positive and negative ions [34]. The following governing equations are the foundations for the direct current discharge (DCD) and the dielectric barrier discharge (DBD). Often the system of equations has been referred to as the Drift-diffusion theory by virtual of the fact the charged particle mass flux is the product of the combined drift and diffusion motions of an electrically charged particle, $\Gamma_e = -D_e \nabla n_e - n_e \mu_e E$ and $\Gamma_+ = -D_+ \nabla n_+ + n_+ \mu_+ E$.

$$\partial n_e/\partial t - \nabla \cdot \left(n_e \mu_e E + d_e \nabla n_e\right) = \alpha(|E|)|\Gamma_e| - \beta n_+ n_e - \nu_a n_e + \kappa_d n_n n_e \tag{10.5b}$$

$$\partial n_+/\partial t + \nabla \cdot \left(n_+ \mu_+ E - d_+ \nabla n_+\right) = \alpha(|E|)|\Gamma_e| - \beta n_+ n_e - \beta_i n_+ n_- \tag{10.5c}$$

$$\partial n_-/\partial t - \nabla \cdot \left(n_- \mu_- E + d_- \nabla n_-\right) = \nu_a n_e - \kappa_d n_n n_- - \beta_i n_+ n_- \tag{10.5d}$$

In Equation (10.5b), the depletion model of electron number density is accomplished by the recombination and attachment processes. Dissociative recombination is the fastest mechanism of the bulk recombination of a weakly ionized gas and is a simple binary chemical reaction. The typical value of coefficient β ($\beta n_+ n_e$) has been assigned a value of 2×10^{-7} cm³/second, and the characteristic decay time scale is less than 10^{-3} second. The rate of ion-ion recombination is molded by collision kinetics and the rate constants β_i ($\beta_i n_+ n_-$) between O_2^- and O_4^+, as well as, NO^+ and NO_2^-, are on the orders of magnitude of 10^{-25} and 10^{-26} cm⁶/seconds. The maximum value of the ion-ion recombination model in one atmosphere has a value of 10^{-6} cm³/seconds [1].

The electron attachment and detachment processes are the generation and depletion processes of a negatively charged electron in partially ionized air. The chemical reaction rates ν_a ($\nu_a n_e$) are uncertain but have a wide range of values from 10^{-8} to 10^{-30} cm⁶/seconds, and especially for the triple-collision reactions, and a strong electronic temperature dependence is noted [35]. Finally, the negative-ion ionization κ_d ($\kappa_d n_n n_e$ and $\kappa_d n_n n_-$) by splitting a portion of the electron generation as the electron attachment process [36]. In short, the physics-based formulation for charge attachments by inelastic collision kinetics has been widely adopted for most DBD simulations [33].

Figure 10.3 displays the computational result for the detailed electromagnetic field generated by a DCD over a side-by-side electrode arrangement. The high-resolution numerical solution is produced by the residual diminishing delta algorithm [16]. The separation distance between electrodes is 1.0 cm and the cathode is located to the left of the anode. The breakdown voltage of 441.7 V and an electrical current of 5.19 mA at the ambient pressure of 1.0 Torr (1 mm Hg) correspond to a high altitude rarefied air condition. The computational simulations are verified by experimental observations including the cathode fall of the electrical field potential across the cathode layer. A shortened positive discharge column over the cathode is also noted in comparison with the classic parallel electrode arrangement [37]. The electric current vectors are aligned perfectly with the electric field intensity, and the zero current over the dielectric material between the metallic electrodes is also predicted until it approaches the anode where a slip velocity is developed by the strong electric field intensity. The discharge domain diminishes rather rapidly from the numerical simulation to cease <1.67 cm in height over the electrodes which is much smaller than the glow region observed by the experiment. The numerical simulation illustrates the detailed DCD structure that has been always obscured by the ordinarily glowing discharge.

Apply DCD for hypersonic flow control is based on the intrinsic characteristic of the similitude parameter $M_\infty \tau$, for which any flow deflection in a hypersonic stream will automatically trigger an oblique shock generating a pressure rise downstream. This feature is especially pronounced near a sharp leading known as the Mach wave interaction. When a DCD is activated, the Joule heating over the cathode creates a low-density domain over the electrode which leads to a sudden increasing displacement thickness of the shear layer, in turn, induces oblique shock over the electrode and elevates the surface pressure distribution downstream like a virtual deflected control surface.

This phenomenon is easily observed in Figure 10.4. The numerical simulation is generated at the identical flow condition in a hypersonic MHD (Hydrodynamic) channel at a Mach number of 5.15 and a Reynolds number of 2.56×10^5, based on the model length. The DCD is maintained by an applied electric voltage of 1.20 kV, and

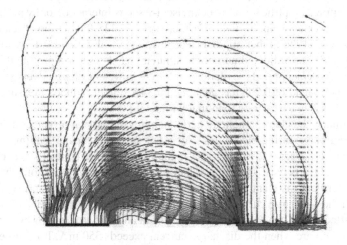

FIGURE 10.3   Electromagnetic field over a side-by-side DCD.

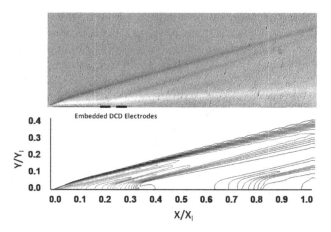

FIGURE 10.4   DCD over a $M = 5.15$ hypersonic sharp leading plate.

the separation distance between the embedded electrodes is 2.80 cm. The cathode is placed upstream of the anode at a distance of 2.48 cm from the sharp edge. From the computational result, the intensive heat transfer at the edges of the discharging electrodes is clearly and correctly displayed. The Joule heating is concentrated most over the cathode and is fully substantiated by the experimental observation [16,29].

The DCD applying as a virtual leading edge plasma actuator has been investigated in the same hypersonic MHD channel. The over plasma is generated by a pair of embedded electrodes in a sharp leading edge flat surface by a total electric currents in a range from 50 to 350 mA by an applied electric field intensity up to 1.2 kV. The maximum electron number density of the plasma is $3 \times 10^{11} / cm^3$, and the electrode temperature is estimated to be 460 K. At the freestream pressure of 0.59 Torr and a temperature of 43 K, the air number density is $1.33 \times 10^{17} / cm^{17}$, thus the degree of ionization is merely 2.25×10^{-5}. The weakly ionized air has an electric conductivity on the order of 1 mho/m. Meaningful effectiveness flow control effectiveness requires amplification by the inviscid-viscous interaction or by an externally applied transverse magnetic field. Furthermore, visible fluctuating discharges are often observed at the higher applied electric voltage and with an externally applied magnetic field. The validated numerical simulations also show a profound influence of a relatively weak applied magnetic field for the flow control effectiveness. A transverse magnetic field of 0.2 T has generated a 34% higher induced pressure level on the virtual leading edge strale than that without it [38].

In summary, the electromagnetic-aerodynamics interaction produces a high-pressure plateau between the electrodes over a flat surface near the leading edge. The higher surface pressure distribution over the immobile flat control surface with an activated DCD acts as if the surface had executed a pitching motion: the aerodynamic performance is identical to a movable leading-edge strake. A series of calculations with the discharge current from 50 to 350 mA yields a range of the equivalent angle of attack from one to five degrees of virtual angle of attack. However, the DCD transits into an arc when the discharge current exceeds 350 mA. It must be pointed out that the elevated surface pressure of the virtual leading edge is only sustained

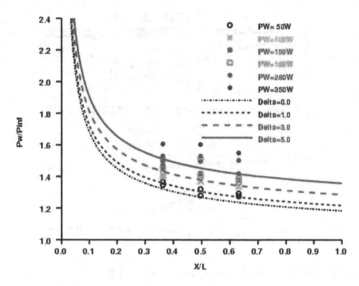

FIGURE 10.5 DCD power supple scaling for virtual leading edge strake.

over the limited DCD domain and decay quickly downstream [37,39–41]. The computational results are substantiated by experimental data as depicted in Figure 10.5.

The entire virtual variable geometry cowl adopts a DCD actuator that combines the electromagnetic and viscous-inviscid interaction [42,43]. In a side-by-side experimental and computational investigation, a pair of DCDs again are applied as the plasma actuator on square inlet sidewalls. The electrodes of the DCD are embedded in the sidewalls of a constant cross-section area inlet; this arrangement is nonintrusive when deactivated. Once the discharge is actuated the increased slope of the displacement thickness through the electromagnetic-aerodynamic interaction consistently generates an oblique shock like the side compression of a convergent inlet cowl. A similar performance is also observed for a circular cylindrical inlet but at a reduced applied voltage and increased electric currents (775 versus 480 V, 80 versus 150 mA). The compression is controlled by the plasma generation power input and the induced compression near the leading edge of the inlet emulates a variable geometry cowl.

Figure 10.6 depicts the numerically simulated square virtual variable inlet cowl. The incoming air is characterized by a Mach number of 5.15 and a Reynolds number of 2.57×10^5. The experiments for creating a virtual variable cross section inlet are also carried out with a lower amount of energy input by two pairs of DCD actuators flushed mounted on the sidewalls by an applied DC current of 775 V and with a current of 80 mA. The oblique shocks that originate from the leading edge coalescing with the DCD actuator induce a steepened oblique shockwave angle.

The numerical simulation captures the physics in that the Joule heating and the convective electrode heating release a significant amount of thermal energy into the air stream. The elevated temperature by DCD near the inlet significantly increases the displacement thickness of the shear layer. In addition, the total Joule heating over the cathode is found on the same order of magnitude as the convective electrode

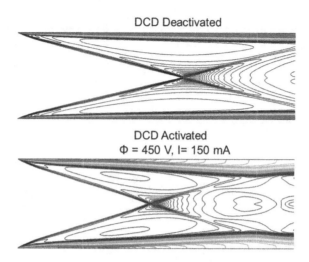

FIGURE 10.6 Virtual variable geometry inlet cowl by DCD.

heating, which is 7.4 versus 6.6 W. Whereas the magnitude of the computed electrostatic force is merely 430 dyne/cm³ (4.30×10^{-6} N/m³) and this force is exerted mostly downward toward the cathode. In a shear layer over a flat surface, this force is not supported by the shear stress and is directly transmitted to the solid surface. Therefore, the computed result substantiates the fact that the thermal effect of the DCD is dominant over that of the electrostatic force for the ensuing viscous-inviscid interaction.

From computational investigations and verified experimental data on the virtual rectangular inlet cowl, the magnetic-aerodynamic interaction has produced an additional compression of 11.7% compared to the inactivated counterpart with a plasma generation power of 64 W [30,42,43]. The DCD in the cylindrical inlet produces less compression, 6.7% under identical free-stream conditions at a plasma generation power of 2.8 W/cm². The lower compression has two distinct contributors: first, the electric current density is much lower in the electrode configuration for the cylindrical inlet. Second, the three-dimensional relief effect or the Mangler scale reduces the outward displacement thickness of the electromagnetic perturbation. It should be reminded that the degree of ionization by DCD exists only at the trace amount; the flow control effectiveness must be both by increasing the degree of ionization or amplifying by an external transverse magnetic field and remaining as an interesting basic research area.

The dielectric barrier discharge (DBD) is another electron impact ionization process but has the advantage of its self-limiting ability from transition to arc [44,45]. In an alternative current cycle, the exposed metallic electrode and the other electrode are embedded within a dielectric filament of a few mime meter thickness. In the positive polarity when the electric breakdown happens, the secondary electron emission from the anode leads to an avalanche and forms a random group of micro discharge or streaks. All positively charged ions during the charge separation are expelled from the exposed anode and attached to the dielectric surface. Hence, the accumulated ions reduce the electric potential across the electrodes and thus help to prevent the discharge transition to the arc. In the negative polarity of the AC cycle, the exposed metallic electrode acts as a cathode the surface charge accumulation just reverses. During the period of charge separation, the heavy ions collide with neutral particles to create a wall-jet-like flow field over the dielectric surface, the induced wall-jet has a velocity <10 m/s [46]. Although the periodic electrostatic forces during each AC cycle are opposite to each other in microwave frequency, due to different electric permittivity of electrodes, a net gas motion results from the exposed electrode toward the dielectric, and the magnitude of the so-called electric wind becomes the main mechanism for using DBD for flow control.

In Figure 10.7, the fundamental self-limiting characteristic of the DBD is convincingly displayed by the computational results at the externally applied electrical potential of 3.0 and 4.0 kV across the overlapped electrodes. At the quiescent atmospheric condition, the discharge breakdown takes place at a voltage of 2.8 kV before the externally applied electrical field reaches its peak values for both polarities. The lower and constant electric potential is maintained by the conductive current and the surface charge accumulation on the electrodes within the discharge. In the positive polarity phase, the discharge occurs when the positive-going potential exceeded the breakdown voltage and continues until the negative-going external field falls beneath the breakdown threshold. The identical behavior is also observed for the negative polarity phase. During the initial breakdown process, a sudden drop in the electric potential with respect to time induces a surge of the displacement current in the discharge. In experimental measurements, this surge indicates the existence of multiple micro discharges and also knows to lead to a train of pulsations in the electrical circuit [44,47]. Similarly, the sudden drop of the electrical potential has also generated non-physical oscillations known as the Gibbs phenomenon in some numerical simulations [40,48]. After the DBD is ignited, the discharge appears as continuous glowing but it is an aggregated micro discharge or streak by high-speed photographic recordings [44].

The force and momentum of DBD are difficult to measure because the discharge is characterized by multiple random micro discharges or streamers in space and time.

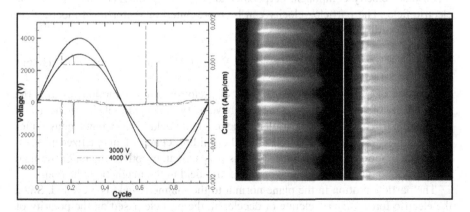

FIGURE 10.7 Electromagnetic characteristics of DBD.

The net force of DBD is time-average information over a period of microseconds due to charge separation. Nevertheless, based on inelastic collision models, independent computations have been performed [40,48,49]. The numerical results show the maximum force is within the plasma sheath, and equally important that the periodic electrostatic force is a push-and-pull during the complete AC cycle of the DBD to nearly negate each other. The net time-average balance of the counteracting force has a value of just a few Newtown per cubic meter.

In any event, the first remarkable flow visualization using DBD for flow control is recorded by Post [50]. It illustrates the effectiveness of the DBD actuator in controlling leading edge separation on a generic NACA 66_3-018 airfoil by energizing the flow in the stagnation region to overcome the adverse pressure gradient. The DBD has also been shown to be a reliable propulsion system for micro-air-vehicle but has not been applied for hypersonic flows.

10.6 HALL CURRENT

The drift velocity of an electrically charged particle is defined as the rate of velocity change along the electric field between collisions.

$$u_d = E \times B \tag{10.6a}$$

One notices immediately that the drift velocity is independent of the mass and the polarity of the charged particle. This velocity component is referred to as the plasma drift velocity in the presence of an electromagnetic field.

The particle in the transverse plane to the magnetic field has a circular motion with a cyclotron frequency and moves with cyclotron or Larmor radius

$$u_\perp = \omega_b \times r_b \tag{10.6b}$$

Therefore, the particle motion in a steady uniform electrostatic and magnetic field is a superposition of a circular motion in the plane normal to the magnetic field and a constant velocity component perpendicular to both the magnetic field B and the electric field. The particle velocity can be represented in vector form independent from the coordinate system as

$$u = \omega_b \times r_b + E_\perp \times B + (q/m)E_\parallel t + u_o \tag{10.6c}$$

The motion of a charged particle consists of cyclotron circular motion which is the drift velocity of the mass center, the constant velocity by the parallel electric field, and the initial particle velocity u_o parallel to the magnetic field. The instantaneous center of gyration is commonly referred to as the guiding center of magnetized plasma. A vector component of the drift velocity is perpendicular both to the magnetic field and the perpendicular component of an electric field to the magnetic induction.

The particle motion in the plane normal to the magnetic field is a cycloid. Since the electric force, $qE_\perp$ accelerate or decelerate the particle based on the polarity of the charge carried, the radius of curvature of the particle's trajectory will change according to the normal electric field component. Thus, the Larmor radius of the ion will be greater than that of the electron, and the Larmor frequency of the ion will be lower than the electron. However, the drift velocity must be identical to the paired ion and electron, as the consequence, the curvature of the ion trajectory is greater than that of the electron. Thus the collision frequency of ion-neutral is greater than the electron-neutral collision, and the ion motion will be retarded by the more frequent collisions with neutral species. Now the drift velocity of ion and electron will be differed and an electric current is produced normal to both electric and magnetic field and flows in the opposite direction of the drift velocity to be identified as Hall current [51]. In essence, the curvature of the electron trajectory in a magnetic field is responsible for a transverse current. In other words, the Hall current leads a voltage difference across plasma due to interaction of charged particles motion with an external applied magnetic field that transverse both to the electric and magnetic fields.

From the definition of a conductive electric current, the Hall current is a component of the current driven by the drift velocity;

$$J_h = n_e e (E \times B) \tag{10.6d}$$

The Hall current occurs in solid, liquid, and ionized gas; it depends on the ratio of electron cyclotron frequency to electron collision frequency. The Hall parameter is then

$$b_e = \omega_b / v_{ec} = eB/mv \approx \lambda_e / r_b \tag{10.6e}$$

where λ_e denotes the mean free path of electron and r_b is the Larmor radius of the electron. The ratio is actually the number of gyrations by an electron between collisions and increases with increasing magnetic flux density B with a linear relationship. From the pure physics viewpoint, the Hall effect offers the first concrete proof that electric currents in metal are carried by moving electrons. On the other hand, the Hall current in ionized gas is an important mechanism for aerospace engineering applications in ion thruster for space propulsion and magneto-hydrodynamic (MHD) electric generator, as well as the MHD accelerator.

10.7 JOULE HEATING

The electromagnetic field of plasma always produces a conductive electric current which is related to the diffusion velocities of the electrically charged particle species. Therefore, the electrical conductivity must be evaluated as microscopic properties together with the thermodynamic state of the plasma. The individual particles move between collisions by the $E \times B$ drift and diffusion velocities which are the driving mechanisms for the charged particle motions. The moving particles need to overcome the electrical resistivity to sustain the collision process. By introducing a mean or drift velocity of the electrons in the gas u, the electric conductive current density is defined as

$$J = nqu = nq^2 E / mf_e \tag{10.7a}$$

In Equation (10.7a), m is the mass of the electron and f_e is the average electron collision frequency which is related to the atomic scale momentum transfer cross section and the random thermal motions of the individual particles [1]. The current density is related empirically to the applied electric and magnetic fields by a bulk electric conductivity $J = \sigma(E + u \times B)$. The proportional constant between them is known as electrical conductivity.

$$\sigma = nq^2 / m\, f_e \qquad (10.7b)$$

The electrical conductivity is the reciprocal of the electric resistance with the SI unit of ohm-meter, which is a function of the medium temperature. Therefore, the electrical conductivity is expressed in the reciprocal of ohm-meter or mho. For example, the conductivities of glass and silver have the values of 10^{-12} and 6.1×10^7 mho/m respectively. The generalized Ohm's law at a point is $J = \sigma E$, J and E have the same direction in an isotropic medium in the absence of a magnetic field [20,51]. Under this special circumstance, electrical conductivity is a scalar quantity. However, in the presence of a magnetic field, collisions play a similar role in determining the drift velocity of a charged particle. The magnetic field through the Lorentz acceleration generates a gyro motion and the helical trajectory of the charged particles has a wide and varying range of orientations in the collision; the electrical conductivity is no longer a scalar quantity but is a tensor of rank two.

Joule heating is also known as Ohmic or resistive heating, the process of releasing heat by an electric current passes through a conductive medium. In other words, the electromagnetic field does work to overcome the resistive force which equals the dissipated heat in response to resistance. The amount of energy released is proportional to the square of the current. This relationship is known as Joule's first law and the IS unit of the energy was subsequently named after Joule by the symbol J ($J = 1$ Newton-meter, $J = 10^7$ erg). Joule heating is the consequence of the interaction of a charged particle in an electric circuit accelerated by an electric field but must give up some of its kinetic energy every time it collides with ions of the conductor. The increased kinetic or vibrational energy of the ions manifests itself as heat and an increase in the temperature of the conductive conductor.

The most general and fundamental formula for Joule heating is

$$q = E \cdot J = I^2 / \sigma = \sigma E^2 \qquad (10.7c)$$

The electric current in a circuit is the volumetric integral of the electric current density

$$I = \iiint J\, dv \qquad (10.7d)$$

The total energy dissipated in the conductor in time is then $W = \int E \cdot I\, dt$ known as Joule's law. An identical formula can also be used for AC power, except the parameters are averaged over one or more cycles. In these cases of a phase difference between I and E, a multiplier $\cos\phi$ must be included.

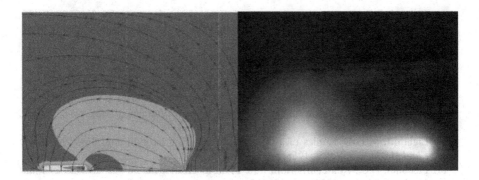

FIGURE 10.8   Joule heating of direct current discharge.

As an example of the Joule heating in a direct current discharge, a side-by-side DCD electrode configuration is presented in Figure 10.8. The composite graph includes the computational simulation and the photograph observation under identical conditions; the gap between the cathode (on the left) and the anode is 1.5 cm. Each electrode has a length of 0.5 cm and a width of 0.25 cm. The discharge is maintained by an electric potential of 439.0 V, and a discharge current of 5.2 mA at an ambient air pressure of 5.0 Torr. The experimental observation shows intensive glow over the closest edges of electrodes. As can be seen from the numerical simulation, the electric field intensifies at the sharp edge of the electrodes. The electron number density has a maximum value of $1.7 \times 10^{10}/cm^3$ locally, and this value is 3.5 times greater than the infinite parallel electrodes DCD. At the inner edge of the anode, the anode layer is completely distorted by the stronger local electric field. From the contour presentation, the cathode layer can be clearly discerned even through the discharge structure over the cathode and is no longer uniform. The plasma sheath is clearly displayed in the cathode layer, and the current density is highly concentrated over the cathode. Thus, the Joule heating is concentrated along the inner edge of the cathode layer closest to the anode, at a total applied direct current electric power of 21.52 J/cm^2s, the Joule heating of the glow discharge at the pressure of 5 Torr is estimated to be 1.30 J/cm^2s [39]. This rate of Joule heating amounts to 6.04% of the total plasma generation power input.

Joule heating is one of the two additional energy transfer mechanisms from the ionized air, which introduces the aerodynamics consideration. It must be reminded that at low ambient pressure, Joule heating is small in comparison to stochastic heating. However, Joule heating is dominant at high pressure due to the high collision frequency such as the laser generation process.

REFERENCES

1. Raizer, Y.P., *Gas discharge physics*, Springer-Verlag, Berlin, 1991.
2. Hansen, C.F., Approximations for the thermodynamics and transport properties of high temperature gas, NACA TN 4150, 1958.
3. Hansen, C.F., Dissociation of diatomic gases, *Journal of Chemical physics*, Vol. 95, 1991, pp. 7226–7233.

4. Chernyi, G.G., *Introduction to hypersonic flow*, Academic Press, New York, 1961.

5. Cox, R.N. and Crabtree, L.F., *Elements of hypersonic aerodynamics*, Academic Press, 1965, New York.

6. Jones, L.J. and Cross, A.E., Electrostatic probe measurements of plasma parameters for two reentry flight experiments at 25,000 feet per second, NASA TN D 66-17, 1972.

7. Moore, W.J., *Physical chemistry*, Prentice-Hall Inc., Englewood Cliff, NJ, 1963.

8. Kuo, K.K., *Principles of combustion*, 2nd Ed., John Wiley & Sons, Hoboken, NJ, 2005.

9. Saha, M.N., Ionization in the solar chromosphere, *Philosophical Magazine*, Vol. 40, No. 238, 1920, p. 472.

10. Lighthill, M.J., Dynamics of a dissociating gas: Part I equilibrium flow, *Journal of Fluid Mechanics*, Vol. 2, 1958, pp. 1–32.

11. Freeman, N.C., Non-equilibrium flow of an ideal dissociation gas, *Journal of Fluid Mechanics,* Vol. 4, 1958, pp. 407–425.

12. Hilsenrath, J. and Klein, M., Tables of thermodynamics properties of air in chemical equilibrium including second viral corrections from 1500 k to 15,000k, AEDC, Tullahoma, TN Repprt TDR-63-161, 1964.

13. Josyula, E. and Shang, J.S., Numerical study of hypersonic dissociated air past blunt body, *AIAA Journal,* Vol. 29, 1991, pp. 704–711.

14. Fay, J.A. and Riddell, F.R, Theory of stagnation point heat transfer in dissociated air, *Journal of the Aeronautical Sciences*, Vol. 25, 1958, pp. 73–85 & 121.

15. Kemp, N.H., Rose, R.H., and Detra, R.W., Laminar heat transfer around blunt bodies in dissociated air, *Journal of Aerospace Sciences*, Vol. 26, No. 7, 1959, pp. 421–430.

16. Shang, J.S., *Computational electromagnetic-aerodynamics*, John Wiley & Sons, Hoboken, NJ, 2016.

17. Shang, J.S. and Surzhikov, S.T., *Plasma dynamics for aerospace engineering*, Cambridge University Press, Cambridge, United Kingdom, 2018.

18. Jahn, R.G., *Physics of electric propulsion*, McGraw-Hill, New York, 1968.

19. Park, C., *Nonequilibrium hypersonic aerodynamics,* John-Wiley & Sons, New York, 1989.

20. Howatson, A.M., *An introduction to Gas discharge*, 2nd Ed., Pergamon Press, Oxford, England, 1975.

21. Meyerand, R.G. and Haught, A.F., Gas breakdown at optical frequencies. *Physical Review Letters*, Vol. 11, No. 9, 1963, pp. 401–403.

22. Raizer, Y.P. Breakdown and heating of gas under the influence of laser beam, *Soviet Physics Uspekhi*, Vol. 8, No. 5, 1966, pp. 650–673.

23. Surzhikov, S.T., *Fundamental of radiative gas dynamics*, Von Karman Lecture Series, von Karman Institute for Fluid Dynamics, Rhode-ST-Genese Belgium, 2005.

24. Adelgren, R.G., Yan, H., Elliott, G.S., Knight, D.D., Beutner, T.J., and Zheltovodov, A.A., Control of Edney IV interaction by pulsed laser energy deposition, *AIAA Journal*, Vol. 43, 2005, pp. 256–269.

25. Langmuir, L., The interaction of electron and positive ion space charge in cathode sheath, *Physical Review*, Vol. 33, No. 6, 1929, p. 954.

26. Riemann, K.U., The Bohm criterion and sheath formation, *Journal of Physics D*, Vol. 24, 1991, pp. 493–518.

27. Jackson, J.D., *Classic Electrodynamics*, 3rd Ed., John Wiley & Sons, New York, 1999.

28. Gallimore, A., Meyer, R., Kelley A., and Jahn, R., Anode power deposition in applied-field segmented anode MPD thruster, *Journal of Propulsion & Power*, Vol. 10, 1994, pp. 262–268.

29. Shang, J.S., Surzhikov, S.T., Kimmel, R., Gaitonde, D., Menart, J., and Hayes, J., Mechanisms of plasma actuators for hypersonic flow control, *Progress in Aerospace Sciences*, Vol. 41, No. 8, 2005, pp. 642–668.

30. Shang, J.S., Chang, C., and Surzhikov, S.T., Simulating hypersonic magneto-fluid dynamic compression in rectangular inlet, *AIAA Journal,* Vol. 45, No. 11, 2007, pp. 2710–2720.

31. Corke, T.C., Enloe, C.L., and Wilkinson, S.P., Plasma actuators for flow control, *Annual Review of Fluid Mechanics*, Vol. 42, 2010, pp. 505–525.
32. Surzhikov, S.T. and Shang, J.S., Two-component plasma model for two-dimensional glow discharge in magnetic field, *Journal of Computational Physics*, Vol. 199, Issue 2, 2004, pp.437–464.
33. Shang, J.S. and Huang, P.G., Surface plasma actuators modeling for flow control, *Progress in Aerospace Sciences*, Vol. 67, 2014, pp. 29–50.
34. Alfven, H, *Cosmical electrodynamics*, Clarendon Press, Oxford, United Kingdom, 1950.
35. Gibalov, V.I. and Pietsch, G.J., Dynamics of electric barrier discharges in coplanar arrangement, *Journal of Physics D: Applied Physics*, Vol. 37, 2004, pp. 2082–2092.
36. Unfer, T. and Boeuf, J.P., Modeling of a nanosecond surface discharge actuator, *Journal of Physics D: Applied Physics,* Vol. 42, 2009, pp. 194017-1-12.
37. Shang, J.S., Surface direct current discharge for hypersonic flow control, *Journal of Spacecraft & Rockets*, Vol. 45, 2008, pp. 1213–1222.
38. Kimmel, R.L, Hayes, J.R., Menart, J.A., and Shang, J.S., Effect of magnetic fields on surface plasma discharges at Mach 5, *Journal of Spacecraft & Rockets*, Vol., 2006, pp. 1340–1346.
39. Shang, J.S., Huang, P.G., Yan, H., and Surzhikov, S.T., Computational simulation of direct current discharge, *Journal of Applied Physics*, Vol. 105, 2009, pp. 023303-1-14.
40. Shang, J.S. and Huang, P.G., Modeling AC dielectric barrier discharge, *Journal of Applied Physics*, Vol. 107, No. 11, 2010, pp. 113302-1-7.
41. Shang, J.S., Roveda, F., and Huang, P.G., Electrodynamic force of dielectric barrier discharge, *Journal of Applied Physics*, Vol. 109, No. 11, 2011, pp. 113301-8.
42. Shang, J.S., Electrostatic-aerodynamic compression in hypersonic cylindrical inlet, *Communication in Computational Physics*, Vol. 4, No. 4, 2008, pp. 838–859.
43. Shang, J.S., Kimmel, R., Menart, J., and Surzhikov, S.T., Hypersonic flow control using surface plasma actuator, *Journal of Propulsion and Power*, Vol. 24, No. 5, 2008, pp. 923–934.
44. Enloe, C.L., McLaughlin, T.E., Van Dyken, R.D., Kachner, K.D., Jumper E.J., and Corke, T.C., Mechanisms and responses of a single dielectric barrier plasma actuator: Plasma morphology, *AIAA Journal,* Vol. 42, 2004, pp. 589–594.
45. Elisson, B. and Kogelschatz, U., Nonequilibrium volume plasma chemical processing, *IEEE Transactions on Plasma Science*, Vol. 19, 1991, pp. 1063–1077.
46. Corke, T.C., Post, M.L., and Orlov, D.M., SDBD plasma enhanced aerodynamics: Concepts, optimization and applications, *Progress in Aerospace Sciences*, Vol. 43, 2007, pp. 193–217.
47. Moreau, E., Airflow control by non-thermal plasma actuators, *Journal of Physics D: Applied Physics*, Vol. 40, 2007, pp. 605–636.
48. Huang, P.G., Shang, J.S., and Stanfield, S.A., Periodic electrodynamic field of dielectric barrier discharge, *AIAA Journal,* Vol. 49, No. 1, 2011, pp. 119–127.
49. Boeuf, J.P. and Pitchford, L.C., Electrohydrodynamic force and aerodynamic flow acceleration in surface dielectric barrier discharge, *Journal of Applied Physics,* Vol. 97, 2005, pp. 103307-1-10.
50. Post, M.L., Plasma actuators for separation control on stationary and oscillating wings, Ph.D. dissertation, University of Notre Dame, 2004.
51. Mitchner, M. and Kruger, C.H., *Partially ionized gases*, John Wiley & Sons, New York, 1973.

11 Radiative Heat Transfer

11.1 FUNDAMENTAL OF RADIATION

Quantum mechanics germinates from studying the radiation and radiative transition phenomenon which always presents in high-enthalpy hypersonic flows. Radiation emerges as an essential energy transition mechanism and one of the unique features of hypersonic flows. Traditionally radiation has been described as a particle phenomenon for energy transmission by photons or light quanta, but in the framework of an electromagnetic field, radiation is described by wave dynamics not only for its energy content but also by its optical behavior. The radiative field in time and space must be described by the distribution of its intensity with respect to the frequency and direction of the energy transfer. The theory of electromagnetics governs the radiation and radiative transitions, and the rule of quantum electrodynamics provides a precise description of electromagnetic radiation and its interactions, including both the macroscopic limit of continuous waves and the microscopic limit of quanta.

The interaction between a charged particle and the radiation field is described by the potential of the Hamiltonian function of the charged particle moving in an electromagnetic field. The interacting potential can be constructed by the conjugate momenta P and a vector quantity R associating with radiation; $V = -(e/m)p \cdot R + \left(e^2/2m\right)R \cdot R$. The first term leads to the physical process for the change of radiation gains or loss of one quantum per time, or simply the emission and absorption of radiation. The second term of the potential is the process of two quanta which describes an incidence quantum modified by both its frequency and direction of motion, or in the general term of scattering [1]. In atomic spectroscopy, the most common radiation transitions are of the electric-dipole types. The transition probability depends on the dipole moment of the atom or molecule averaged between the lower and upper quantum states [2]. Therefore, radiation transitions are analogous to the classic oscillating magnetic dipoles, electric or magnetic quadrupoles, and higher multipoles. Sometimes, they may account for some of the most intense atomic spectrum lines.

The radiative energy transfer occurs when a molecule, atom, or ion is in a highly excited energy state at the instant of a quantum jump. The energy depleted or gained is transmitted by photons or light quanta by the classic terminology. An instant radiation field is always described by the distribution of the intensity for radiation by its frequency, location, and direction of the energy transfer; the intensity is a unique function of the photon particle distribution $f(v, s, \Omega, t)$. The propagation of radiating energy in a medium, therefore, is best described by an electromagnetic wave.

All quantum transitions for energy transfer have been divided into three groups according to the continuity criterion of its energy spectrum by the initial and final states of the transition as the bound-bound, bound-free, and free-free [3]. The bound-bound transitions take place in atoms and molecules from one discrete quantum

DOI: 10.1201/9781003212362-13

state to another. These transitions emit or absorb energy according to the atomic and molecular structure that follows the permissible quantum jumps. For molecules, the transitions are followed by changes in the quantum states of the internal degrees of freedom. In case the transition does not involve electronic states will occur in the infrared spectrum. The energy transfer is limited to the unit of $h\nu$ or the product of the Planck constant and radiating frequency ($h = 6.626 \times 10^{-34}$ m^{2}kg/s), and the frequency is restricted to an extremely narrow range. The amount of the transferred energy is the difference between two quantum energy levels of an atom and is referred to as selective absorption. The cross sections of absorption are associated with short photon mean free paths, at the standard atmospheric condition the mean free path of a photon is around 10^{-10} cm.

The bound-free transition is resulting from the amount of energy transfer that exceeds the bounding energy of an electron to an ion, atom, or molecule; the electron becomes free. If the input energy is greater than the bounding energy, the excess energy is transformed into the kinetic energy of the free electrons. The energy of a free electron can receive more energy than the binding potential and thus can have continuous absorption and emission spectra. In the case of the reverse transition, the bonded electron will release the energy by photons.

The free-free transition is also known as the bremsstrahlung because the motion of free electrons is slowing down in the field of the ion and loses a part of its energy by the radiation process. Since bremsstrahlung is associated with a free electron, its emission and absorption have a continuous spectrum. This energy transition is the key element of ionization via radiation. For free-free energy transition, the absorption must be accomplished by collision with an ion. Therefore, the coefficient of bremsstrahlung absorption is proportional to the number of ions and the number of free electrons in a unit volume. The coefficient of free-free transition is approximately an order of magnitudes smaller than the bound-free counterpart [3]. The scattering cross section of the electron is also known as the Thomson scattering cross section which is relatively small to have a value $\sigma_s = 6.65 \times 10^{-25}$ cm^2. In partially ionized plasma, the scattering mean free path of a photon resides in a continuous spectrum which is always greater than the corresponding absorption mean free path. Therefore, the scattering becomes significant only in an extremely rarefied and fully ionized gas.

The radiation spectrum covers a wide range from radio waves, infrared, visible, ultraviolet, X-ray, and gamma-ray frequencies from 10^3 to 10^{22} Hz, by wave lengths it varies from 10^3 to 10^{-12} m. The portion of spectra associated with the high-enthalpy hypersonic flows with quantum transitions for thermal energy transfer however is mostly concentrated in the infrared, visible, and ultraviolet spectra (from 4×10^{14} to 7.9×10^{17} Hz). Within each spectrum bands, the internal degrees of freedom transition zones also are identified. Thermal radiation is strongly dependent upon the temperature of its propagating medium. For visible radiation, the characteristic medium temperatures are in the order from 7,000 to 13,000 K, as shown in Figure 11.1.

In view of the far-ranging radiative transfer phenomena, radiating energy transfer involves the extremely wide intertwining scientific disciplines from quantum physics, quantum chemistry, optical physics, and thermodynamics to aerodynamics. The

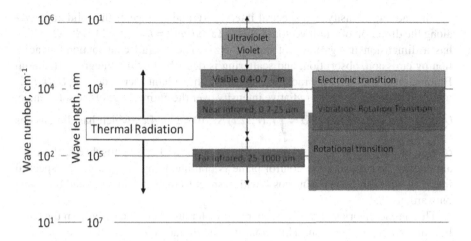

FIGURE 11.1   Spectrum of radiation.

aspects of radiation are concentrated on the formulations of the radiation energy transfer and the methods of analysis for theoretical and computational simulations for aerospace engineering applications. Therefore, we are focusing the discussions on the variations of multi-flux formulations, the method of ray tracing, and the Monte Carlo method for photon dynamics. The specialized techniques such as the narrow band, wide band, and view-factor approaches are not included.

11.2 CLASSICAL THEORIES

A radiation field is always described by the distributions of absorption, emission, and scattering intensities of radiation by its frequency, position, and direction of the energy transfer. These mechanisms are keys for the attenuation of a radiative emission from the absorption of the medium and the colliding interaction with participants. According to the kinetic theory, the entire radiative field is described by the distribution of the intensity of radiation by its frequency v, location r, and instance in time t, as well the direction Ω of the energy transfer, the radiation intensity is a unique function of the photon particle distribution $f(v,r,\Omega,t)$. When the radiative energy emits from a medium into another, the energy flux frequently has a different strength in a different direction due to, attenuation, diffraction, reflection, and refraction. The energy flux may have infinitely many directions and traditionally is described as the vector in terms of spherical coordinates. The total surface area of a unit radius is enclosed by the solid angle above the control surface and the polar angle θ, which is measured from the surface outward normal, and the azimuthal angle φ from the axis of the vector projects onto the surface. The domains of the two angulars measured for a hemisphere are $0 \le \theta \le \pi/2$ and $0 \le \varphi \le 2\pi$, and thus the solid angle $\bar{\Omega}$ is the project area of the vector. The direction of the position vector is associated with an infinitesimal solid angle defining the infinitesimal area on the hemispherical surface, $d\bar{\Omega} = \sin\theta d\theta d\varphi$.

The radiant intensity in a spectral interval dv and a range of the solid angle $d\bar{\Omega}$ along the direction of a unit vector Ω is $I_v(r,\Omega,t)dvd\bar{\Omega} = hvf(v,r,\Omega,t)dvd\bar{\Omega}$, which has a dimension in cgs units of $cm^{-2}s^{-1}Hz^{-1}$. The overall contribution to radiation by emission, absorption, and scattering is described by the integro-differential Boltzmann equation for the radiation intensity. The radiant energy density, $U_v(r,t)$, is obtained by integrating the radiative intensity over the entire range of the solid angle;

$$U_v(r,t) = hv\int_{4\pi} f(v,r,\Omega,t)d\bar{\Omega} = \int_{4\pi} I_v(r,\Omega,t)d\bar{\Omega},\ (0 < \bar{\Omega} < 4\pi),\ \text{which has dimensions}$$

of erg/cm^3. By dividing the domain into the right and left hemispheres, the radiant energy flux passing the control plane is obtainable by integrating the spectral intensity with direction cosine $\cos\vartheta$ with respect to the solid angle $\bar{\Omega}$ and the plane outward normal.

The classic theories of radiative energy transfer are derived mostly from the equilibrium radiative condition. This condition is characterized by that the radiation emitted by the medium per unit time and volume at a given frequency in a given solid angle $\bar{\Omega}$ that is identical to that absorbed by the medium. The radiating field is then isotropic; the field is a universal function of the frequency and temperature. The allowed energy levels are 0, hv, $2hv$, $3hv$, etc. The average energy is therefore.

$$\bar{\varepsilon} = hv\left(\Sigma ie^{-ir}/\Sigma e^{-ir}\right) = hv/(e^r - 1) = hv/\left(e^{hv/kT} - 1\right) \tag{11.1a}$$

The final form of Equation (11.1a) is achieved by an identity of infinity series summation to a compact expression $\Sigma\exp(-hv/kT) = \left[1-\exp(-hv/kT)\right]$. According to Equation (11.1a), the mean energy of an oscillator whose fundamental frequency is v approaches the classical value of kT when hv becomes much less kT. The functional form of the radiative energy density follows immediately from Bose-Einstein quantum statistics [4]. Planck derived an energy distribution formula using this approach in place of the classic equipartition of energy and in excellent agreement with experimental data for black-body radiation [2]. The energy density $E(v)dv$ is

$$E(v)dv = \left(8\pi hv^3/c^2\right)dv/\left(e^{hv/kT} - 1\right) \tag{11.1b}$$

From the quantum statistics, the spectral energy density in the hohlraum has been obtained by Planck [5,6]. It gives explicitly the amount of energy per unit volume at equilibrium at a frequency v.

$$U_v = \left(8\pi hv^3/c^3\right)/\left(e^{hv/kT} - 1\right) \tag{11.1c}$$

The spectral intensity in the isotropic radiation condition is given by the Planck distribution function or simply the Planck's function in terms of the wavelength.

$$I_v = \left(2hv^3/c^2\right)/\left(e^{hv/kT} - 1\right) \tag{11.1d}$$

The Planck function distribution over the wavelength from 10^{-2} to $10\ \mu m$ in the temperature range of 2,000 to 20,000 K is depicted in Figure 11.2. The magnitude of the

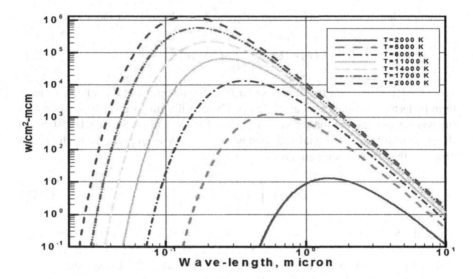

FIGURE 11.2   Planck's distribution function.

Planck function reaches its maximum around the wavelength of one-tenth of a micron at the temperature of 20,000 K, and shifts to more than one micron at the lowest temperature condition. In other words, the equilibrium point of spectral absorption and emission of a medium moves to a shorter wavelength radiative spectrum at an increasing temperature condition. The strong dependence of the radiation intensity is clearly displayed and the wavelength of the peak intensity shifts with the medium temperature.

   Planck's law is valid for thermal radiation over the entire range of frequency under thermal equilibrium conditions. In the low-frequency region of $h\nu \ll kT$; the Planck formula reduces to the Rayleigh-Jeans law [7,8]. The radiation in this low-frequency region does not frequently encounter in most engineering applications, and the spectral radiative energy density becomes

$$U_v = \left(8\pi kT/c^3\right)v^2 \tag{11.1e}$$

At the opposite limit of the high frequency of $h\nu \gg kT$; the spectral energy density, Equation (11.1c) yields Wien's displacement law:

$$U_v = \left(8\pi h v^3/c^3\right)/e^{hv/kT} \tag{11.1f}$$

Planck's law provides the description of an idealized black-body radiation under the equilibrium condition. In order for a body to be in equilibrium with its environment, the emission of radiation must be equivalent to absorption in wavelength and energy. It is possible to conceive of objects that are perfect absorbers of radiation, the ideal black bodies. However, no substances approach very closely to the ideal black body over an extended wavelength range. Only a cavity or a small orifice consisting of a standing wave with excellent insulating walls could become practically ideal blackbody radiation [2].

Stefan's law is developed by integrating the energy density for radiation.

$$U_v(T) = \left(8\pi h/c^3\right)\int_0^\infty v^3 dv \Big/ \left(e^{hv/kT} - 1\right) \tag{11.2a}$$

From the Riemann Zeta function of order four, the integrand of the infinity series is term by term to get the result as $\pi^4/15$ [9]. The equilibrium radiant density becomes the well-known expression for equilibrium radiation density. Stefan's law states in thermal equilibrium, the total radiation energy density is proportional to the fourth power of the absolute temperature.

$$U(T) = \int_0^\infty U_v dv = 4\sigma T^4/c \tag{11.2b}$$

where the Stefan-Boltzmann constant is $\sigma = 8\pi^5 k^4/15h^3 c^2$. In the cgs system, it has the value of 5.67×10^{-5} erg/(cm$^2 \times$ second $\times$ K^5). The amount of radiant energy is given by the amount per unit time and area over all frequencies is therefore σT^4.

An outstanding characteristic of the propagating radiating ray in a gas medium is its opaqueness, which is caused by the amplitude attenuation of electromagnetic waves by the collision and non-collision processes. The latter is incurred by the scattering of photons by diffraction or by penetration of photons into particles and by changing direction from refraction. As the consequence, the medium of radiation propagation has been classified as optically thick and optically thin.

The absorbing and scattering coefficients, k_v and σ_v have a strong dependence on the composition and thermodynamic state of the transmitting medium. It represents the reciprocal of the radiation mean penetration distance of the medium. The optical thickness of a medium is obtained by integrating the combined absorption and scattering coefficient along a beam path to becoming a measure of the penetration distance of radiation; $\tau_v = \int_0^t \beta_v(r) dr$. When $\beta_v(r) = k_v(r) + \sigma_v(r)$ has a constant value $\beta_v = k_v + \sigma_v$, then it represents the reciprocal of the radiation mean penetration distance of the medium. If $\tau_v \gg 1$, the beam path in the medium is considered to be optically thick. Whereas the dimensionless parameter is very small in comparison with unity, $\tau_v \ll 1$, the beam path in the medium is referred to as optically thin which means that the absorption along the beam path length is negligible over the radiating domain. In fact, it's the most common occurrences in aerospace engineering applications.

An important relationship between thermal emission and absorption coefficients for radiative transfer exists only under a strict thermodynamic equilibrium condition. In an adiabatic process, radiation equilibrates at a uniform temperature and isotropic field. Kirchhoff's law is the basic theory of thermal radiation, which determines the amount of energy that can be emitted by any object at any fixed temperature and wavelength. In fact, it defines the communication processes between emission and absorption. Kirchhoff's law states: "For a body of any arbitrary material emitting and absorbing thermal electromagnetic radiation at every wavelength in thermodynamic equilibrium, the ratio of its emissive power to its dimensionless coefficient of absorption is equal to a universal function only of radiative wavelength and temperature." In essence, the emissivity is equal to absorptive in thermodynamic equilibrium.

The universal function describes the perfect or black-body emissive power $U_v = \alpha_v \pi I_{b,v}(T)$. This law for radiation applies to an isotropic and the non-polarized medium is homogeneous. As the consequence, the spectral and integrated directional absorptive power of a surface is equal to a spectral and integrated directional emissivity; $\alpha_v(r,\Omega) = \varepsilon_v(r,\Omega)$.The absorption must be exactly equal emission in all frequencies and directions; it's the critically important Kirchhoff-Planck relation [4,7].

For the following discussions of radiative energy transfer, the concept of the local thermodynamic equilibrium (LTE) condition is very important. The equilibrium state is in the adiabatically isolated condition from its environment, and the density of energy at any point of the region does not vary with respect to time, and the thermodynamic parameters defining a state of any two points are equal. However, the region still can contain radiative activity. If the system is in local thermodynamic equilibrium and there will not be any net heat exchange between any parts of the region. At the thermodynamic equilibrium condition, the respective temperatures of all radiating objects are equal. Therefore, the absorption and emission with other objects are in a dynamic balance and all have the same temperature among these objects.

The LTE condition is defined as when the radiation at each point of a medium with a uniform temperature is close to equilibrium; then the radiation intensity can be described by the Planck function at the temperature of the point in consideration. In fact, the Planck spectral distribution of emissive power is derived from quantum mechanics; it describes the equilibrium between spontaneous emission and absorption. Meanwhile, the Kirchhoff-Planck relation alleviates the difficulty of determining the emission coefficient by solving the quantum chemical physics. It is of great importance because the Planck distribution provides quantitative values for spectral radiation from a blackbody to be used as a benchmark reference.

11.3 RADIATION RATE EQUATION

In most hypersonic flights in ionized gas with a temperature around 10,000 K or higher, and with an electron number density of around 10^{13} per cubic centimeter, the radiative pressure is negligible. However, the radiative energy transfer manifests itself by exchanging with the surrounding gas medium or to free space can be quite appreciable. The mechanism of radiation is remarkably different from conductive and convective heating by virtue of its transfer rate by the light speed in comparison with the sonic velocity of the gas. Therefore, the radiation heat transfer is generally considered to be instantaneous and independent of time in comparison with gas motion. For this reason, the radiant energy and the flux of energy have resembled an isotropic flux that can be given by the divergence of the radiation flux.

The total energy transfer per unit volume and time can be found by integrating the difference between spontaneous emission and absorption over the retired solid angle and spectral frequencies; $Q = \int dv \int k_v \left(I_{v,o} - I_v \right) d\Omega$. Since the total energy transfer is nearly independent of the time scale in gas motion. The total energy transfer is equal to the divergence of the integrated radiation flux. $q = \int \cos\theta I_v(\Omega) d\Omega$ where θ is the angle between the radiation propagation and the outward normal of the control surface.

Radiative energy transfer is emitted and absorbed during quantum jumps from one energy state to another, the absorption of a photon is always happened by excitation of an atom or molecule. The radiation transfer is by electromagnetic wave with the speed of light $c = 2.998 \times 10^{16}$ cm/s, and by a unit of hv. The scattering of light or electromagnetic waves under terrestrial conditions can always be neglected in comparison with absorption. However, the condition is reversed in astrophysics.

The amount of energy spontaneously emitted per unit volume per unit time is $\varepsilon_v I_v(r,\Omega,t) dv d\Omega$ and ε_v is the emission coefficient of the propagating wave medium. The radiating energy in the same frequency interval dv and in the element of solid angle $d\Omega$ is $k_v I(r,\Omega,t) dv d\Omega$ and k_v denotes the absorption coefficient. From the quantum mechanics, the total emission per unit volume and time becomes $\varepsilon_v \left(1 + c^2/2hv^3\right) I_v(r,\Omega,t) dv d\Omega$. The first term is called the spontaneous emission and the second term is the induced emission [4]. It is also important to know that Kirchhoff's law via the general principle of detailed balancing allows the emission and absorption coefficients to be closely related, namely one can always calculate from one to another.

The radiation interacts with its propagating medium is a function of its atomic cross section of collision and population number density that are measuring by absorption, emission and scattering. As the radiative wave passes through a medium, energy is generally removed to be known as total absorption by the medium's opacity, the absorption coefficient is defined as k_{va}. In a moving medium, by the result of Doppler shift, a photo moves with a frequency as $v(1 - \Omega \cdot u/c)$.

The emission coefficient $e_v(r,t,\Omega,v)$ is defined by the medium depending on the collision cross section and traveling length at a solid angle. The emissivity can be isotropic in the stationery frame but not in a moving medium. The scattering of radiation is rather complex by including optical refraction and diffraction and the detailed inelastic collision process between photons and particles of the medium. Therefore, it must be determined by the collision integral analogous to that of the gas kinetic theory.

From the previous discussion, it is noticed that absorption and emission are important mechanisms of radiation energy transfer mechanisms together with the scattering process. The coefficient of bound-bound and bound-free absorptions is proportional to the number of absorbing atoms per unit volume of gas which is the unique property of an absorbing atom. The controlling parameter is the absorption cross section defined as $\sigma_a = k_{va}/N$, in which k_{va} is the absorption coefficient and is the reciprocate of the absorption mean free path, and the number of atoms per unit volume is designated by N. In fact, the absorption coefficient is defined by the spectral radiation intensity, I_v with a frequency v; $dI_v = -\kappa_v I_v dr$. The collision cross section of an atom or any other particle for the scattering phenomenon can be defined identically as that for absorption.

There are really five detailed basic mechanisms in the radiation heat transfer process, such as spectral absorption, emission, scattering, reflection from media interface, and re-absorption. The spectral emission is more complex, traditionally it has been categorized as spontaneous, induced, or medium emissivity because it depends on the state and properties of the emitting medium. The spectral radiation scattering coefficient $\sigma_v(r)$ gives the scattering rate per unit volume for radiation propagated in an elemental solid angle $d\Omega$, therefore it follows; $\sigma_v(r)I_v(r,\Omega,t)d\Omega$. Attenuation of radiative energy in a medium can be caused by absorption, reflection, and also by scattering, the opacity of a medium is a measurement by the attenuation of the transmitted radiation energy. By definition, the spectral scattering indicatrix $p(r,\Omega',\Omega,v',v)$ is a function that characterizes the probability of scattering to the direction from Ω to direction Ω' and to the frequency v from a frequency v'. The rate of energy transfer between two states will have a value of $(1/4\pi)\sigma_v(r)I_v(r,\Omega',t)p(r,\Omega',\Omega,v',v)d\Omega'dv'$.

In short, the radiating mechanism is generated by the kinetics of the photon or light quanta which is treatable by the integro-differential Boltzmann equation for photon distribution function in phase space. We have the radiation rate equation as [7,10]

$$(1/c)\partial I_v(r,\Omega,t)/\partial t + \partial I_v(r,\Omega,t)/\partial r + \left[k_v(r)+\sigma_v(r)\right]I_v(r,\Omega,t)$$

$$= I_v^{em}(r,t) + \left[\sigma_v(r)/4\pi\right]\int_{v=0}^{\infty}\int_{\Omega=4\pi} p(r,\Omega',\Omega,v',v)I_{v'}(r,\Omega,t)d\Omega'dv' \tag{11.3a}$$

The two leading terms on the left-hand side of the Equation (11.3a) are the total derivative of the radiative intensity, and the convective velocity is the speed of light along the direction of the solid angle. The third term represents the spectral intensity associated with the absorption and scattering with the coefficient of $k_v(r)$ and $\sigma_v(r)$ respectively. The first term on the right-hand side of the equation represents the amount of spectral emission in the medium. The total emission rate per unit volume at LTE conditions $I_v^{em}(s,t)$ is obtained by invoking *Kirchhoff's law*, which expresses the general principle of detailed balancing applied to the emission and absorption of light. Therefore, the absorption coefficient is the emission coefficient of a medium $I_v^{em}(s,t) = \kappa_v(s)I_{b,v}[T(s)]$, and $I_{b,v}[T(s)]$ is the spectral intensity of Planck's function. In this case, the spectral intensity does not depend upon the solid angle Ω. However, all the other coefficients are dependent only on the local thermodynamic and optical properties of the gas medium. Therefore, they are the sole functions of the space variable r to be determined as soon as the thermodynamic variables of the flow field are known. The last term describes the contribution to the rate of change for the radiative intensity by scattering of the radiative wave.

It is intuitively obvious that the time-dependent term of the radiative rate equation is scaled by the speed of light, therefore is negligible in comparison with the spatial derivative. The time-independent radiative rate equation acquires the following form:

$$\partial I_v(r,\Omega,t)/\partial r + \left[k_v(r)+\sigma_v(r)\right]I_v(r,\Omega,t)$$

$$= I_v^{em}(r,t) + \left[\sigma_v(r)/4\pi\right]\int_{v=0}^{\infty}\int_{\Omega=4\pi} p(r,\Omega',\Omega,v',v)I_{v'}(r,\Omega,t)d\Omega'dv' \tag{11.3b}$$

The time-independent radiation rate equation, Equation (11.3b) is the fundamental governing equation for the rate of radiative transfer. The basic equation is derived from the Boltzmann equation of the photon probability distribution function. There are widely adopted simplifications to the radiative heat exchange rate equation neglecting the dispersion of radiation over the frequency of electromagnetic wave propagation. This is the so-called approach of coherent dispersion. Under this condition that photons of radiation incident on the elemental physical volume are diffused on the same frequency of electromagnetic radiation. In most hypersonic flows, the scattering of the radiative wave is neglected by virtue of very low degree of ionization, or the radiation is considered optically thin, then the radiation rate equation is reduced to the partial differential equation.

$$\partial I_v(r,\Omega,t)/\partial r + k_v(r,t)I_v(r,\Omega,t) = I_v^{em}(r,t) \tag{11.3c}$$

The governing integro-differential or the simplified differential of the radiation rate equations must be closed by appropriated initial value and boundary conditions, but the far field boundary conditions for equation are rather complex. In general, they are categorized into four classes, such as the vanishing flux, plane-parallel flux, diffuse radiation flux, and the most general reflection indicatrix or the transmission indicatrix. These types of well-posed initial values and boundary conditions are problem dependent and will be deferred until later.

The equally challenging issue in solving the radiative rate equation is the determination of the absorption and scattering coefficients because of the interlocking of nonequilibrium thermodynamics and optical properties of the gas medium in different quantum states. The quantum transitions accompanying the energy absorption and emission are subdivided into three types: The bound-bound transition has a discrete spectrum in an atom that produces line spectra, and in the molecule, it appears as band spectra that contain a large number of rotational lines. The bound-free transition leads to photoelectric absorption. The free-free transition is referred to as the bremsstrahlung emission and absorption. Therefore, the spectral absorption coefficient distribution will reflect distinctive characteristics at different thermodynamic states of a radiating medium. Theoretic derivation of the spectral absorption coefficient can only be accomplished by approximations to the quantum mechanics and statistic quantum mechanics through a perturbation to the quantum mechanics or by the *ab initio* approach [7,8]. These spectral optical properties are uniquely related to the thermodynamic state and chemical composition of the radiating medium and must be derived from the nonequilibrium flowfield. In general, the optical data generation process can be accomplished by the *ab initio* application of quantum mechanics or some physical-based approximation from the molecular and atomic spectral structures. However, the most commonly accepted databases under the local thermodynamic equilibrium (LTE) approximation have been provided by NASA [11], NIST [12], and by Russian Academy of Sciences [7].

Surzhikov has demonstrated the unique behavior of radiation at two different temperatures of 5,000 and 20,000 K and at the same pressure of one atmosphere [7]. It is clearly shown that the magnitudes and distributions of absorption spectral coefficient as a function of wavenumber are entirely different from each other. The difference is definitely not merely a shifting of the distribution shape in the spectrum but with drastic change in intensities to reflect the different excited internal energy states of the air mixture.

Additional understanding of the closely coupled relationship between radiative energy transfer to the local thermodynamic state in the high-enthalpy environment can be gained by numerical simulations. The Stardust reentry probe with ablation is simulated including the phenolic-impregnated carbon ablator (PICA) [13]. The radiative emissions have been detected by the measurements using Echelle spectrograph [14,15]. The active emission is concentrated around the wave number from 10^4 to 10^5/cm which corresponds to the wavelength from 1,000 to 100 nm. This wide range of wavelengths or spectra contains the Schumann-Runge band of O_2, the β and γ systems of NO, the first and second positive of N_2, the fourth positive band of CO, and the violet and red systems of CN. Some ionized species may also include the first negative band of O_2^+, as well as the first negative N_2^+ band but it cannot be clearly identified.

In Figure 11.3, the radiative heat fluxes on the surface are depicted at the reentry altitude of 65.4 km which is characterized by a velocity of 12.413 km/s, a density of 1.06×10^{-4} kg/m^3, and a temperature of 221.4 K. The intensity of radiative energy emission in time over the spectral measured by wave numbers from 10^3/cm to 2.0×10^5/cm, corresponds to wavelength range from 10,000 to 196 nm. At the earlier stage of reentry, the high intensity of surface radiative flux is concentrated in the upper range of 10^4/cm ~ 2.0×10^5/cm and shifts toward the middle wave number range. The lower wave number range contains the first positive band of nitrogen (740–870

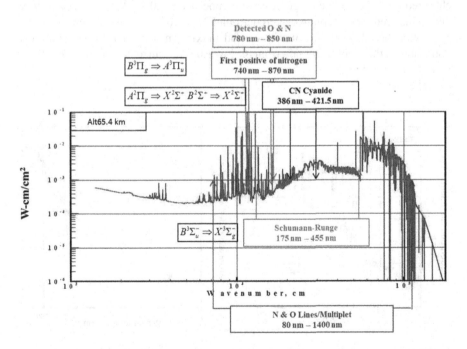

FIGURE 11.3 Planck's distribution function.

nm), the Schumann-Rung system of oxygen (780–850), and the Violet system of CN (386–421.5 nm). The atomic nitrogen N, and oxygen O, in fact, have lines and multiples (80–1,400 nm) at the higher wave number range up to 1.25×10^5/cm. The radiative energy exchange in the studied spectrum remains similar and sustained from the attitude of 51.2 km, including the peak heat transfer condition on the Stardust capsule, 59.8 km. At the later stage of reentry at the altitude of 46.5 km, the radiative heat flux on the surface decreases over the entire studied spectral. In the lower wave number domain from 10^3/cm to 10^4/cm, the magnitude of intensity reduces by nearly two orders of magnitude. The radiation heat transfer ceases beyond the spectral value of 10^5/cm. The downward shift of radiative wave number is reflected by the lower temperature in the shock layer according to the Wein displacement law for spectral energy density function.

The spectral absorption coefficients near the surface of the Stardust capsule from the altitudes 65.4 ($t = 48$ seconds) to 46.5 km ($t = 76$ seconds) are presented in Figure 11.4. The radiative energy transfer is calculated based on the local flow field condition to show the dependence of radiative intensity on the propagating medium. The displayed absorption coefficients consist of two results generated within the boundary layer and the maximum value near the surface of the Stardust capsule to show the influence of the near-surface temperature of the emitter. At the initial and the maximum forebody heating condition, at an altitude of 59.8 km, the calculated absorption coefficient in terms of wave number range is from 10^4/cm to 1.8×10^5/cm. All the absorption coefficients have a similar pattern; the spontaneous pikes reflect the random quantum transitions. And the high absorption coefficients on the surface are clustered near the middle range of wave number from 5.0×10^4/cm to 1.2×10^5/cm, and a clustered spike domain centered around 2.0×10^4/cm. The difference between absorption based on the local surface and maximum near-surface temperature is noticeable in the spectral domain from the low 3.0×10^4/cm to the middle 5.0×10^4/cm wave number range.

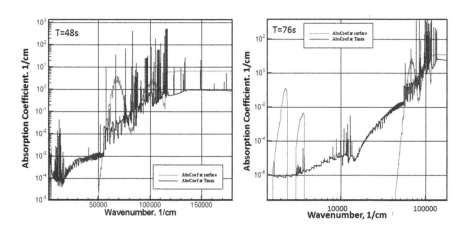

FIGURE 11.4   Radiative absorption coefficient distribution on Stardust reentry probe.

Finally, the results at the altitude of 46.5 km ($t = 76$ seconds) are given in the spectral range of 10^3/cm to 2.0×10^5/cm. The absorption coefficients indicate spectral intensity centered on the wave number of 10^4/cm and 10^5/cm. The difference between coefficients of the local and maximum near-surface temperature is very pronounced. This discrepancy can be understood by which the maximum near-surface temperature in the shock layer is the highest at the initial reentry stage and decreases steadily over time. The accurate near-surface temperature is shown to be paramount for predicting the emission and absorption of radiative heat transfer.

11.4 MULTI-FLUX METHODS

Three traditional methods for solving the radiation rate equation are built on the multi-flux approach based on the intrinsic characteristic of the approximated governing equation. These methods are developed for solving the rate equation of radiation transfer and the radiative intensity is approximated with a series of moment equations. The underlying principle of this approach is based on the Milne-Eddington approximation for evaluating spectral quantities [16,17]. In gist, the fundamental idea is considered the radiation transmitting in one direction, and all directional components have a value independent of the angle; thus the local radiation in each coordinate direction is isotropic. All multi-flux methods modify the rate equation of radiative transfer equation (11.3b) by multiplying an increasing power of the direction cosine to the radiative intensity to form a set of moment equations. At the same time, they also incur a weakness for the closure issue by the system of equations, namely there are always more unknowns than the numbers of governing equation [18].

The method of spherical harmonics is one of the most widespread methods for solving the radiation transfer rate equation. The procedure is based on an approximation of angular dependence of emittance and a scattering indicatrix by a series of spherical harmonics [6]. The spherical harmonics are the orthogonal Legendre polynomials deriving from a solution of the Laplace equation in spherical coordinates. The continuous solutions of the Laplace equation are known as potential functions, these basis functions are often called spherical harmonics. A major advantage of the method of spherical harmonics is derived from its high efficiency to solving the radiation transfer equation in light-scattering mediums for elementary and complicated geometries. The specified efficiency is achieved on the one hand by representing the angular dependence of emittance and a scattering indicatrix by a series of spherical harmonics. On the other hand, the orthogonality of the Legendre polynomials allows transforming of the initial integro-differential equation for radiative transfer to a system of the differential equations. At the same time, the accuracy of solution to radiating transfer and the improvement to computing efficiency increase with growth of albedo or the reflectance of the plasma for the first-order scattering phenomenon [8,19].

The approach to the spherical harmonic method is first expanded the radiation intensity into spherical harmonics series:

$$I_v(r,\Omega) = \sum_{n=0}^{\infty}(2n+1)\sum_{m=-n}^{n}\frac{1}{2\pi(1+\delta_{0,m})}\frac{(n-|m|)!}{(n+|m|)!}\varphi_{nm}(r)Y_n^m(\mu,\psi) \quad (11.4a)$$

where $\varphi_{nm}(r)$ are position-dependent coefficients to be determined by the problem. The functions $Y_n^m(\mu,\psi)$ are angular-dependent normalized spherical harmonics. The series of approximated solutions are constituted by spherical orthogonal functions, which are expressed in terms of associated Legendre polynomials of the first kind

$$
\begin{aligned}
Y_n^m(\mu,\psi) &= P_n^{(m)}(\mu)\sin m\psi, & m &> 0, \\
Y_n^{-m}(\mu,\psi) &= P_n^{(m)}(\mu)\cos m\psi, & m &\leq 0,
\end{aligned}
\qquad -1 \leq \mu < 1, \quad 0 < \varphi < 2\pi \qquad (11.4b)
$$

The infinite series approximation at the upper limit of $n \to \infty$, the spherical harmonics approximation is the exact solution to the rate equation of radiation transfer. The function $\varphi_{nm}(r)$ is a location-dependent coefficient and will be determined by the solution that satisfied the boundary conditions.

In practical applications, the odd-order number of the Legendre polynomial series is preferred over the even-order number, because the even-order polynomials are difficult to resolve for boundary conditions implementation. Therefore the truncated finite N terms are mostly odd numbers for engineering radiative transfer evaluation, such as the P_1 and P_3 approximations which have been found to be adequate for most applications. The higher-order polynomial is more accurate for the approximation, but the added complexity frequently makes them impractical.

Typical boundary conditions of an arbitrary geometry configuration for the P_1-approximation usually consist of two possibilities. The first physical meaningful condition is that if there is no external radiation at boundary surfaces at $(\Omega \cdot \hat{n}) < 0$, $I_v(r,\Omega) = 0$. The second permissible boundary condition is for a cylindrical geometry with axial symmetry or radial symmetry for spherical geometry; $\hat{r} \cdot W_v(r) = 0$.

The next higher-order spherical harmonic method is the P_3-approximation, the method has been frequently applied to analyze the radiation heat transfer over axial symmetric configurations. The series solution to the radiation rate equation can be systematically derived by substituting Equation (11.4b) into (11.3b) with the aid of the recursive formula of the Legendre polynomials. The resultant series become a boundary-value problem relating to the boundary conditions for the radiative intensity [6,10].

The half-moment method is separating the incoming and outgoing radiative fluxes at a point in space, thus it has also been called the two-flux approximation. The original idea assumes the radiation is isotropic by splitting the direction of propagation into two pieces from its source. By splitting the direction of propagation in terms of solid angle direction Ω into two disjoint components Ω^+ and Ω^-, such that $\Omega = \Omega^+ \cup \Omega^-$ and $\Omega^+ \cap \Omega^- = 0$. The only coupling between the positive and the negative components is based on the intensity in the plane perpendicular to the radiative flux, and the total flux is the sum of the two halves. The major advantage of this approximation is that it allows independent incoming and outgoing boundary conditions which are always difficult with the full moment formulation.

For a flat radiative gas layer, the half-moment method first splits the flux vector into two opposite components; $I_v^+ = I_v(s,\Omega,t); 0 \leq \theta \leq \pi/2$ and $I_v^- = I_v(s,\Omega,t); \pi/2 \leq \theta \leq \pi$. The angle θ defines the domains between the direction of radiation propagation and the reference coordinate. The final result will be obtained by integrating the split radiating characteristics over the entire domain. Take the different order moments for the radiative intensity with respect to direction cosine; $\mu = \cos\theta$ to get spectral half-moment characteristics $M^\pm_{n,v}$; $M^+_{n,v} = 2\pi \int_0^1 I^+_v \mu^n \, d\mu$ and $M^-_{n,\mu} = 2\pi(-1)^n \int_0^1 I^-_v \mu^n \, d\mu$.

The Equation (11.3b) is now approximated by an equivalent system of the following differential equations for spectral half-moment characteristics $M^\pm_{n,v}$.

$$\frac{dM^+_{n+1,v}}{d\tau_v} = -M^+_{n,v} + \frac{2\pi}{n+1} I_{b,v}$$

$$\frac{dM^-_{n+1,v}}{d\tau_v} = -M^-_{n,v} + \frac{2\pi(-1)^n}{n+1} I_{b,v} \quad n = 1, 2, \dots$$

(11.4c)

In Equation (11.4e) the symbol $d\tau_v$ is just the scaled spatial coordinate for the absorption coefficient $d\tau_v = \kappa_v dx$.

The approximation to the radiative rate equation becomes the sum of a large number of the characteristic equations. In order to limit the number of the characteristic equations system (11.4c) to be solved, it is necessary to use a truncation series technique or a closure condition; such as by putting all high-order half-moment functions equal to zero [7]. The integrated half-moment characteristics over the spectral range Δv can be obtained by using $M^+_{0,v}(x)$, $M^+_{1,v}(x)$, $M^-_{0,v}(x)$, and $M^-_{1,v}(x)$ as the integrand over the entire spectrum. Then the transformation of these integrals to the physical coordinate will give the radiation heating for the plane-parallel, non-uniform layers.

The discrete ordinates method is related to the multi-flux methods, which may be carried out to any arbitrary order of accuracy as needed. The basic idea is transforming the coordinate of the radiation rate equation for a gray medium from the Cartesian frame of reference onto the coordinates of the propagating radiative wave path by the chain rule of differentiation. The orientation of the radiation intensity is further discretized into many directions in the transformed coordinates [20,21] into a set of simultaneous partial differential equations. A solution to the radiation rate equation is then found by solving the rate equation by a set of discrete directions spanning the total solid angle range of 4π.

$$dI_v/dr = (\partial x/\partial r)\partial I_v/\partial x + (\partial y/\partial r)\partial I_v/\partial y + (\partial z/\partial r)\partial I_v/\partial z \quad (11.4d)$$

The metrics of coordinate transformation, $\partial x/\partial s, \partial y/\partial s$ and $\partial z/\partial s$ are direction cosines (α, β, γ) that link the transformed coordinates to the Cartesian frame. The transformed rate equation of radiation including the scattering mechanism acquires the following form:

$$\alpha \, \partial I_v/\partial x + \beta \, \partial I_v/\partial y + \gamma \, \partial I_v/\partial z = k_v I_{b,v} - (k_v + \sigma_v) I_v$$

$$+ (\sigma_v/4\pi) \int_0^{4\pi} I_v(s,\Omega) p(s,\Omega,\Omega') d\Omega \quad (11.4e)$$

Equation (11.4e) represents the general transfer relations for the radiative intensity through which the angular average is projected over each transformed ordinate direction. The integral results over the incident angular directions are approximated by the sum of phase functions, $p(s, \Omega, \Omega')$. The outgoing and incoming angular directions are designated by the superscript m and m', meanwhile the integral of the scattering can be approximated by the summation over the directional cosines with a weight function w_m. The equation of transfer in the m direction becomes

$$\alpha_m I_\Omega^m + \beta_m I_\Omega^m + \gamma_m I_\Omega^m = k_v I_b - (k_v + \sigma_v) I_\Omega^m + (\sigma_v/4\pi) \sum_{m'} w_{m'} I_\Omega^{m'} p^{mm'} \quad (11.4f)$$

Along the propagation path in the direction m, the initial value at the source of emission must be specified. The equations system (11.4f) is constituted by m number of discretized first-order differential equations, requiring only one boundary condition for $\hat{n} \cdot \Omega^m > 0$. When the scattering is presented ($\sigma_v \neq 0$), then the equations are coupled and an iterative procedure becomes necessary. In the absence of scattering and with an initial emitting condition, the solution of the discrete radiation intensity will be the exact solution.

The boundary condition written in terms of incident intensities is the sum of the intensity leaving and reflecting from the surface along each ordinate direction m or m' as $I_0^m = \varepsilon I_{b,v} + \dfrac{1-\varepsilon}{\pi} \sum_{m'} \alpha_{m'} w_{m'} I_v^{m'}$ and $\alpha_{m'}$ is the direction cosine between the m' direction and the outward normal $\hat{n}$ of the emitting surface and ε is the emissivity of the surface.

In the computational domain of Cartesian coordinates, the discrete ordinates method is divided into finite numbers of non-overlapping tetrahedral cells. On a discrete ordinate, the radiation heat transfer equation without the scattering mechanism can be given as

$$\alpha_m \partial I_v^m/\partial x + \beta_m \partial I_v^m/\partial y + \gamma_m \partial I_v^m/\partial z = \kappa_v \left(I_{b,v} - I_v^m \right) \quad (11.4g)$$

In Equation (11.4g), I_v^m is the spectral radiation intensity, which is a function of position and direction Ω_m, and $I_{b,v}$ is the spectral blackbody radiation at a given temperature of the medium. The boundary condition on the outermost control surface Γ is the vanishing radiative intensity.

$$I_v^m = 0, r = \Gamma \quad (11.4h)$$

The discrete ordinates methods have been developed for three-dimensional rectangular and cylindrical polar coordinates for radiative rate equation [7,20]. A modified discrete ordinates solution of three-dimensional radiative transfer in enclosures with localized boundary loading conditions [22]. Another discrete ordinate method has been adopted for thermal radiations for a free-burning argon arc [21]. A similar method has also been applied to study radiative heating of the internal surface for the air and hydrogen laser-supported plasma generator [7].

11.5 MULTI-SPECTRAL GROUP APPROXIMATION

Among the most efficient numerical algorithms for solving the radiative rate equations, the multiple-spectral group technique applying to the optical property is a sensible engineering technique. The approach divides the entire spectral domain into multiple spectral subgroups according to its frequency and sums over all the constant frequency approximations to become the complete solution. The numerical accuracy can be improved by an increasing number of spectral subgroups. However, this approach is unable to accurately simulate a medium of rapidly varying opacity.

The radiation heat transfer of the ablating Stardust probe contains optically active components CO_2, H_2O, CH_4, N_2, O_2, NO, N_2^+, C_2, CO [23–25]. Numerical simulation demands the development of effective numerical simulation methods for prediction of spectral radiation fluxes both within the shock layer and on the space vehicle surfaces. The first simplification for numerical simulation is focused on the input from spectral optical data. The Multi-group models use the averaged absorption coefficients in the limits of each spectral group, instead of the data of each individual frequency. The optical data from a given thermodynamic state are generated by the wide-and, narrow-band, and even utilizing the line-by-line integration of a spectrum. All of these models use similar computational algorithm, by restricting the spectral range of $\Delta v \in (v_{min}, v_{max})$ and divide into finite spectral regions or the spectral groups of a selected representative frequency. The equation of radiation heat transfer, Equation (11.3b) is then solved by the frequency-independent spectral coefficients of emission and absorption. Integrated in the full spectral range is determined by summation of the results of the divided spectral groups of Δv.

The approximation is presented in Figure 11.5 for the absorption coefficient distribution of twenty chemical species of the Stardust probe. The probe has a heat sheath of the phenolic-impregnated carbon ablator (PICA). A group of four solutions is generated for the spectral region of $1,000 < \Delta v < 4,000$ cm^{-1}. The progressive divisions of the spectral region by 10, 37, 100, and 500 groups. The radiation heat transfer is calculated by the half-moment method [25].

The unique feature of the multi-spectral group approach for analyzing the nonequilibrium radiation heat transfer is that the basic pattern of radiating fluxes can be captured by the sparse grouping of the spectrum. The definition is sharpened as the number of groups is refined. The multi-group spectral approach has also applied to the Fire-II flight simulation at a higher and wider spectra domain in the wave number from 1,613 to 50,015 cm^{-1} [10,26]. For this numerical simulation, these radiative fluxes reveal a weak dependence on the spectral group number. Nevertheless, it should be stressed that the influence of spectral group number on radiation fluxes can be significant in other circumstances.

Radiation heat transfer in the air has one peculiarity because the rotational line structure of diatomic and multi-atomic spectra has an average transmission of the optically thick path, which is a non-linear function versus physical coordinate. However, the multispectral method is valid only for an optically thin medium. Unfortunately, very in-shock layers and wakes can be characterized as the optical thick or an intermediate case between the optically thick and optically thin cases.

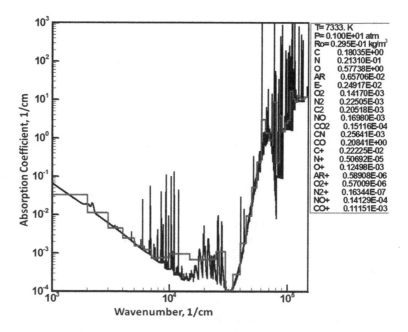

FIGURE 11.5 Approximated absorption coefficient by sub-spectral group method.

Therefore, the multi-spectral group method needs additional precaution in applications for high-enthalpy hypersonic flow.

11.6 RAY-TRACING TECHNIQUE

Radiative energy transfer in hypersonic flow is carried out by the electromagnetic wave at the optic frequency spectrum. The ray tracing technique is established on the optical physics and the procedures of geometric optics and applies to the radiative exchange process. The radiative energy flux is evaluated on each single radiating ray which is different from the flow field coordinate on which the optical properties are determined from the local thermodynamic condition. Whenever the beam path encountered multiple reflective surfaces; the reflecting ray will carry the property of the reflector. For thermal radiation study, the reflecting surfaces have a finite absorptive and thus will attenuate the radiative energy meanwhile they will also emit energy. After a sufficient number of rays are sampled, the completely radiating energy intensity is summing over the entire field of interest. The physics are complex but the ray tracing method is widely adopted in aerospace engineering because the method is naturally suitable for parallel computational simulation. In fact, the tracing ray technique is embarrassingly easy to map on a high performance concurrent or massively parallel computing architecture and the computational efficiency is scalable to tens of thousands of multi-computers [27].

The view-factor approach is equally applicable in the optical domain for optically transparent gas medium [28–30]. The view-factor approach is a zonal method for evaluating radiative energy transfer between isothermal elements and the key idea is establishing the parameters of visibility in the radiating domain. Therefore, it eliminates the interpretive process between coordinates of the ray-tracing method and can be very computationally efficient. However, in a three-dimensional space, this approach degenerates into a semi-analytic procedure.

The critical heat loading of most interplanetary reentries takes place in the upper atmosphere and under a rarefied gas environment [25]. The local thermodynamic equilibrium (LTE) condition and a non-scattering medium are the accepted engineering practice for all radiative heat transfer simulations for interplanetary entry [23,31,32]. The appropriate radiation transport equation simplifies to the most widely adopted formulation in applications [6]:

$$dI_v(r,\Omega)/dr = \kappa_v I_{b,v} - k_v(r,\Omega) \tag{11.5a}$$

The general solution to the above first-order partial differential equation is

$$I_v(r) = I_v(r_0)e^{-\tau_v(r_0)} + \int_{r_0}^{r} \kappa_v(r')I_{b,v}(r')e^{-\tau_v(r')}dr' \tag{11.5b}$$

Recall, the integral of the absorption coefficient along the distance of the beam path is an approximate optical penetration depth, $\tau_v(r') = \int_{s'}^{s} \kappa_v(r'')dr''$, which is a dimensionless parameter for measuring the opacity of a medium excluding the scattering mechanism at the location r'.

For solving Equation (11.5a), the absorption coefficient of radiation, κ_v, along the position vector is needed and must be derived from an emitting/absorbing medium locally. These spectral optical properties are uniquely related to the thermodynamic state and chemical composition of the radiating medium and must be derived from the nonequilibrium flowfield. These macroscopic spectral optical databases are adopted for most radiative simulations [7,11,12,33]. In the process to interpret the optical property of the medium, the effectiveness of the nearest neighbor search plays a pivotal role in computational efficiency.

The ray tracing technique replicates electromagnetic wave propagation by issuing a group of rays from a designated point on the radiating surface covering the entire range of the solid angles. The radiation flux density is obtained by integrating the radiative spectral intensity over the solid angle and over all radiating spectra. Typically a few hundred surface points are processed in a cross-flow plane of an earth reentry simulation, and a total of several hundred points along each tracing ray are calculated. Finally, at least, one hundred spectral subgroups of the spectral data are integrated over the entire spectrum; as the consequence, a hundred million calculations must be performed within an internal iteration. The iterative coupling process demands a substantial amount of computational resources in addition to a large number of optical data shipments through the nearest neighbor search process for the ray tracing calculations.

For the ray tracing method, the thermodynamic state and gas composition are generated first by the solutions of nonequilibrium aerodynamic conservation laws on a body-oriented coordinate, and then the optical property is interpreted along the tracing rays. Figure 11.6 depicts the selected rays on the body-oriented coordinate either on over a blunt forebody of a reentry space vehicle. A typical coupling coordinates system along the radiating ray from a surface point is characterized by a set of solid angles overset on the frame of the nonequilibrium flow field.

The radiative rays are emitted from a solid surface either uninterrupted toward outer space or bouncing from other reflectors and then exiting the control volume. The coupling process between data on different coordinates is the nearest neighbor search (NNS) or commonly referred to as the *proximity search*. It is an optimization problem for locating the closest discretized points among the two coordinates in metric space.

For the proximity search process, a set of data points of r_b and the query points r_r on an emitting ray are both located in the metric space M. The objective of this search process is to find the closest point r_b to r_r. Here, M is restricted to a Euclidean space with a dimension of three. The space partition algorithm is introduced to achieve the most efficient search algorithm to locate the query points [34]. The fundamental relationship between the two independent coordinates is the metric tensor for coordinate transformation which is formally described as [35].

$$ds^2 = g_{m,n}dq^m dq^n$$
$$g_{m,n} = \left(\partial x/\partial q^m\right)\left(\partial x/\partial q^n\right)+\left(\partial y/\partial q^m\right)\left(\partial y/\partial q^n\right)+\left(\partial z/\partial q^m\right)\left(\partial z/\partial q^n\right)$$

(11.5c)

Here $\left(x_b, y_b, z_b\right)$ is designated as the coordinate of the body conformal system for solving the interdisciplinary governing equation of the nonequilibrium hypersonic flows. On the other hand, the query points along the emitting ray are described by the coordinates (x_r, y_r, z_r).

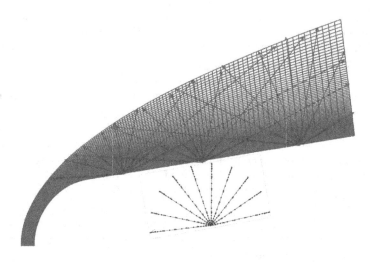

FIGURE 11.6   Radiating rays overlay the body-oriented flowfield grid.

The discretized points between the two distinctive coordinate systems rarely coincide. Therefore, the chemical species concentrations, nonequilibrium thermodynamic conditions, and the internal energy states of the gas mixture must be interpreted from one coordinate system to another by a nearest neighbor search process. The first step in the NNS process is trying to put the two coordinates on a common fundamental frame $\vec{r}$. The distance between a point $\vec{r}_b$ and a query point $\vec{r}_r$ in the Euclidean space can be given, and the criterion for the NNS is clearly defined:

$$ds^2 = \left[\left(x_r - x_b \right)^2 + \left(y_r - y_b \right)^2 + \left(z_r - z_b \right)^2 \right] = \min.$$

The body-conformal coordinate for shock capturing is always generated over a given configuration by a constant-value parametric line, η for the solid surface. Therefore, the surface outward normal establishes itself as the unique reference for the position vector of a radiating ray; $\eta = \eta(x,y,z) = \eta(r,\theta,\varphi)$.

The outward normal of any surface point on the parametric line η actually defines the ranges of latitudinal, θ and azimuthal, φ angular placements of an emitting ray, and can be given as $\vec{n} = \nabla\eta/\|\nabla\eta\| = \left(\eta_x\vec{i} + \eta_y\vec{j} + \eta_z\vec{k} \right) / \sqrt{\left(\eta_x^2 + \eta_y^2 + \eta_z^2 \right)}$. The tracing rays originating from any surface point are usually described by a local spherical polar coordinates system. Therefore, the query points along the radiating ray, q^n (x_r,y_r,z_r) can be easily mapped onto the body-conformal system (x_b,y_b,z_b). The coordinate transformation of the query point is completed by a simple superimposing of the local spherical system onto the body-conformal coordinate:

$$x_r = x_b + r\sin\theta\cos\varphi$$

$$y_r = y_b + r\sin\theta\sin\varphi \qquad (11.5d)$$

$$z_r = z_b + r\cos\varphi$$

The radial vector $\vec{r}$ is the emitting ray originated from a chosen surface point. The latitudinal and azimuthal angular displacements (θ,φ) in reference to the orientation of the body conformal coordinate are simply; $\theta_b = \tan^{-1}\left(\eta_y/\eta_x \right)$, $\varphi_b = \tan^{-1}\left(\eta_z/\eta_x \right)$. The range of latitudinal angle, θ of the emitting ray is defined as; $0 \le \theta \le \pi$ by a tangential plane on the surface. Therefore when the outward normal is not a bisector of the latitudinal angle, the range of the ray in θ becomes $\pi/2 - \theta_b \le \theta \le \pi/2 + \theta_b$. Similarly, the range of the azimuthal angular, φ can also be given as $\pi/2 - \varphi_b \le \varphi \le \pi/2 + \varphi_b$. In fact, the combined latitudinal and azimuthal angular displacements define the solid angle of radiation.

A major portion of efforts in using the ray tracing method is focused on the optical data interpolation from the nonequilibrium flowfield and projects onto the line-of-the-sight tracing rays. The numerical resolution and computational efficiency improvement have been demonstrated by the space partition algorithm of the nearest neighbor search between the two coordinate systems for optical property determination. The basic idea is a kin to both the principal axis trees (PAT) [36] and the binary trees (K-D trees) [37] by reducing the data search domain. The more recent proximity search algorithm is put forward by Shang et al. [34], the uniqueness of the space partition algorithm (SPA) is built on the vector projection of a tracing ray with respect to its surface outward normal. The optimizing technique is expected to reduce an order of magnitude of computational resources required for the optical data interpolation. An efficient ray-tracing (Monte Carlo) method can be realized by jointly applying the line-by-line spectral data integration through the Gauss-Lobatto quadrature via the local resolution refinement.

In a conventional ray tracing method, a group of tracing rays is designated by its solid angle $\Omega_{m,n}$ [25,32]. The spectral flux density is often obtained by a simple first-order approximation as

$$w_\nu(s) = \int dv \int_0^{2\pi} d\varphi \int_0^{\pi/2} I_\nu(s,\Omega)\cos\theta\sin\theta d\theta$$

$$= \int dv \sum_1^m (\varphi_{m+1} - \varphi_m) \sum_1^n I_\nu(\Omega_{m,n})[\sin\theta_{n+1}\cos\theta_{n+1} - \sin\theta_n\cos\theta_n](\theta_{n+1} - \theta_n)$$

$$(11.6a)$$

The integration over the solid angle, $\Delta\Omega$ in a uniformly angular displacement space, can also be obtained by the product sum with a weight function. In fact, the three-point Simpson's rule generates a more accurate result than the above first-order approximation, and can be described as the following:

$$w_\nu(s) = \int_\infty dv\Delta\Omega \sum_n w_n I_{\nu,n}(s_n)$$

$$= \int_\infty dv \sum_m \Delta\varphi \sum_n (\Delta\theta/3)\big[I_{\nu,n}(s_n)\cos\theta_n\sin\theta_n + 4I_{\nu,n+1}(s_{n+1})\cos\theta_{n+1}\sin\theta_{n+1}$$

$$+ I_{\nu,n+2}(s_{n+2})\cos\theta_{n+2}\sin\theta_{n+2} + (\Delta\theta^5/90)I_{\nu,n}^{(4)}(s)\big]$$

$$(11.6b)$$

There are also known formulations for even higher order Simpson's rule, however one needs to break down the finite number of the discrete data by the groups of three-point or more segments and perform an algebraic sum. This integrating rule also has restrictions for an even number of intervals or an odd number of base points over a constant incremental value of $\Delta\Omega$. Although higher-order integration is desirable the choice of the integrating method should be compatible with the accuracy of the interpolated radiation intensity $I_\nu(\Omega_{m,n})$.

As the radiating beam propagates through an inhomogeneous optical path, the spectral intensity needs to take into account the opaque effect along the beam path of the wave motion. In general, it can be accomplished by including the integration formulation as a modifier [7]

$$I_\nu(\Omega_{m,n}) = \int_{r_b}^r \left\{ I_{b,\nu}(r)\kappa_\nu(r)\exp\left[-\int_0^r \kappa_\nu(r')dr'\right]\right\} dr \qquad (11.6c)$$

In Equation (11.6c); the integration limits; r_b and r are the beginning and ending locations along the ray, and the $I_{b,v}$ is the source term of the emission from the body surface, which is the classic black body radiation intensity.

Finally, the radiative flux density by the product sum with a weight function acquires the form:

$$w_v(r) = \int_\infty dv \sum_1^m \Delta\varphi_m \sum_1^n \Delta\theta_n w_n \cos\theta_n \sin\theta_n \left\{ \int_{r_0}^r I_{b,v}(r)\kappa_v(r)\exp\left[-\int_0^r \kappa_v(r')dr'\right]\right\}$$

(11.6d)

In some practices, the local radiation spectral intensity $I_{b,v}(r)$ and the absorption coefficient $\kappa_v(s)$ have been approximated by the averaged values between the start and the end points of a tracing ray. Surzhikov [7] has also recommended additional simplification by performing the integral as

$$I_v\left(\Omega_{m,n}\right) = \sum_{k=1}^k I_{b,v,k+\frac{1}{2}}\left(\tau_{k+1} - \tau_k\right)$$

(11.6e)

In the above equation, the radiation spectral intensity $I_{b,v,k+\frac{1}{2}}$ is obtained by an average temperature along the tracing ray between the positions of r_b and r. This simple approximation is essentially the trapezoidal rule and is fully compatible with the Mean-value theorem. The symbol τ_k in Equation (11.6e) denotes the optical depth at these locations, $\tau_k = \exp\left[-\int_{s_0}^{s_k} \kappa_v(s')ds'\right]$.

In short, the divergence of the radiation heat transfer rate in the global conservation energy equation, Equation (9.4c) appears as either a heat sink or source, is now can be given as

$$\nabla\cdot q_r = \int_0^\infty \kappa_v\left[4\pi I_{b,v} - \int_0^{4\pi} I_v(r,\Omega)d\Omega\right]dv$$

(11.6f)

On the relatively small databases of 15,416 and 60,636 discrete points over the reentering RAMC-II probe, the optical properties are obtained by the ray tracing technique. The magnitudes of radiation heat transfer in the stagnation region at five different locations on the probe surface are depicted in Figure 11.7. The computational results at different locations from the stagnation point are indicated by the inset and over the wave number in the range of $10^4 \le w \le 10^5$ cm^{-1}. The surface heat transfer distributions along the probe surface are generated at the altitudes of 61, 71, and 81 km with a constant surface temperature of 1,000 K. The radiative energy exchanges are calculated by a total of 3,721 tracing rays. It is clearly shown that the radiative heat transfer essentially ceased from a distance of 0.2 m along the probe surface originating from the stagnation point. At the higher altitudes, 71 and 81 km, the radiating heat transfer

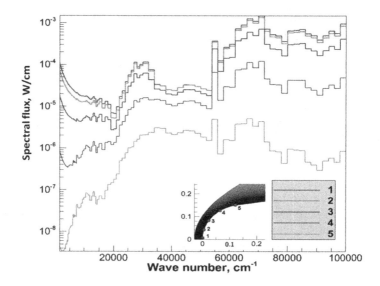

FIGURE 11.7   Radiative intensity over RCMC-II probe at altitude of 61 km.

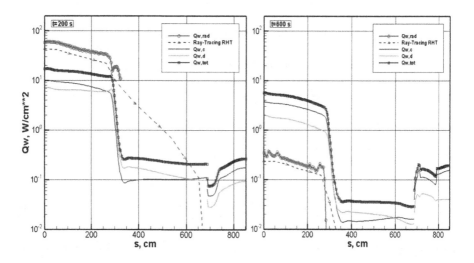

FIGURE 11.8   Heat transfer distributions over the Orion capsule at two reentry stages.

rates are 36.38 and 29.13 W/cm². At the altitude of 61 km, the rate increases to a value of 38.61 W/cm² to become the peak radiative heating condition.

The ray tracing technique has also been applied to ORION space capsule [38] on the catalytic, non-catalytic and. pseudo-catalytic surfaces. Hypersonic flow conditions for International Space Station Crew Exploration Vehicle (ISS CEV) are simulated according to the trajectory data [39,40]. The configuration of all the reentry capsules is similar to the Stardust probe which consists of a large radius frontal surface connecting to an inverted conical after the body. The common feature is designed to cut down the conductive heat transfer at the stagnation point.

Different trajectory points are analyzed for distributions of conductive, convective, and radiation heat fluxes by the half-moment and ray tracing methods along the surface from the forward until the rear stagnation point. In Figure 11.8 two re-entry stages are presented at 200 and 600 seconds from a reference point in time, at the altitude of 78.2 and 42.6 km, and velocities of 7.7 and 5.2 km/s respectively. It is not surprising that radiative heating on the front shield exceeds both conductive and convective heat transfer because of the extremely large blunt-nose radius. The half-moment method gives the total radiation heat fluxes inside the shock layer by the infinite plane layer approximation but is uncertain for the radiation heat transfer results beyond the front shield. The ray-tracing method also uses to predict the radiative heating over the vehicle and reach an excellent agreement with the half-moment method in the shock layer, but the instantaneous solution is not acceptable for the radiation prediction with unsteady aerodynamic problems.

Nevertheless, ray tracing together with the Monte-Carlo method is very appealing because these procedures intrinsically closely mimic the physical process of radiation [25].

11.7 MONTE CARLO SIMULATION

The Monte Carlo method is based on a statistical sample technique and can be applied to simulate the dynamics of photons or light quanta. For most thermal radiation applications, the collisions between particles and photons occur over many mean free paths of the photon. Therefore solving the radiation rate equations, Equation (11.3b) is really tracing the history of a statistically meaningful random sample of photon for the past encounter or its source of emission. Computational simulations by Monte-Carlo methods have great potential for developing the universal computational procedure for the prediction of spectral emissivity and absorption of radiation over different aerospace vehicles. The Monte-Carlo methods have been applied to solve the radiation rate equation, as well as, used to simulate directly the photon trajectories within ion engines [10].

In solving especially the radiation rate equation, the acceleration of the convergence by the Monte-Carlo algorithm has been enhanced through coupling the forward and adjoint radiation transport equations. The solution of the radiation intensity is expressible in terms of a volume Green's function, which leads to an adjoint integro-differential equation [41,42],

$$\Omega \cdot \nabla I(r,\Omega) + k_v I(r,\Omega) = \sigma_v(r) \int_\Omega I(r,\Omega') p(r,\Omega,\Omega') d\Omega' + I_v^{em}(r,\Omega)$$

$$(11.7a)$$

$$-\Omega \cdot \nabla I'(r,\Omega) + k_v I'(r,\Omega) = \sigma_v(s) \int_\Omega I'(r,\Omega') p(r,\Omega,\Omega') d\Omega' + I_v'^{em}(r,\Omega)$$

The boundary conditions for $I'(r,\Omega)$ will be dual to that of for $I(r,\Omega)$, because they must satisfy a reflecting boundary condition as the case for light transport where Fresnel and Snell's law applied [43], we have then

$$\int_\Omega I(r,\Omega) I_v'^{em}(r,\Omega) dr d\Omega = \int_\Omega I'(r,\Omega) I_v^{em}(r,\Omega) dr d\Omega \qquad (11.7b)$$

The approach of applying the Monte Carlo method to solve the radiation transport equation directly is mathematically rigorous and can automatically adapt itself to the specific needs of individual radiation problems.

The integro-differential radiation rate equation, Equation (11.3b) is very difficult to solve and the equation evaluates the probability or the distribution function, the Direct Simulation Monte Carlo (DSMC) technique is a perfect solving algorithm. The Monte Carlo method for solving the distribution equation is based on the Lagrangian frame of reference which tracks directly a large group of particles in a control volume, instead of observing these particles from a fixed location like that of the Eulerian formulation. The basic approach relies on multiple and repetitive random samples to obtain numerical results. This approach duplicates the randomness of particle chaotic motion by collisions; the computational results are deterministic in principle.

However, there is no unique way to implement the DSMC method for solving the Radiation rate equation. The essential repetitive procedure includes the definition of the selected sampled particles, developing a tracking algorithm for these samples with the required boundary conditions, and then performing the random binary collisions with a series of random numbers. When the result met a preset convergent criterion, an average process is used to get the macroscopic results [44]. The key steps of the DSMC method are to define the stimulated representative particles and impose boundary conditions, and the next step will be to set the index and track these particles. The random collision process is simulated by a pseudo-random numbers generator (PRNG) [45]. The final step samples the macroscopic properties of the electromagnetic wave dynamics.

The heart of any DSMC algorithm is a random number generator, which can be described as $p(\varepsilon < x) = \int_{\infty}^{x} p(x)\,dx$, which is the random quantity distribution function (commutative function). It can be either provided by a tabular, random transmitter, or by a pseudo-random numbers (PRNG) generator. The probability density within a solid angle Ω is $\int_{\Omega} p(\Omega)\,d\Omega = (1/4\pi) \int_{0}^{2\pi} d\Omega \int_{0}^{\pi} \sin\theta\,d\theta$. All DSMC schemes must process a huge number of data and always produces a noticeable amount of statistic noises. In order to achieve a statistic convergence or a sufficient amount of data has been sampled. The statistical error estimate becomes crucial and can be determined as; $p[|\,p^* - p\,| < \varepsilon] = 2\Phi\left[ef / \sqrt{p^*(1-p^*)} \right]$ in which $P^* = n/f$ estimates the frequency of occurrence, $\Phi(x) = (1/2\pi) \int_{\infty}^{x} e^{-x^2/2}\,dt$ is the probability integral, and ε is an arbitrarily selected error criterion.

The fundamental approach using Monte-Carlo simulation for radiative energy transfer is tracking photon groups over an averaged spectral band [46,47]. The direct computational simulations analyse the energy transfer over thousands of spectral bands in which the optical thickness always varies by orders of magnitude. Perform simulations associated with complex configurations require the high performance parallel computers with distributed memory and concurrent calculations can reduce dispersion of numerical simulation results. However, this method always incurs statistical error after tracing N number photon bundles of the sampling results $S(N)$. The exact answer is theoretically possible by sampling infinitely many energy bundles $S(\infty)$, but it will not be practical. The error estimate is achievable by the mean square root deviation for the number of bundles and the number of subsamples of radiative intensity $\varepsilon^2 = \sum \left[S(N_i) - S(N) \right]^2 / N(N-1)$.

The advantage of the Monte Carlo technique is unique from which the statistical and discretization errors can be isolated and quantified. Therefore from the pure computational consideration, it may limit the number of photon bundles that can be used in practice. In order to reduce the required computing resource the accuracy of spatial discretizing can be restricted to a lower-order scheme. The Monte-Carlo method for solving the radiative rate equation is in good agreement with either the tracing method or the discrete ordinate method, but the computational resource requirement is greater than the formers.

A basic Monte-Carlo imitative algorithm for radiation energy of heat transfer in an arbitrary inhomogeneous, light-scattering media has been considered by Surzhikov and Howell [47]. As an example to calculate the radiative energy in the infrared spectrum (from 500 to 10,000 cm⁻¹), the resolution of the half-width of a rotational line and the width of the vibrational band of air is important for predictive accuracy. The half-width of the rotational band spans a range from $0.01 < \gamma < 1$ cm⁻¹, and the width of the vibration band Δv varies from 200 to 500 cm⁻¹. From this estimate, it is easily observed that the accurate simulation depends only on the number of statistical samples chosen and the required number of samples is huge. In short, the line-by-line approach integrates the radiative heat transfer equation on the spectrum of rotational lines, as well as, the line-by-line application to trajectories for limited number of photons must be conducted.

In the basic approach, the given spectral range is divided into N number of spectral sub-regions Δv_g in terms of wavenumber. The following averaged absorption coefficient is introduced in the limits of each sub-region as

$$\kappa_g = \left(1/v_g \right) \int_{\Delta v} \left[\kappa^p (v) + \kappa^g (v) + \sum_i^{N_l} \kappa_i^l (v) \right] dv \qquad (11.7c)$$

where $\kappa_i^l (v)$ is the spectral volumetric absorption coefficient of i-th line in terms of wavenumber; N_l is the number of lines located in the sub-region Δv_g. The spectral volumetric absorption of media either of solid or gas, or liquid are designated as $\kappa^g (v)$ and $\kappa^p (v)$ respectively.

Numerical dispersion of direct statistical simulation results is noticed which has been accumulated in the solution of similar problems in "grey" medium. Nevertheless, a satisfactory accuracy simulation is obtainable with moderate optical thickness $(\tau < 1)$. On the contrary, for multi-scattering medium of greater optical thickness, this number shall be further increased. In general, the computational simulation can achieve a reasonable accuracy by simulation trajectories not to exceed 10^4 photon groups (as in the regular line-by-line method), and usually only up to 10–15 photon groups should be sufficient. It must be stressed that it is impossible to determine the spectral characteristics inside Δv_g by this manner, but to achieve only the averaged characteristics in the spectral region Δv_g with an acceptable accuracy.

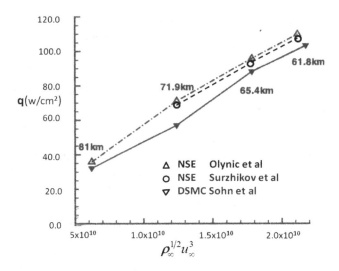

FIGURE 11.9   Comparison of Stardust probe stagnation point heat transfer with DSMC results.

Figure 11.9 depicts the comparison of the stagnation point heat transfer of the Stardust re-entry probe between the coupled radiation rate equations with Navier-Stokes equations [25,48] and DSMC method [49]. The comparison is presented on a correlation ordinate to exhibit a linear radiative heat flux. An excellent accord is consistently maintained among solutions generated by the coupled equations with the Navier-Stokes equation. A reasonable agreement between all simulations is reached at the early stages of the re-entry trajectory where the convective heat transfer is dominant over the conductive and radiative mechanisms. The observation echoes an earlier study on the DSMC approach to radiation, which still requires sustaining basic research in this arena.

In summary, the Monte Carlo method for solving the radiative rate equation can realistically represent the nature of radiation energy transfer via photon dynamics. The random sampling technique has opened a new avenue to study and understand radiation on a microscopic scale that is not achievable by any existing method known to us.

REFERENCES

1. Leighton, R.B., *Principles of modern physics*, McGraw-Hill, New York, 1959.
2. Moore, W.J., *Physical chemistry*, 3rd Ed., Prentice-Hall, Inc, Englewood Cliffs, NJ, 1963.
3. Zel'dovich, Y.B. and Raizer, Y.P., *Physics of shock waves and high-temperature hydro-dynamic phenomena*, Dover Publication, Mineola, NY, 2002.
4. Mihalas, D. and Mihalas, B., *Foundations of radiation hydrodynamics*, Oxford University Press, New York, 1984.
5. Modest, M.F., *Radiative heat transfer*, McGraw-Hill, Inc., New York, 1993.
6. Siegel, R. and Howell, J.R., *Thermal radiation heat transfer*, 4th Ed., Taylor & Francis, New York, 2002.

7. Surzhikov, S.T., *Radiation modeling and spectral data*, Von Karman Lecture Series, von Karman Institute for Fluid Dynamics, Rhode-ST-Genese Belgium, 2002.
8. Surzhikov, S.T., Radiative modeling in shock-tubes and entry flows, RTO-AVT-VKI Lecture Series, Von Karman Institute for Fluid Dynamics, Rhode-ST-Genese Belgium, 2008.
9. Bonwein, J., Bradley, D.M., and Crandall, R., Computational strategies for the Riemann Zeta function, *Journal of Computational and Applied Mathematics*, Vol. 121, 2000, pp. 247–296.
10. Shang, J. and Surzhikov, S., *Plasma dynamics for aerospace engineering*, Cambridge University Press, Cambridge, United Kingdom, 2018.
11. Whiting, E.E., Park, C., Liu, Y., Arnold, J.O., and Paterson, J.A., NeqAIR96, Nonequilibrium and equilibrium radiative transport and spectral program: User manual, NASA Report 1389, December 1966.
12. Ralchenko, Y., et al., NIST atomic spectra data base, Version 3.10, July 2006.
13. Desai, P.N., Lyons, D.T., Tooley, J., and Kangas, J., Entry, descent, and landing operations analysis for the Stardust entry capsule, *Journal of Spacecraft and Rockets*, Vol. 45, No. 6, 2008, pp. 1262–1268.
14. Jenniskens, P., Observations of the Stardust sample return capsule entry with a slit-less Echelle spectrograph, AIAA 2008-1210, Reno, NV, January 2008.
15. McHarg, M.G., Stenbaek-Nielsen, H.C., and Kanmae, T., Observations of the Stardust sample return capsule entry using a high frame rate slit-less Echelle spectrograph, AIAA 2008-1210, Reno, NV, January 2008.
16. Eddington, A.S., *The internal constitution of the stars*, Dover Publication, Mineola, NY, 1959.
17. Milne, F.A., *Thermodynamics of the stars, Handbuch der Astrophysics*, Vol. 3, Springer-Verlag, Berlin, 1936, pp. 65–635.
18. Ripoll, J.F. and Wray, A.A., A half-moment model for radiative transfer in a 3D gray medium and its reduction to moment model for hot, opaque source. *Journal of Quantitative Spectroscopy & Radiative Transfer*, Vol. 93, 2005, pp. 473–519.
19. Andrienko, D., Surzhikov, S., and Shang, J., Spherical harmonics method applied to the multi-dimensional radiation transfer equation, *Journal Computer Physics Communications*, Vol. 184, No. 10, 2013, pp. 2287–2298.
20. Fiveland, W.A., Three-dimensional radiative heat transfer solutions by the discrete-ordinates method, *Journal of Thermophysics and Heat Transfer*, Vol. 2, 1988, pp. 309–316.
21. Menart, J., Radiative transport in a two-dimensional axisymmetric thermal plasma using the S–N discrete ordinates method on a line-by-line basis, *JQSRT*, Vol. 67, 2000, pp. 273–291.
22. Ramankutty, M.A. and Crosbie, A.L., Modified discrete-ordinates solution of radiative transfer in three- dimensional rectangular enclosures, *JQSRT*, Vol. 60, 1998, pp. 103–134.
23. Liu, Y., Prabhu, D., Trumble, K.A., Saunders, D., and Jenniskens, P., Radiation modeling for the reentry of the Stardust sample return capsule, *Journal of Spacecraft & Rockets*, Vol. 47, No. 5, 2010, pp. 741–752.
24. Shang, J., and Surzhikov, S.T., Simulating Stardust reentry with radiation reentry, *Journal of Spacecraft and Rockets*, Vol. 48, No. 3, 2011, pp. 385–396.
25. Surzhikov, S.T. and Shang, J.S., Coupled radiation-gasdynamic model for Stardust earth entry simulation, *Journal of Spacecrafts & Rockets*, Vol. 49, No. 5, 2012, pp. 875–888.
26. Surzhikov, S.T. and Shang, J.S., Numerical rebuilding of Fire-II Flight data with the use of different physical-chemical kinetics and radiation models, AIAA 2013-0190, Grapevine, TX, January 2013.
27. Shang, J.S., *Computational electromagnetic-aerodynamics*, IEEE Press Series on RF and Microwave Technology, John Wiley & Sons, Hoboken, NJ, 2016.

28. Andrienko, D., Surzhikov, S., and Shang, J., View-factor approach as a radiation model for the reentry flowfield, *Journal of Spacecrafts & Rockets*, Vol. 53, 2016, pp. 74–83.
29. Bose, D. and Wright, M. J., View-factor based radiation transport in a hypersonic shock layer, *Journal of Thermophysics and Heat Transfer*, Vol. 18, No. 4, 2004, pp. 553–555.
30. Glassner, A.S., *An introduction to ray tracing*, Morgan Kaufmann, San Francisco, CA, 1989, pp. 1–32.
31. Feldick, A.M., Modest, M.F., and Levin, D.A., Closely coupling flowfield: Radiation interactions during hypersonic re-entry, *Journal of Thermophysics and Heat Transfer*, Vol. 25, No. 4, 2011, pp. 481–492.
32. Shang, J.S., Surzhikov, S.T., and Yan, H., Hypersonic nonequilibrium flow simulation based on kinetics models, *Frontiers in Aerospace Engineering*, Vol. 1, No. 1, 2012, pp. 1–12.
33. Surzhikov, S.T., Capitelli, M., and Colonna, G., Spectral optical properties of nonequilibrium hydrogen plasma for radiation heat transfer, AIAA Preprint, 2002-3222, 2002.
34. Shang, J.S., Andrienko, D.A., Huang, P.G., and Surzhikov, S.T., A computational approach for hypersonic nonequilibrium radiation utilizing space partition algorithm and Gauss quadrature, *Journal of Computational Physics*, Vol. 266, 2014, pp. 1–21.
35. Langwitz, D., *Differential and Riemannian Geometry,* Academic Press, New York, 1965.
36. NcNames, J., A fast nearest-neighbor algorithm based on a principal axis search tree, *IEEE Transaction on Pattern analysis & Machine Intelligence*, Vol. 23, No. 9, 2001, pp. 964–976.
37. Freidman, J.H., Bently, J.L., and Finkel, R.A., An algorithm for finding best matches in logarithmic expected time, *ACM Transactions on Mathematical Software*, Vol. 3, No. 3, 1977, pp. 209–226.
38. Surzhikov, S. and Shang, J., Radiative and convective heating of Orion space vehicles at Earth orbital entries, AIAA 2011-0251, Orlando, FL, January 2011.
39. Hayden, S., Oza, H., Mah, R., Mackey, R., Narasinha, S.S., Karsai, G., Poll, S., Deb, S., and Shirley, M., Diagnostic technology evaluation report for on-board Crew launch vehicles, NASA-TM-2005-21402, 2005.
40. Ried, R.C., Ruchelle, W.C., and Mihoan, J.D., Radiating heating to the Apollo Command module, NASA TM-58091, 1972.
41. Kong, R., Ambrose, M., & Spanier, J., Efficient, automated Monte Carlo methods for radiation transport, *Journal of Computational Physics*, Vol. 227, pp. 9463–9476, 2008.
42. Olynick, D.R., Taylor, J.C., and Hassan H.A., Comparisons between Monte Carlo method and Navier-Stokes equations for re-entry flows, *Journal of Thermophysics and Heat Transfer,* Vol. 8, 1994, pp. 251–258.
43. Hecht, E., *Optics*, 4th Ed., Addison Wesley, New York, 2002.
44. Oran, E., Oh, C., and Cybyk, B., Direct simulation Monte Carlo; Recent advances and applications, *Annual Review of Fluid Mechanics*, Vol. 30, 1998, pp. 403–441.
45. Von Neumann, J., Various techniques used in connection with random digits, *NBS Applied Mathematics Series*, Vol. 12, 1951, pp. 36–38.
46. Lee, P.Y.C., Hollands, K.G.T., and Rathby, G.D., Reordering the absorption coefficient within the wide band for predicting gaseous radiative exchange, *Journal of Heat Transfer*, Vol. 118, No. 2, 1996, pp. 394–400.
47. Surzhikov, S.T. and Howell, J.R., Monte-Carlo simulation of radiation in scattering volumes with line structure, *Journal of Thermophysics and Heat Transfer*, Vol. 12, No. 2, 1998, pp. 278–281.
48. Olynick, D., Chen, Y.K., and Tauber, M.E., Aerothermodynamics of the stardust sample return capsule, *Journal of Spacecraft and Rockets*, Vol. 36, No. 3, 1999, pp. 442–462.
49. Sohn, I., Levin D.A., and Modest, M.F., Coupled DSMC-PMC radiation simulation of a hypersonic reentry, *Journal of Thermophysics and Heat Transfer*, Vol. 22, 2012, pp. 22–35.

12 Multi-Disciplinary Governing Equations

12.1 MAGNETOHYDRODYNAMICS EQUATIONS

The interdisciplinary governing equations for high-enthalpy hypersonic flows in the continuum domain must be developed by integrating the time-dependent compressible Navier-Stokes and Maxwell electromagnetic equations with nonequilibrium quantum chemical physics. The integrating of electromagnetics into aerodynamics follows the identical train of thought as that for the theory of Magnetohydrodynamics (MHD) [1]. However, a significant improvement to physics fidelity is by removing the simplification treatment of the ionized gas as a homogenous multi-fluid plasma medium. Instead, the individual ionized gas species from nonequilibrium chemical reactions, including the internal degrees of freedom for vibrational and electronic excitations are explicitly included [67].

The ionized air increases a physical dimension to hypersonic flows through the quantum mechanics and electromagnetic phenomena [58]. All electromagnetic physics is governed by the Maxwell equations, and these equations in the time domain consist of four fundamental laws of electromagnetics. The Maxwell equations were established by James Clerk Maxwell in 1873 and verified experimentally by Heinrich Hertz in 1888. Albert Einstein's special theory of relativity further affirmed the rigorousness of Maxwell's equations in 1905 [40,41]. For these reasons, the Maxwell equations occupy the widest range of applicability and unique position in plasma physics.

$$\partial B/\partial t + \nabla \times E = 0$$

$$\partial D/\partial t - \nabla \times H + J = 0$$

$$\nabla \cdot B = 0$$

$$\nabla \cdot D = \rho_e$$

(12.1a)

The first equation is Faraday's law relates the time rate of change by the magnetic flux density B (Wb/m^2) with the curl of the electric field intensity E (V/m). The second equation is the generalized Ampere's circuit law determining the time rate of the electric displacement D (C/m^2) with the curl of the magnetic field strength H (A/m) and electric current density J (A/m^2). The most significant fact is that the generalized Ampere's law introduces the displacement current $\partial D/\partial t$ by Maxwell, so there is no ambiguity that electromagnetic waves can propagate in a vacuum. The last two equations are the Gauss' laws for magnetic and electric fields, which simply require that

DOI: 10.1201/9781003212362-14

the divergence of B must vanish in any electromagnetic field to preclude the existent of a magnetic dipole. The second law of Gauss defines the relationship between the electric displacement D and the electric charge density ρ_e (C/m^3).

From a purely mathematical point of view, Maxwell's equations have a closure issue, namely, the number of dependent variables has exceeded the number of equations. This dilemma is removed by the constitutive relations. These constitutive relationships are derived from the relativistic and described by tensors' consideration but in an isotropic media, the required constitutive properties are scalar, $D = \varepsilon E$ and $B = \mu H$. The scalar electric permittivity ε in the free space has a value of $4\pi \times 10^{-7}$ H/m, and the scalar magnetic permeability has a value of approximately 8.85×10^{-12} F/m.

On a moving boundary, because the partial temporal derivatives are no longer commuted with the surface integral, the boundary conditions are required to formulate by the integral form of the Maxwell equations. Additional components via the normal velocity are required to append on the tangential formulation of the electric and magnetic field strength [40].

$$n \times (E_1 - E_2) - (n \cdot u_b)(B_1 - B_2) = 0$$

$$n \cdot (B_1 - B_2) = 0$$

$$n \cdot (D_1 - D_2) = \rho_s \tag{12.1b}$$

$$n \times (H_1 - H_2) + (n \cdot u_b)(D_1 - D_2) = J_s$$

The symbol n designates the surface outward normal and u_b is the velocity of the moving boundary in equation (12.1b).

On a stationary frame, the boundary condition becomes simpler. The tangential electric field E is continuous across the interface boundary. Conversely, the normal component of the magnetic flux density B is also permitted to pass uninterrupted through the media interface. The normal component of the electric displacement D must be balanced by the surface charge density. It is important to note that the surface charge is the actual electric charge separated by a finite distance between charged particles but not the surface charge due to polarization. Finally, the discontinuity of the tangential components of the magnetic field strength H is equal to the surface current density J_s.

The ideal magnetohydrodynamics (MHD) equations are the integrating result of the Maxwell and Euler equations [1,70]. These equations are originally developed for the study of the astrophysics and geophysics. An important feature in MHD is that the motion of collective charged particles is described by an electrically conducting fluid with the usual fluid dynamic variables of velocity, density, and pressure or temperature. In the basic formulation for classic MHD, the electrostatic force is dropped by virtue of the global electric neutrality of plasma, and another leading approximation is that the displacement current $\partial D/\partial t$ of Ampere's circuit law is omitted [79]. Thus, the electric current density is only associated with the curl of the magnetic field strength $J = \nabla \times H$. In essence, the Ampere equation reduces to the original time-independent circuit law. The controlling mechanisms for the temporal change of magnetic flux density are approximated as $\partial B/\partial t = -\nabla \times E \approx \nabla \times \left[u \times B - \nabla \times (B/\mu\sigma) \right]$, in which the leading and consecutive terms represent the production and diffusion of the magnetic flux density [82].

In the process, the generalized Ohm's law is adopted to directly relate the electric field strength with current density. The simplified generalized Ohm's law can be written as $J = \sigma(E + u \times B)$ at a much slower MHD time scale than the plasma frequency [47]. This approximation greatly simplifies the formulation at the expense of overlooking the detailed composition of the plasma [12,35]. As the consequence, the electric current density depends upon a prescribed value of electric conductivity but not the thermodynamic state of the plasma. However, it should be cautioned that in the presence of an applied magnetic field, the electric conductivity becomes a tensor of rank two. Ohm's law is subjected to a constant driving field of E and B can still be constructed by assigning a simpler scalar electrical conductivity as

$$E = \left[J + \beta_e (J \times B) + sB \times (J \times B) \right]/\sigma \qquad (12.1c)$$

In Equation (12.1c), the second term is identified as the Hall current and the third term is referred to as the slipping current unique to weakly ionized plasma [36,47,67]. The MHD formulation gives an accurate description of a wide range of astrophysics and geophysics phenomena. At extremely high temperature and low-density environments when the degree of ionization changes vigorously and with an externally applied magnetic field, then the accuracy of MHD approximation becomes unreliable.

The full magnetohydrodynamics equations are the generalized classic MHD equations by including the viscous dissipation, Joule heating, and the work done by the ionized medium to the system [23,62,65]. It essentially integrates the reduced Maxwell equations with compressible Navier-Stokes equations. The resultant equations extend the ideal MHD formulation to the finite Reynolds number regime [10].

$$\frac{\partial \rho}{\partial t} + \nabla \cdot (\rho u) = 0$$

$$\partial B/\partial t + \nabla \cdot (uB - Bu) = \nabla \cdot (\nabla B)/\mu\sigma$$

$$\partial \rho u/\partial t + \nabla \cdot \left[\rho u u - \overline{\overline{\tau}} - BB/\mu + (p + B^2/2\mu)\overline{\overline{I}} \right] = 0 \qquad (12.1d)$$

$$\partial \left(\rho e + B^2/2\mu \right)/\partial t + \nabla \cdot \left[\left(\rho e + p + B^2/2\mu \right)u - \kappa\nabla T - u \cdot \overline{\overline{\tau}} - (B \cdot u)B/\mu \right]$$

$$= \left(\nabla \times B/\mu \right)^2/\sigma + (B/\mu) \cdot \left[(\nabla \cdot \nabla B)/\mu\sigma \right]$$

By reducing the full MHD equation to a dimensionless form with respect to the unperturbed free stream values, two well-known similitudes are revealed [38,67]. One of them is the interaction parameter or the Steward number, $S_m = \sigma B^2 L/\rho_\infty u_\infty$ and the other is the magnetic Reynolds number, $R_m = \sigma\mu u_\infty L$ [47,79]. The former similitude is a measure of the relative importance of the magnetic flux density to the inertia of the fluid dynamics or the ratio of Lorentz force to the inertia of the fluid motion. It may be interesting to point out that the Lorentz force is a first-order relativistic effect.

The dimensionless similarity parameter R_m becomes a very important indicator to highlight the relatively dominant force in an interacting ionized gas flowfield. For most hypersonic flows, R_m is much less than unity, $R_m < 1$, the induced magnetic flux density is negligible in comparison with the externally applied magnetic field. The applied magnetic fields are usually implemented for electric propulsion, flow control by plasma actuators, and remote energy deposition [68]. On the other extremes, when the magnetic Reynolds number is much greater than unity, $R_m \gg 1$, the strong magnetic field dominates the charged particle motion. The electric conducting particles can still move freely along the magnetic line but are unable to cross the magnetic field. In literature, it often describes this phenomenon as the charged particles are frozen in the magnetic field line [47]. This particular feature of plasma offers a unique capability for flow control and even suppresses the transition to turbulent flow.

12.2 HYPERSONIC FLOW IN AN APPLIED MAGNETIC FIELD

The effect of an applied magnetic field on hypersonic flow simply cannot be understated. The magnetic flux density increases mechanisms for electromagnetic effects by introducing the Lorentz acceleration, Larmor gyration, and Hall current to the ionized hypersonic flows which allow the possibilities for plasma confinement, altered plasma sheath thickness, and amplification of electromagnetic-aerodynamics interactions.

The most common adopted magnetic generator is the helical electric induction coils or a solenoid winding around a ferrite core and operating by a pulsing or steady electric current with water cooling. If the length of the solenoid is much greater than its radius, the magnetic flux density at the center of the solenoid is proportional to the permeability of the medium, number of turns of the coil, current through solenoid, and inversely proportional to the length of the solenoid; $B = \mu_m nI/l$ [41]. An externally applied magnetic fields generated by a solenoid coil does not produce a uniform distribution as displayed in Figure 12.1. On the axis of symmetry, the theoretic magnetic field decay rate for a dipole is inversely proportional to the cubic power of the distance from the pole. According to a steady-state experimental measurement [84], the magnetic field decay rate of a solenoid is found to behave as $B = B_o \left(r_b/r \right)^{3.61}$. This empirical formulation has also been confirmed by other independent measurements [39].

The most powerful steady-state magnetic field is generated by the superconducting material for its winding coil because it permits an enormous electric current flow. The superconductivity is a quantum mechanical process known as the Meissner effect [4], and under this condition, the electric resistance vanishes. Theoretically, an infinite amount of electric current can be realizable by applying a low applied electric potential. The most widely adopted superconducting material is either niobium-titanium (NBTI) or niobium-tin (NBSN), and the critical temperatures of superconductor for NBTI and NBSN are 10 and 18 K, respectively. The required operational condition must keep the wire temperature at the cryogenic temperature of 4.2 K in

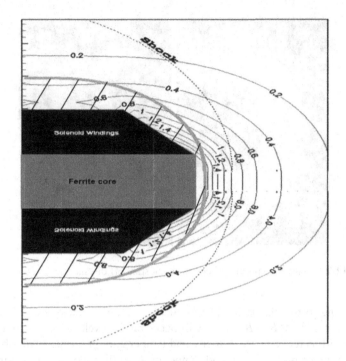

FIGURE 12.1   Magnetic field distribution of solenoid coil.

an insulated container. The cryostat requires the wire temperature to be maintained at least around 40 K. The highest known magnetic flux density generated by a super magnet is around 45 T. These super magnets have not been practical for most aerospace engineering applications (Figure 12.2).

The most vivid demonstration of an applied magnetic force to modify aerodynamics is displayed by Ziemer's indisputable experiment in 1959. The bow shock wave standoff distance is substantially increased by the magnetic pressure. The experiment is conducted with a hemispherical cylinder immersed in an electromagnetic shock tube and the partially ionized is moving at supersonic speed. The electric conductivity of the ionized air is estimated in the range of 28–55 mho/cm. A magnetic field is generated by a magnetic coil located coaxially in the hemispherical cylinder with a maximum magnetic field strength, which is 4.0 T. The activated magnetic field generates a magnetic pressure that combines with the static hydrostatic pressure and maintains a nearly constant value in the stagnation region. The magnetic pressure attains the maximum value at the stagnation point over the magnetic pole and the hydrodynamic pressure will decrease proportionally as the flow approaches the solid surface. As a consequence, a large stream tube cross section is required to ingest the same amount of mass flow in the shock layer to satisfy the continuity condition if the magnetic field ceased. An outwardly displaced standoff distance of the bow shock is recorded by the photograph, but the quantification of the standoff distance is not ascertained due to optical distortion by the emitting glow.

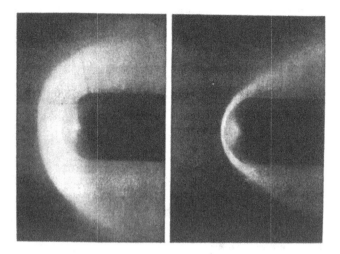

FIGURE 12.2   Magnetic pressure effect on shock standoff distance.

According to the classic MHD formulation, Lorentz acceleration always has a component $\sigma(E + u \times B) \times B$, which will decelerate the velocity of ionized gas [58]. The net result is reducing the velocity and temperature gradients of the shear layer. The flattening of both velocity and temperature profiles reduces both skin friction and heat transfer to the body surface. This unique feature of an applied magnetic field at the stagnation region creates heat transfer mitigation for hypersonic flows. Under some conditions, the eddy current induced by the turbulent fluid particles crossing lines of the magnetic field tends to dampen the turbulent oscillations. Even at a moderate value of Hartmann number $(w/2)B\sqrt{\sigma/\mu_m}$, the wall shear of a turbulent flow will be modified [47]. Therefore, the multi-disciplinary hypersonic flow formulation needs to make provisions to accommodate an externally applied magnetic field.

Several preliminary experiments have provided some concrete data by conducting measurements in an arc-heated facility [28]. The flow medium is the ionized argon generated by a maximum power of 1.4 mW to achieve a stagnation pressure and a specific enthalpy of 37.5 kPa and 2.0 mJ/kg respectively. A freestream velocity of 2,350 m/s is measured by diode laser absorption spectroscopy, and the partially ionized argon has a range of electron number densities from 2.50 to 2.75×10^{11}/cm^3. On quartz covered hemispheric cylinder and a flat frontal surface cylinder, the externally applied magnetic field revealed a reduction of surface temperature as high as 16% and 44% at different applied magnetic flux densities.

Another combination of experimental and computational investigations has also been performed over hemispherical cylindrical models with both a conductive and insulated surface in an arc-heated tunnel [48]. The applied magnetic field generates a weakly ionized supersonic flow with a Hall parameter $(\beta_e = eB/mv \sim 300)$. Their major findings substantiate an outward displacement of the bow shock standoff distance by an externally applied magnetic field [84]. In addition, their data also show a dependence on the model surface conductivity. However, quantification of a reduced heat transfer in the stagnation region is obscured by the electromagnetic wave interference to their spectral measurement techniques. In all, the quantifica-tion for mitigation of the stagnation point heat transfer is extremely difficult, even by the most sophisticated measurement techniques. This formidable challenge is extravagated by interferences created by an applied magnetic field to measuring instrumentations.

Innovative computational simulations for hypersonic blunt bodies with the applied magnetic field have also been conducted. One of the simulations applies the non-ideal MHD equations at a Mach number of 16.34 [26], the other investigation actu-ally solved the inter-disciplinary computational fluid dynamics equations at a Mach number of 5.0 [55]. The externally applied magnetic field duplicates a solenoid coil with a pole aligned with the body axis. However, both formulations do not including the nonequilibrium chemical reaction process but adopt a power law of temperature dependence of the variable electric conductive to determine the electric current den-sity. Nevertheless, their results show indeed the applied magnetic field reduced heat flux in the stagnation region. However, the most critical issue of the scalar relation-ship between the electric current and electric field intensity by the generalized Ohm's law remains unresolved.

In the later 1990s, the scramjet MHD bypass concept for improved propulsive per-formance created significant interest. The MHD bypass engine is proposed as a part of the AJAX vehicle concept [22,29]. The fundamental idea is manipulating electro-magnetic energy to extract and bypass a portion of the energy in form of electricity from the inlet upstream to the combustor, thereby acting as an MHD electric genera-tor. The extracted energy in form of electric current is reintroduced to the scramjet downstream flow path to accelerate the exhaust gas. A schematic of the MHD bypass engine with added components is presented in Figure 12.3. The placement of MHD electric generator is embedded between the isolator and the combustor. The MHD accelerator is combined with the thruster to generate a greater jet exhausting speed.

The operational principle is based upon the generalized Ohm's law in that a mov-ing conductor like plasma through a magnetic field will generate an electric current. The electromagnetic force in the momentum conservations law for plasma motion is approximated by the Lorentz force. The pertaining basic electromagnetic equations are:

$$J = \sigma(E + u \times B) - b_e(J \times B)$$

$$F = J \times B$$

(12.2a)

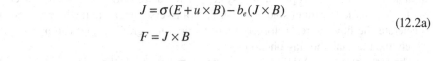
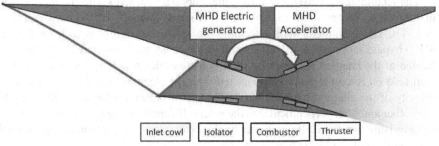

FIGURE 12.3   Schematic of the flow path of MHD bypass scramjet.

Two major parameters for an effective MHD bypass engine are the electrical conductivity σ and the Hall parameter b_e ($b_e = eB/mv = \lambda_e/r_b$, the electron mean free path versus Larmor radius). It is essential to maintain a high electric current density and the ratio of Larmor frequency and characteristic collision frequency of plasma to be minimized.

The electrical power delivered by an MHD generator per unit volume is $P = J \times B$ and the rate of energy that can be extracted from the plasma per unit volume is $u \cdot (J \times B)$. The efficiency of an MHD generator is given as

$$\eta_e = J \cdot E/[u \cdot (J \times B)] \tag{12.2b}$$

The electric generation efficiency of an MHD generator is the ratio between the energy that can be delivered by the plasma from the possible total energy possesses the flowing plasma. Both the velocity of the ionized gas motion and the applied magnetic flux density are important factors together with the electric current density of the generated plasma.

The electromagnetic accelerator is operated on an arrangement often referred to as the Hall current accelerator. The Lorentz force along the axial direction is given by [79]

$$F = \left[\sigma u B_z^2/\left(1+b_e^2\right)\right]\left(Kb_e^2 - 1\right) \tag{12.2c}$$

In Equation (12.2c) B_z is the transverse component of the magnetic field perpendicular to the applied electric field intensity and K is known as the accelerator loading factor, $K = E_y/uB$. The optimized acceleration is found to have a value of the loading factor K to be

$$K_{\text{opt}} = \left(1+\sqrt{1+b_e^2}\right)\Big/b_e^2 \tag{12.2d}$$

The arrangement of this device is feasible in a coaxial or annular geometry which is suitable for a high-speed propulsive system. This idea is attractive because according to the classic Hartman solution [31]; the transverse magnetic field force tends to accelerate the flow more uniformly than the pressure gradient and can avoid completely the thermal choking phenomenon.

The technology for MHD electric generators has reached sufficient engineering maturation in recent years to achieve the efficiency needed for practical application. The Hall parameter will determine the optimal design configuration for the MHD bypass scramjet engine. In general, the Faraday field associated with the bulk motion of the charged particles for the Lorentz acceleration is given by $J \times B$. The Hall field on the other hand is associated with the electromotive force by the drift velocity of the charged particles $u \times B$. These two induced fields are orthogonal to each other and both perpendicular to the externally applied magnetic field. It may not be surprising that the Hall parameter finally determines the appropriate operational configuration of the MHD bypass engine.

The rule of thumb for the upper limit of the axial electric field is no more than 100 V/m, which is imposed on a generator that operates on a high electrical resistance medium. It shall be pointed out that the MHD electric generator actually slows down the incoming airflow thus associated with it an adverse pressure gradient along the inlet ramp compression, the adverse pressure perturbation will propagate upstream through the subsonic portion of the boundary layer as the upstream-feeding to increase the workload for the isolator.

Assessments on the MHD bypass conclude that the thermodynamic analysis reveals some realizable engineering benefits of the new concept; in principle, the innovation also shows the basic scientific feasibility of implementing the MHD bypass for propulsive efficiency improvements [8,24,52]. However, the overall system performance is considered to be extremely sensitive to any non-isentropic losses in the flow path through which the favorable operational characteristic will vanish if these losses become significant.

In particular, Gaitonde utilizes the full three-dimensional interdisciplinary MHD equations coupling to the compressible Navier-Stokes equations [24]. The governing system is the essential low magnetic Reynolds magnetohydrodynamics formulation but without the detailed nonequilibrium chemical reactions process. The electric generator and accelerator are simulated by four pairs of Faraday coarsely segmented electrodes that are mounted on both sides of a constant-area isolator/combustor components. The computational simulation reveals complex aerodynamic-electromagnetic interactions with separated flow regions and vortical structures with an eddy to interfere with the mainstream electric current thus the electromotive force field. The separated flow in the electric generator has a major impact on the electric current pattern downstream. One of the more pronounced effects is the formation of eddy current patterns within the flow path. His conclusion for the performance of MHD scramjet bypass components is consistent with nearly all known assessments. In view of the margin of improvement being extremely narrow by evaluating the relative merits of conventional and the scramjet with MHD energy bypass; an accurate analysis is recommended utilizing physically realistic partially ionized gas composition and turbulent flow modeling.

12.3 CONSERVATION EQUATIONS FOR HIGH-ENTHALPY HYPERSONIC FLOW

The high magnetic Reynolds number environments such as the ionosphere and solar atmosphere are the consequence of the astronautic length scales; $R_m = \sigma \mu u_\infty L$ [47]. In most hypersonic flows at suborbital fights, the degree of ionization in a shock layer is in the order of magnitude around 10^{-5}, or the number density of the ionized components is $<10^{13}/cm^3$, thus the electric conductivity of the ionized air seldom exceeds 100 mho/m. The high-enthalpy hypersonic flows meet the low magnetic Reynolds number criterion as indicated by Table 12.1. The electrostatic force has appeared essentially as a small perturbation in a globally neutral air plasma dynamic. Any strong aerodynamic and aerodynamics-electromagnetic interactions must be derived from an externally applied magnetic field.

TABLE 12.1

Typical electric conductivity, magnetic Reynolds number in applications

Environment	σ (mhos/m)	R_m	S_m
High enthalpy hypersonic flows	10^2	10^2	10^5
Glow discharge	10^2	10^3	10^1
MHD generator	10	10^3	10^1
Arc heater	5×10^3	6×10^2	5×10^2
MHD thruster	5×10^3	3	10
Thermonuclear reaction	10^5	10^2	10^6
Ionosphere	10	10^3	10^3
Solar atmosphere	10^3	10^8	10^8

For nonequilibrium hypersonic flows in the presence of a given externally applied magnetic field and the pertaining characteristic time scales are within the microwave spectrum. These conditions focus the study of electromagnetic-aerodynamic interaction onto aerodynamics rather than plasma wave dynamics. Under this low magnetic Reynolds number condition and by restricting the temporal scale of the dynamics to within the microwave spectrum, the induced magnetic flux density $\partial B/\partial t$ is negligible in comparison with the externally applied field. Faraday's induction equation now can be decoupled from the governing equations system. The approximation includes Ampere's circuit law by neglecting the displacement current. The constitutive relation between magnetic flux density and magnetic field intensity is also retained, and Gauss' law is invoked in equation derivations but not explicitly imposed. However, the conductive electric current density is determined directly from the detailed charged particles' composition and their dynamic motion. In short, the difference between the full MHD equation and the multi-disciple hypersonic governing equation lies in the incorporations of species composition, transport properties for the heterogeneous gas mixture, and radiative energy transfer.

The complete conservation equations of the multi-species reacting gas medium for the high-enthalpy hypersonic flows can now be written. The species conservation equation base on the law of mass action is

$$\partial \rho \alpha_i/\partial t + \nabla \cdot \left[\rho \alpha_i \left(u + u_i \right) \right] = \partial w_i/\partial t \qquad (12.3a)$$

In Equation (12.3a), the symbol α_i is the mass fraction of species i and u is the globally averaged mass velocity of the gas mixture. The rate of species production or depletion for the nonequilibrium chemical reactions is denoted by $\partial w_i/\partial t$. For the N number species system, only $N - 1$ numbers of the species conservation equations are needed. Any one of the N number equations may be replaced by the global continuity equation of the mixture. In practice, all species conservative equations are computed by the nonequilibrium chemically reacting forward and backward formulation, Equation (8.2e). The global continuity equation can be obtained by summing over all species conservation equations by noting that the sum of all species diffusion velocity must be vanished by definition; $\Sigma\rho\alpha_i u_i \equiv 0$ and $\Sigma \partial w_i/\partial t = 0$. The global continuity equation acquires the following form:

$$\partial\rho/\partial t + \nabla \cdot \rho u = 0 \qquad (12.3b)$$

The conservation momentum equation is the result of integrating Ampere's electric circuit law but decoupling Faraday's induction law. The variables of the conservation momentum equation are vectors in multiple-dimensional spaces, and the momentum equations are the only vector equation of the system. The electrostatic force and the Lorentz acceleration are appending to the momentum conservation equation as the source term.

$$\partial\rho u/\partial t + \nabla \cdot [\rho uu + p - \tau] = J \times B + \rho_e E \qquad (12.3c)$$

The role of an externally applied magnetic field can be singled out by expressing the Lorentz acceleration in terms of the magnetic flux and by invoking Gauss's law for magnetic field and by vector identity [67].

$$J \times B = \left[\nabla \cdot BB - \nabla(B \cdot B)/2\right]/\mu_m \qquad (12.3d)$$

The conservation of momentum equation after substituting the identity yields the following form for high-enthalpy hypersonic flows with an externally applied magnetic field B.

$$\partial\rho u/\partial t + \nabla \cdot \left[\rho uu - BB/\mu_m + \left(p + B \cdot B/2\mu_m\right) - \tau\right] = \rho_e E \qquad (12.3e)$$

Even though the induced magnetic flux density is completely neglected, the provision of an externally applied magnetic flux density is included in Equation (12.3e). An important remark must be made here. The externally applied electromagnetic field produces the magnetic stress matrix BB and the magnetic pressure $B \cdot B/2\mu_m$ to the conservation momentum equation. The magnetic stress matrix is the degenerate Maxwell stress tensor without the contribution from the electric field intensity and singles out the normal stress components $B \cdot B/2\mu_m$ as the magnetic pressure. The added pressure is the result of the Lorentz acceleration in the conservation momentum equation. The magnetic stress matrix is a second-rank tensor similar to the dyadic of the velocity uu. On the Cartesian coordinates, the tensors in matrix forms are

$$uu = \begin{bmatrix} uu & uv & uw \\ vu & vv & vw \\ wu & wv & ww \end{bmatrix} \quad \text{and} \quad BB = \begin{bmatrix} B_x B_x & B_x B_y & B_x B_z \\ B_y B_x & B_y B_y & B_y B_z \\ B_z B_x & B_z B_y & B_z B_z \end{bmatrix} \qquad (12.3f)$$

The consistent conservation energy equation of the multi-disciplinary hypersonic flow assumes that the electrostatic force according to Coulomb's law is negligible in comparison with the Lorentz acceleration in globally neutral plasma. The work done by the Lorentz force to the thermodynamic system is also negligible due to the fact that the electric current is nearly parallel to the mass-average velocity of the flow field; $[\rho_e E + (J \times B)] \cdot (u + u_i) \approx 0$. The only contribution of the electromagnetic effect is Joule heating.

$$\partial \rho e/\partial t + \nabla \cdot \left[\rho e u - \kappa \nabla T + \sum \rho_i u_i h_i + q_{rad} + u \cdot p + u \cdot \tau \right] + Q_{vt} - Q_{et} = E \cdot J \qquad (12.3g)$$

In Equation (12.3g), the convective heat transfer $\Sigma \rho_i u_i h_i$ is represented by the summation of each chemically reacting species carrying by the species diffusion velocity including the standard heat of formation, ∇h_i^o. The energy transferred from quantum transitions by vibration and electron modes is characterized by the Q_{vt} and Q_{et} respectively. The modeling of the quantum transition has been discussed in Chapter 6, thus will not be repeated here.

The explicit dependence of an externally applied magnetic field can be highlighted. The Joule heating on the right-hand side (RHS) of Equation (12.4a) is rearranged by substituting the electric current by the Ampere's circuit law, $J = \nabla \times H$ and the electric field intensity by the generalized Ohm's law, $E = \nabla \times H/\sigma - u \times B$ to appear as [67].

$$E \cdot J = \nabla \times (B/\mu)^2 /\sigma - \nabla \cdot \left[uB^2 - (B \cdot u)B \right]/\mu_m - \left(1/2\mu_m \right) \partial (B^2)/\partial t$$
$$+ (B/\mu_m) \cdot \left[(\nabla \cdot \nabla B)/\mu_m \sigma \right] \qquad (12.4a)$$

Substituting Equation (12.4a) into Equation (12.3g), the conservation energy equation becomes [67].

$$\partial \left(\rho e + B^2/2\mu \right)/\partial t + \nabla \cdot \left[\left(\rho e + p + B^2/2\mu_m \right) u + u \cdot \tau - k\nabla T + \Sigma \rho_i u_i h_i + q_{rad} \right.$$
$$\left. - (B \cdot u)B/\mu_m \right] + Q_{vt} + Q_{et} = \nabla \times \left[(B/\mu_m)^2 /\sigma + (B/\mu_m) \right] \cdot \left[(\nabla \cdot \nabla B)/\mu_m \sigma \right] \qquad (12.4b)$$

The momentum and energy conservation equations (12.3e) and (12.4b) are derived explicitly for the multi-discipline hypersonic flow with an externally applied magnetic field.

The complex governing equations for high-enthalpy hypersonic flows in the continuum regime consist of the compressible Navier-Stokes equations, the reduced Maxwell equations of electromagnetics, and the nonequilibrium chemical reaction from quantum chemical-physics. The transport properties of the flow medium are also determined from the gas kinetic theory by collision integrals. However, from experimental evidences and order-of-the magnitude analyses, the electromagnetic effects appear mostly as a perturbation to the inertia of the flowfield. In most practical hypersonic flow environments, the magnetic Reynolds number $R_m = ul / (\sigma\mu)^{-1}$, usually has a value much less than unity. In this framework, the electrostatic force, the Lorentz acceleration, and the Joule heating appear as the perturbing source terms to the multi-discipline equations system. Therefore, the essential physics of the high-enthalpy hypersonic flow can be effectively described by the governing equation system which includes the energy conservation equations for the internal degree of freedom of vibration and electron, and is summarized as

$$\partial \rho \alpha_i / \partial t + \nabla \cdot \left[\rho \alpha_i (u + u_i) \right] = \partial w_i / \partial t \tag{12.5a}$$

$$\partial \rho u / \partial t + \nabla \cdot [\rho u u + p - \tau] = J \times B + \rho_e E \tag{12.5b}$$

$$\partial \rho e / \partial t + \nabla \cdot \left[\rho e u - \kappa \nabla T + \sum \rho_i u_i h_i + q_{\mathrm{rad}} + u \cdot p + u \cdot \tau \right] + Q_{vt} - Q_{et} = E \cdot J \tag{12.5c}$$

$$\partial \rho_e e_{iv} / \partial t + \nabla \cdot \left[\rho_i (u + u_i) (e_{iv} + q_{iv}) \right] = Q_{v,\Sigma} \tag{12.5d}$$

$$\partial \rho_i e_e / \partial t + \nabla \cdot \left[\rho_i (u + u_i) e_e + (u \cdot p_e I + q_e) \right] = E \cdot J + Q_{e,\Sigma} \tag{12.5e}$$

The definition of internal energy now needs to include the electronic kinetic energy arising from the charge particle dynamics, and the constant volume-specific heat must also include all internal modes of the air mixture $c_{vi} = c_{v,t} + c_{v,r} + c_{v,v} + c_{v,e}$.

$$\rho e = \sum \rho_i \left(\int c_{vi} dT + u \cdot u / 2 \right) + \sum_{i \neq e} \rho_i \nabla h_i^o + \rho_e u_e^2 / 2 \tag{12.5f}$$

where ∇h_i^o is the standard heat of formation for all reacting species.

In Equations (12.5b)–(12.5e), the coupling between aerodynamics and electro-magnetic force/energy appears as the electrostatic force $\rho_e E$, Lorentz acceleration, $J \times B$ and Joule heating $E \cdot J$ [63,69]. The vector field of E and B in general is the sum of the externally applied and the induced electrical and magnetic field intensities. In the absence of an externally applied magnetic field, the charged particles are accelerated by the electrostatic force alone due to charge separation in free space. The electrostatic force $\rho_e E$ thus vanishes except in the plasma sheaths over the cathode and anode layers of electrodes. On the other hand, the electric current always imparts thermal energy to the external stream through Joule heating mostly in the cathode layer in addition to the electrode heating. It is the basic mechanism of plasma actuator by thermal effect [67].

The electric current density J is directly evaluated from the number density of charged species in place of the generalized Ohm's law. It is one of the most distinguished features of the multi-discipline governing equations over the MHD formulation for including all species of a weakly ionized flow medium. The current density J is calculated directly by the definition of the conductive electric current density by the charged number density and the charged particle velocity.

$$J = \Sigma \rho_i u_i e \tag{12.6a}$$

The next level approximation is determined by the drift-diffusion theory for the partially ionized gas for plasma actuators. The process is based on the multi-component plasma model by separating the ionized components into positively and negatively charged species [76].

$$J = e\left[-d_+\nabla n_+ + d_e\nabla n_e + d_-\nabla n_- + \left(n_+\mu_+ + n_e\mu_e + n_-\mu_-\right)E\right]$$
$$+ \left[n_+\mu_+\left(u_+ \times B\right) + n_e\mu_e\left(u_e \times B\right) + n_-\mu_-\left(u_- \times B\right)\right]$$

(12.6b)

In Equation (12.6b), the notations n_+, n_e, and n_- designate the species number densities of the positive, negative ion, and electron. The transport property consists of the drift velocity and the diffusion of electrons and ions. The electron and ion mobilities (μ_e, μ_+, μ_+) which are traditionally evaluated as functions of the reduced electric field intensity E/P, and the diffusion coefficients (d_e, d_+, d_-) of the charged particles can be calculated by the Einstein formula [56]. The diffusion and drift velocity for electrons and both positively and negatively charged ions are used to describe the individual random and forced motions of the charged particles.

The governing equation system, Equations (12.5a)–(12.5e) is derived from the time-dependent compressible Navier-Stokes equations. Therefore the nonlinear, second-order partial differential equations system still retains the characteristic of the incompletely parabolic differential system [67,75]. Therefore, the required initial values and boundary conditions are nearly identical for solving the Navier-Stoke equation, except for the electromagnetic variables which have been discussed in Section 12.1.

12.4 NUMERICAL ALGORITHM

From the previous discussions, the formulation is a multi-disciplinary endeavor that involves compressible fluid dynamics, electromagnetics, and nonequilibrium chemical kinetics, and quantum mechanics. The interacting computational simulations among four independent scientific disciplines form an iterative loop and must be solved by an iterative numerical procedure. The initial value of the solutions starts from an equilibrium condition and by solving Equations (12.5a)–(12.5c); essentially the global conservation equations are obtained from the macroscopic flow field properties. Then the nonequilibrium chemical kinetics determines all chemical species and the chemical reacting thermodynamic conditions of the gas mixture. The chemical reactions in general occur at a much short time scale than that of the gas dynamics, and the eigenvalues structure of the equations also have a wide bandwidth and thus require additional effort to maintain computational stability. An iterative process is frequently required to meet stability requirements.

From these updates species composition and thermodynamic state proceed to solve Equation (12.5d) and (12.5e) to obtain the final energy states for all internal degrees of freedom. The detailed data then feed into the conservative laws to satisfy the global behavior of the complex system of equations including the chemical reacting and electromagnetic effects. The iterative process will continue until the final solution meets a preset numerical tolerance between solutions of consecutive iterations. The interaction among the four scientific disciplines is displayed graphically in Figure 12.4. It is important to point out that in the iterative cycle; the approximation of the Maxwellian distribution connecting microscopic and macroscopic properties has been used because the temperature is a well-known and accepted data based for chemical reaction rate constants and optical properties during the quantum transitions.

A major pacing item for super/hypersonic flow computational simulation is evaluating discontinuous solutions generated by shock waves and slipstreams from shock interceptions. In 1959, Godunov demonstrates a multi-dimensional flow field that contains shock waves and contact surfaces can still be analyzed as the Riemann problem [27]. The underpinning principle is the monotonicity preserving property of the hyperbolic difference equation; namely, temporal increment/decrement of the dependent variable is monotonic, in fact, has been revealed by Lax [42]. Based on this property, the total variation diminishing (TVD) scheme has originated and spans a huge amount of research resulting in a variety of flux limiters for analyzing piecewise discontinuous solutions for computational fluid dynamics [30]. For shock capturing, the interdisciplinary governing equations, Equations (12.5a)–(12.5e) are expressed in vector flux form and split into the inviscid and viscous terms. The former is the Euler equation, which is the nonlinear hyperbolic type. The Cartesian coordinate system can be given as

$$\partial U/\partial t + \partial F(U)/\partial x + \partial G(U)/\partial y + \partial H(U)/\partial z = R$$

$$\partial U/\partial t + \partial[F_i(U) + F_v(U)]/\partial x + \partial[G_i(U) + G_v(U)]/\partial y + \partial[H_i(U) + H_v(U)]/\partial z = R$$

$$(12.7a)$$

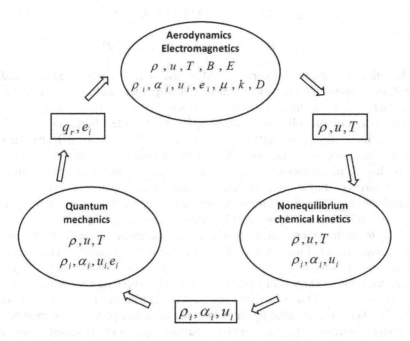

FIGURE 12.4   Coupling and iterating sequence for the high-enthalpy aerodynamics.

From the physics viewpoint, the treatment of shock jump by flux splitting can be easily understood through the concept of the zone of dependence for supersonic flows. According to the distinctive characteristics, the directional wave propagation is constructed according to the phase velocity from the permissible database. It has been shown as a systematic relationship between the real eigenvalue and eigenvector for the split flux formulation. In the process, it also demonstrated the Euler equations together with the equation of state for gas, possessing the homogeneous function of degree one property [73];

$$F_i = \left(\partial F_i / \partial U \right) U = \left[A_x \right] U; \, G_i = \left(\partial G_i / \partial U \right) U = \left[A_y \right] U; \, H_i = \left(\partial H_i / \partial U \right) U = \left[A_z \right] U$$

$$(12.7b)$$

In Equation (12.7b), the partial derivatives of the inviscid flux vector; $\partial F_i / \partial U, \partial G_i / \partial U,$ and $\partial H_i / \partial U$ are the Jacobian matrices of the flux vector or the coefficient matrices. The governing equations in split flux vector form become

$$\partial U / \partial t + \partial \left[\left(F_i^+ + F_i^- \right) + F_v \right] / \partial x + \partial \left[\left(G_i^+ + G_i^- \right) + G_v \right] / \partial y + \partial \left[\left(H_i^+ + H_i^- \right) + H_v \right] / \partial z = R$$

$$(12.7c)$$

The split flux vectors containing all inviscid terms are formed according to the signs of the eigenvalue λ of the coefficient matrices, A_x, A_y, and A_z;

$$F_i = F_i^+ + F_i^- = \left(S_x \lambda_x^+ S_x^{-1} + S_x \lambda_x^- S_x^{-1} \right) U$$

$$G_i = G_i^+ + G_i^- = \left(S_y \lambda_y^+ S_y^{-1} + S_y \lambda_y^- S_y^{-1} \right) U \qquad (12.7d)$$

$$H_i = H_i^+ + H_i^- = \left(S_z \lambda_z^+ S_z^{-1} + S_z \lambda_z^- S_z^{-1} \right) U$$

In Equations (12.7d), the similarity S_i and its inverse matrices S_i^{-1} are designated for the diagonalization of coefficient matrices. The positive and negative eigenvalues associated with the right- and left-running wave are denoted as λ_x^+ and λ_x^-. The solving procedure for the split equation is by applying one-side differencing approximation to achieve the approximate Riemann problem. The basic issue is that the split inviscid flux components are not differentiable at singular sonic points. This behavior is also the peculiarity of the approximate governing equations. The continuous viscous terms of the conservation equations are solved simultaneously by a spatially central scheme. An incisive summary for using the approximated Riemann approximations can be found in the work of Roe [59].

In spite of the rational treatments of the discontinuous numerical solution, the undesirable artifact of numerical oscillation or the Gibbs phenomenon is always presented around a singular point. A series of excellent algorithms for maintaining computational stability and yielding sharp definition of a piecewise continuous numerical solution are the ENO and WENO (weighted ENO) schemes [3,71]. The fifth-order WENO scheme is supported by an overall stencil of five points; the smoothness of a solution is measured by the sum of normalized squares of the scaled L_2 norm for derivatives from the lower-order polynomials. The conditioned information is incorporated into the weights definition to improve the convergence at the critical points. It is revealed that the enhancement of the fifth-order scheme is derived from a large weight assigned to the discontinuous stencils, but not from their superior order of convergence at critical points.

The widely used basic solving scheme for the multi-discipline governing equation is derived from the Beam-Warming algorithm [5]. The gist of this solving scheme is the factored ADI (alternating direction implicit) but solved in a time-consecutive delta form [2]. To illustrate the fundamental character of this algorithm, only the inviscid terms of Equation (12.7c) are included:

$$\left[I + \frac{\theta \Delta t}{1+\xi} \left(\nabla_x A^+ + \nabla_x A^- + \nabla_y B^+ + \nabla_y B^- + \nabla_z C^+ + \nabla_z C^- \right)^n \right] \Delta U^n$$

$$= -\left(\frac{\Delta t}{1+\xi} \right) \left(\delta_x^b F^+ + \delta_x^f F^- + \delta_y^f G^+ + \delta_y^b G^- + \delta_z^f H^+ + \delta_z^b H^- \right)^n + \left(\frac{\xi}{1+\xi} \right) \Delta U^{n-1}$$

$$= R \tag{12.8a}$$

In Equation (12.8a), the windward differencing operator is designated by the symbol ∇ and is dictated by the sign of its associated eigenvalue. The parameters, θ and ξ determine the time-difference approximations, for the trapezoidal formula $\theta = 1/2, \xi = 0$.

The left-hand side (LHS) of the Equation (12.8a) can be factored into the product of two operators as

$$\left[I + \left(\frac{\theta \Delta t}{1+\xi} \right) \left(\nabla_x A^+ + \nabla_y B^+ + \nabla_z C^+ \right) \right] \left[I + \left(\frac{\theta \Delta t}{1+\xi} \right) \left(\nabla_x A^- + \nabla_y B^- + \nabla_z C^- \right) \right] \Delta U_{i,j}^n = R_{i,j}$$

$$\tag{12.8b}$$

In applications, this numerical scheme is always implemented by the temporal sequence.

$$\left[I + \left(\frac{\theta \Delta t}{1+\xi} \right) \left(\nabla_x A^+ + \nabla_y B^+ + \nabla_z B^+ \right) \right] \Delta U^* = R_{i,j}$$

$$\left[I + \left(\frac{\theta \Delta t}{1+\xi} \right) \left(\nabla_x A^- + \nabla_y B^- + \nabla_z C^- \right) \right] \Delta U^n = \Delta U^* \tag{12.8c}$$

$$U^{n+1} = U^n + \Delta U^*$$

When the spatial discretization is based on a semi-discrete finite-volume scheme, the upwind-biasing approximation is similarly applied to the convective and pressure terms. Central differencing is used for the shear stress and heat transfer terms. In the flux-difference splitting procedure, the flux vectors at the control surface are written as an exact solution to the approximate Riemann problem.

$$\nabla F_i = \frac{1}{2} \Big[F(U_L) + F(U_R) - |A|(U_R - U_L)_{i+1/2} \Big]$$

$$- \frac{1}{2} \Big[F(U_L) + F(U_R) - |A|(U_R - U_L)_{i+1/2} \Big] \tag{12.8d}$$

where U_L and U_R are the interpolated values of dependent variables, ρ, ρ_u, ρ_v, ρ_w, and ρ_e on the interface of the control volume. A is the Jacobian matrix of the inviscid or convective term $A = \partial F / \partial U$.

Time advancement is solving implicitly for the complete flow field. A slope limiter is also used to control the discontinuous pressure jumps at the shock front. Specifically, the minmod limiter is adopted for most computations [2]:

$$(U_L)_{i+1/2} = U_i + 1/4 \left[(1-k)\nabla_- + (1+k)\nabla_+ \right]_i$$

$$(U_R)_{i+1/2} = U_{i+1} - 1/4 \left[(1-k)\nabla_+ + (1+k)\nabla_- \right]_{i+1} \tag{12.8e}$$

The one-sided difference operators are $\nabla_+ = U_{i+1} - U_i$ and $\nabla_- = U_i - U_{i-1}$, and the minmod operators are defined as [83]

$$\nabla_- = \min \bmod \left[\nabla_-, (3-k)/(1-k)\nabla_+ \right]$$

$$\nabla_+ = \min \bmod \left[\nabla_+, (3-k)/(1-k)\nabla_- \right] \tag{12.8f}$$

For most investigations in hypersonic flow, temporal accuracy is essential. Therefore a greater stability bound enhancing by an implicit scheme for a larger time step is not necessary, instead, a successful iterative relaxation procedure is selected by the residual diminishing delta formulation in consecutive temporal evolution [33,74]. In essence, only the convective terms are retained on the LHS of the discretized equation and the complete governing equations in time are placed on the RHS by the approximation, The delta formulation of the discretized equation is used as shown in the following to linearize Equations (12.5a)–(12.5e):

$$\frac{\partial(\Delta U)}{\partial t} + \frac{\partial}{\partial x}(\Delta F_x) + \frac{\partial}{\partial y}(\Delta F_y) + \frac{\partial}{\partial z}(\Delta F_z) = \frac{\partial}{\partial x}(\Delta G_x) + \frac{\partial}{\partial y}(\Delta G_y) + \frac{\partial}{\partial z}(\Delta G_z) + R^n$$

where $R^n = \dfrac{\partial(\Delta U)^n}{\partial t} \dfrac{\partial}{\partial x}(F_x^n) - \dfrac{\partial}{\partial y}(F_y^n) - \dfrac{\partial}{\partial z}(F_z^n) + \dfrac{\partial}{\partial x}(G_x^n) + \dfrac{\partial}{\partial y}(G_y^n) + \dfrac{\partial}{\partial z}(G_z^n) + S^n$.

$$\tag{12.8g}$$

where the split vectors $\Delta U = U^{n+1} - U^n$, $\Delta F_i = F^{n+1} - F^n$, and $\Delta G_i = G^{n+1} - G^n$ are identical to Equation (12.8a).

Consequently, it achieves the optimized numerical procedure, the LHS of the discretized equation is required only to maintain the computational stability, and the final solution is determined by the error tolerance of the iterative procedure and the algorithm for solving the RHS of the equation. It should be noted that the RHS of Equation (12.8g) is explicit, and it can be evaluated using an intermediate solution with any high-order scheme such as the compact differencing scheme provided it is numerically stable. In most studies, a central differencing scheme is adopted for the diffusive and source terms. A quadratic upwind differencing scheme is adopted for the convective terms.

In the above formulation, all diagonal terms of the discretization are moved to the LHS of the finite-difference approximation to maintain diagonal dominance and enhance computational stability. For the nonequilibrium chemical reaction with the catalytic surface condition, the steep gradient always occurs near the surface, thus the multigrid method offers the best solving option [9]. In application, only a two-consecutive-different-grids work cycle is sufficient for the purpose. The error of the rapidly varying Fourier component is eliminated by the finer mesh system. In order to improve temporal accuracy, a third-order Runge-Kutta scheme is applied to resolve the rapid variation of the ionizations in the kHz range [33]. By this formulation, the axiom of the optimal numerical simulation is fully realized in that the physical fidelity is controlled by the accurate RHS of the approximation. The LHS of the approximation is pure numerical and is constructed to maintain computational stability.

In Figure 12.5, a comparative study of the implicit solving schemes for solving the charged species conservation law is depicted on a highly stretched grid system. The remarkable convergent acceleration from the conventional tridiagonal dominant scheme to SIP and finally the combination of SIP and multigrid procedure is illustrated. The acceleration in residual reduction by the SIP

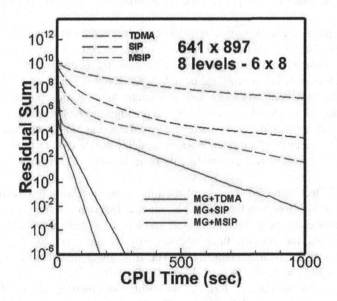

FIGURE 12.5 Iterative convergent acceleration for solving schemes.

algorithm alone exceeds four orders of magnitude. The rate of convergence is further vastly improved by the combined application of the SIP and multigrid techniques. Through these superior computational attributes, the grid-independent solutions are truly realizable.

From the lessons learned, the desired feature of a numerical scheme may be better selected from the optimization in the Fourier space rather than by focusing on the lowest possible truncation error. Along this line of reasoning, a compact-difference scheme becomes a viable method to achieve high resolution. The basic algorithm is an implicit procedure for evaluating derivatives with a small stencil dimension and maintains a lower level of dispersive and dissipative errors than the conventional numerical schemes. The basic formulation is derived from Hermite's generalization of the Taylor series [16]. The compact differencing formulations for evaluating the first-order and second-order derivatives have been given by Lele [44] as

$$\beta U'_{i-2} + \alpha U'_{i-1} + U'_i + \alpha U'_{i+1} + \beta U'_{i+2}$$

$$= c(U_{i+3} - U_{i-3})/6\Delta x + b(U_{i+2} - U_{i-2})/4\Delta x + a(U_{i+1} - U_{i-1})/2\Delta x$$

$$\beta U''_{i-2} + \alpha U''_{i-1} + U''_i + \alpha U''_{i+1} + \beta U''_{i+2}$$

$$= c(U_{i+3} - 2U_i + U_{i-3})/9\Delta x^2 + b(U_{i+2} - 2U_i + U_{i-2})/4\Delta x^2 + a(U_{i+1} - U_i + U_{i-1})/\Delta x^2$$

$$(12.9)$$

In practical applications, only the tridiagonal form is utilized and achieved by setting β and c to zero to get the formulas relating to the pertaining parameters $a = 2/3(\alpha + 2); b = 1/3(4\alpha - 1); c = 0$. The fourth-order Pade formula is obtained by choosing the value of $\alpha = 1/4$, and the sixth-order derivatives are recoverable for $\alpha = 1/3$.

The formulas are a three-point spatially central scheme and require attention on boundaries. Since the scheme is inapplicable on the immediately adjacent grid point next to a boundary, a transitional operator between the boundary and the interior domain is required. The transitional boundary scheme is not only required to transmit data from the boundary to the interior domain but also must preserve stability and accuracy of the global solution. Although the high-resolution scheme is stable in the classic sense the transition operator is one of the sources that contribute to spurious high-frequency oscillations known as time-delay instability. The time-instability is incurred by positive real eigenvalue components which dominate the numerical result.

A very effective remedy to control the time-delay instability has been demonstrated by using a low-pass filter [25]. The spectral function is a symmetric numerical filter that contains no imaginary part and has a low-pass amplitude response. In other words, the low-pass filter modifies only the amplitude but not the phase relation among all Fourier components. The tridiagonal spatial filter is given as
$$\beta U_{i-1} + U_i + \beta U_{i+1} = \Sigma a_n (u_{i+n} + u_{i-n})/2 \quad n = 0,1,2,\ldots,N.$$

The compact differencing formulation ensures the high resolution of optimal numerical simulation for solving the interdisciplinary hypersonic flows.

12.5 EARTH REENTRY SIMULATIONS

The trajectories of interplanetary space exploration are commonly categorized as the suborbital, low earth orbit (LEO, attitude < 2,000 km), the Sun-synchronous orbits (SSO) with the Perigee and Apogee of 6,000 and 8,000 km respectively, and the elliptic Geosynchronous transfer orbit (GTO) has the perigee of 2,000 km and the Apogee of 35,786 km. Hypersonic applications to the earth reentry of space flights are mostly limited to suborbital flight and low earth orbits. The general earth reentry reference timeline is clocked from an altitude of 120 km. It may be interesting to know that the altitude of 100 km is recognized as the Karman line. The initial tracked reentry velocity is generally around 10 km/s and at the altitude near 70 km above the earth to yield a Mach number of about 25. The available initial tracked points on the earth reentry trajectories of the most studied hypersonic flows are listed in Table 12.2.

All the kinetic energy of a reentry vehicle has to be converted into thermal energy when it landed. The amount of kinetic energy is enormous depending on the originated orbits or trajectories of the vehicles, which have values from 6.4×10^7 to 2.3×10^9 J/kg. A part of the energy is deposited on or absorbed by the ablating surface material of the reentry vehicle, and some of it transfers to the passing air stream, another portion radiates into free space. The conductive, convective and radiative heat transfer rates are strongly dependent on the surface material of the reentry vehicle whether it's a non-catalytic or a full-catalytic material. Regardless, thermal protecting materials are necessary to maintain the structural integrity of space vehicles during the re-entry phase. The ablative surface for thermal protection is widely used for its high phase-changing latent heat to absorb the converted thermal energy from the decelerated flow and its low thermal conductivity to minimize the conductive heat transfer to the substrate [14]. The radiative heat transfer strongly depends on the temperature and composition of ionized gas in the shock layer and becomes an increasingly important mechanism for thermal protection for interplanetary exploration.

Applications of the multi-discipline hypersonic governing equations to the Space Shuttle accelerated after the Challenger (STS51-L) accident in 1986. A series of computational analyses were initiated to address discrepancies in the design database

TABLE 12.2
Initial recorded environmental data of reentry vehicles

Vehicle	Altitude (km)	Entry Velocity (km/s)	Ambient Temp (K)	Mach Number
Apollo module	69.80	11.30	NA	NA
FIRE-II probe	76.42	11.36	195.0	40.6
MIRKA	66.18	7.36	228.6	24.3
ORION (CEV)	53.00	9.50	NA	NA
RAMC-II probe	76.42	11.36	195.0	25.9
Stardust capsule	81.02	12.38	217.6	41.9
Space shuttle	120.0	NA	NA	25.0

in regard to both Reynolds number scaling, exhaust-plume-Orbiter interaction, and a complete and more detailed description of the launch vehicle flowfield [11,66]. A more stringent accuracy requirement was also imposed to predict the Orbiter wing root shear to within 5% of the maximum structural capability of the wing. To answer this engineering accuracy requirement for a multiple-component aerospace system, innovation in grid generation and data flow between different computational domains was essential. The earlier numerical solutions for the transonic flow over the vehicle were obtained using the chimera domain decomposition technique, and then logically transitioned into the unstructured grid method to change how the complex and multi-component configurations were simulated [72].

Guiding by the classic theory of reducing the conductive heat transfer at stagnation point transfer, the frontal vehicle face is configured with the maximum radius of curvature. Nearly all reentry capsules shape like an inverted spherical cone. The Stardust sample return capsule consists of a 60° one-half angle spherical cone with a nose radius of 0.229 m as the forebody. The afterbody is a truncated 30° cone with a base radius of 0.406 m. The corner radius at the juncture of the forebody and afterbody is merely 0.02 m [19].

The flow structure around the Stardust capsule is presented by a streamline trace in Figure 12.6. The extremely thin shock layer over the forebody of the capsule stands out. The rapid expansions are also noted at the junction of the forebody and afterbody, as well as, at the base region of the capsule. The vortical formation in the near wake region transits from a counter-flowing, twin recirculation zones to a biased structure as the capsule travels along the trajectory. These near-wake vortices

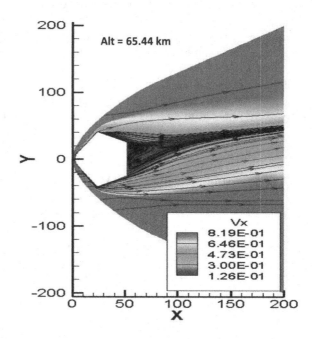

FIGURE 12.6 Streamline traces at three reentry altitudes.

enhance the mixing for the recombining chemical reactions in the base region to affect the radiative heat transfer locally [77].

The assessment of numerical resolution for a complex physical phenomenon is extremely challenging. The computational accuracy is controlled by establishing a uniformly enforced numerical convergent criterion for all dependent variables to the relative error of 10^{-5} and by a series of studies on grid topology. The comparison of the peaking heat transfer rate with results in literature becomes the ultimate assessment of computational simulation. For the reason of clarity and lacking detailed flight data, the interactions of the ablator charring and pyrolysis gas interaction with the external flow have not been taken into consideration in this particular simulation. Therefore the surface blowing contributes to ablation cooling is unaccountable.

The surface temperature and shear flow patterns over the frontal surface and the after body are presented on the LHS of Figure 12.7, the effect of an eight degrees angle of attack is clearly displayed. For the designed capsule configuration, the diffusion heat transfer in the stagnation regime is dominated over the conductive and radiative heat transfers. The heat transfer rate drops drastically over the jointure of the forebody and afterbody into the near wake, and the compression by aligning with the far wake is clearly displayed as anticipated.

The peak heating condition of the forebody occurs at the reentry velocity of 11.137 km/s (altitude of 59.77 km) and can be summarized as the air density which is 2.34×10^{-4} kg/m^3 and the ambient temperature which is 238.5 K [1]. The computational simulation shows a distribution of all components' heat flux on the surface in the plane of symmetry of the Stardust probe, including conductive, diffusive, and radiative components. The conductive heat transfer contributes a value of 320 W/cm^2 to the total heat transfer rate at the stagnation point. The combined conductive and

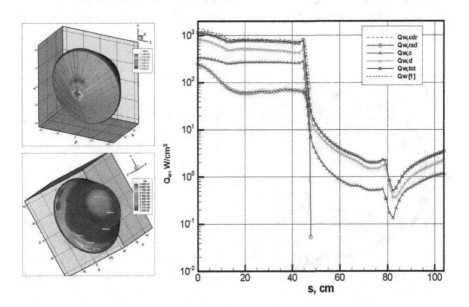

FIGURE 12.7 Heat transfer rate on the Stardust Probe surface.

diffusive heat transfer rates yield a value of 1.19×10^3 W/cm² at the stagnation point of the capsule. The present total heat transfer rate without active ablation is in very good agreement with the non-ablating results by Olynick et al. [50] of 1.2×10^3 W/cm² and agrees equally well with the laminar flow result of Park [51] at 1.189×10^3 W/cm². The radiative heat transfer rate over the forebody however reveals a comparable value of 248 W/cm² with the results by Olynick et al. and Park. In all, the present radiative heat transfer indicates mostly a constant rate over the frontal face of the Stardust capsule and drops sharply at the juncture of the forebody and afterbody of the capsule like that of Olynick et al. At the same instance, the diffusive heating in the wake region begins to rise through energy cascading from vibrational excitation and chemical species recombination as the nonequilibrium flow leaves the computational domain.

The composite presentation of vibration temperatures of molecular nitrogen and oxygen together with heat transfer data for the FIRE II capsule at an altitude of 71.04 km and the reentry velocity is 11.31 km/s is given in Figure 12.8. The only unusual part of the vibrational temperature contour is the heating of nitrogen by the recombination of nitrogen atoms (The upper part of the vibrational temperature contour). The high vibrational temperature of N_2 and O_2 molecules have a specific property of the expanding flows behind any space vehicle at non-equilibrium conditions. It is fully understood that the local heterogeneous kinetics and kinetics of recombination near the leeward surface are questionable. However, such descriptions of thermodynamic variables and assumptions concerning these non-equilibrium regions are commonly used in the literature. Nevertheless, the phenomenon is still significant for radiative and convective heating of the afterbody surface, but it is relatively insignificant in comparison with respect to the front shield of a space vehicle.

Figure 12.9 presents the comparison between most computational with flight test data, and all the computational results are generated by adopting Park's chemical kinetic model [53].

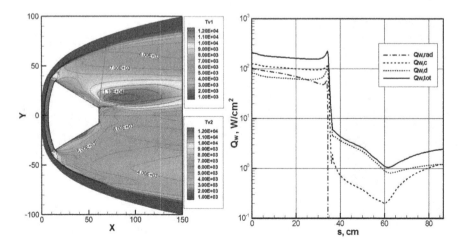

FIGURE 12.8 Vibrational temperature and heat transfer data of the FIRE II capsule.

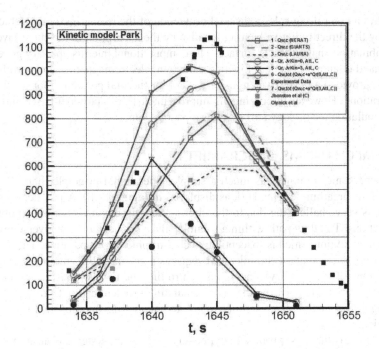

FIGURE 12.9 Comparisons computational results with flight data.

The comparisons are established on the standard databases of FIRE II (designated as the square symbol) and the correlated results [13,37]. The experimental data indicate heat transfer rates of 150 W/cm² at the initial descending stage and a peak heating load of 1,140 W/cm² at $t = 1644.5$ s. The total heat transfer rates reduce to a value <320 W/cm² at the last measured stage, $t = 1,652$ s. The computational results are using the line-by-line approach for spectral properties with data collected by Cauchon at the maximum heat load and follows closely the entire trajectory [78]. The procedure holds independently from the two possible nonequilibrium dissociation models for radiation computations. Equally important, the difference between the total heat transfer rates with/without the atomic spectral lines is 150 W/cm² at the peak heating load. Finally, the predicted heat transfer rates with and without the contribution of the atomic spectral lines bracket the results from the LAURA and GIANTS codes, presented by Olynick et al. [49].

In summary, the critical heat transfer rate in the enveloping shock layer bounded over the earth reentry vehicles has been predicted and verified to be consistent over the temperature around 10,000 K. Under the environment, the air is fully dissociated and partially ionized, and radiation will be produced by quantum transitions from oxygen and nitrogen atoms as well as complex abated compounds. The extremely high heat loading on the earth reentry space vehicle surface is generated in part by the recombination of ionized gas through diffusion from the convective heat transfer process. The converting kinetic energy of the reentry vehicles has effectively dissipated through the heat shield coated with ablating material. Therefore ionized air plays a controlling role in thermal protection of the reentering space vehicle not only by the direct radiative heating but also by the absorption in the shear layer and recombination on the vehicle surface. The computational interdisciplinary governing integrated quantum chemical kinetics, radiation, electromagnetics, and aerodynamics has provided an effective engineering tool for thermal protection for outer space explorations. However, a continuous improvement to the physical-based modeling and simulation technology still remains a pacing item.

12.6 MECHANISMS OF SCRAMJET

The hypersonic scramjet has much a simpler operational principle than any other type of air-breathing engine [17], as displayed in Figure 12.10. A practical scramjet design is very challenging due to the interlocking viscous-inviscid and combustion interactions. The flow path within a scramjet always encounters an adverse pressure gradient condition which is susceptible to catastrophic flow separation. The state-of-the-art effort to create a scramjet still requires a considerable amount of time and effort. One of the primary concerns is controlling the fuel combustion in the combustion chamber. Especially, the very demanding problem arises from the efficient mixing of fuel and oxidizer, ignition delay, and maintaining combustion stability for scramjet.

Scramjet is the most promising hypersonic propulsion system because of its simplicity in construction and is devoid of any moving components in comparison with other systems. The basic concept evolving from interactions between combustion and aerodynamics is known as thermal compression, which was propounded by Ferri in 1959 [20]. The required components of the flow path are the compression inlet, isolator, fuel-injection system, combustor, and exhaust nozzle. The hypersonic flow within the inlet contains complicated shock-boundary-layer interactions, as well as shock wave interactions, which are generated from the leading edge interaction with

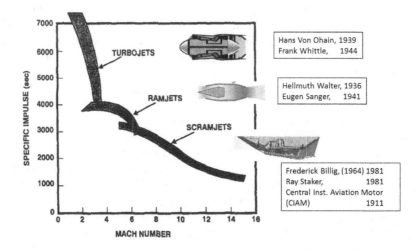

FIGURE 12.10   Operational range and efficiency of air breathing propulsion system.

the incoming flow are determined by the geometry of the compression inlet cowl. In the combustion chamber, the interaction is further intensified by turbulent mixing and by diffusive combustion. A complex pressure field inside the scramjet channel also exists in localized regions of heat release, due to combustion processes, resulting in adverse pressure gradient propagating along the flow path that leads to flow separations and finally the inlet unstart condition. For this reason, the lower bond of the operation envelope of a Scramjet is not set by the thermal and structure limitations but by the onset of shear layer separation in the flow path. The upper bound of the envelope is set by the combustion stability (Figure 12.11).

To demonstrate that the leading-edge induced shock impingement can trigger flow separation; a series of calculations of the hypersonic flow within the cylindrical isolator of a scramjet was carried out for a range of Reynolds numbers. At a Mach number of 5.15, the examined cases have a Reynolds number range from 1.07×10^5 to 2.90×10^5 based on the running length measured from the leading edge. The pressure distribution at the Reynolds number of 1.26×10^5 is fully verified by the experimental data [64]. It is observed in Figure 12.2 that lower Reynolds numbers result in static pressure increases along the axis of the isolator generated by the shock-boundary-layer interaction and the apex of the conical shock moves upstream. At a higher Reynolds number, 1.51×10^5, the oblique shock downstream to the shock apex exits the diffuser without incidence. As the Reynolds number decreases the strengthened conical shock originating from the entrance of the isolator has a steeper conical shock angle and eventually impinges on the sidewall to produce a second shock-boundary-layer interaction. The resultant, stronger, adverse pressure gradient within the constant area

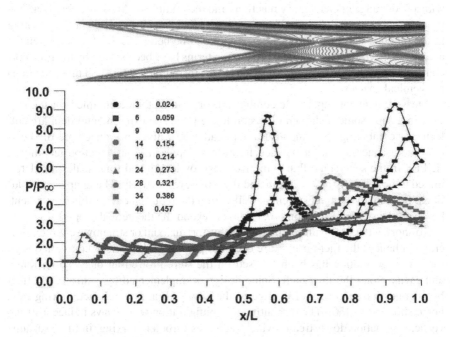

FIGURE 12.11   Transient pressure pulse in scramjet isolator.

duct induces flow separation. Once the flow is separated, the value of the Reynolds number has a relatively small effect on the location of the separated flow.

In order to neutralize an oscillatory back pressure from the combustion chamber that can cause a complete collapse of the internal flow, a special buffer zone, i.e. the isolator between the inlet and combustor, is required. The critical component for a scramjet is the inlet isolator [6]. Consistent with the scramjet configuration, this device typically consists of a constant-area supersonic diffuser serving to contain a complex shock wave system for decelerating the oncoming flow from the inlet to the entrance of the combustor. The major function of the isolator reduces the upstream back-feeding by the rising pressure within the combustion chamber. In gist, the isolator decouples the inlet flow field from the downstream combustion chamber. This decoupling is critical to prevent the inlet from unstart at a moderate flight Mach number. The possibility of unstart diminishes when the scramjet operates at Mach numbers above eight [17].

The heart of a scramjet configuration is the complicated combustor and fuel injector coordination for ignition and combustion stability. The supersonic combustion in a scramjet requires the residence time within the combustor as short as a few milliseconds accompanied by a high mass flow rate. Furthermore, the narrow margin for sustained combustion is crucially dependent upon the issues of mixing reaction rate and the radiative heat transfer. Two fundamental issues in supersonic, turbulent, finite-rate chemical reactions are presently analyzed by physics-based models. The turbulent closure is achieved mostly by a wide class of approximations to the Reynolds averaged Navier-Stokes equations or by the large eddy simulations (LES) [54]. The non-equilibrium chemical reactions are formulated by the law of mass actions with a wide range of elementary reactions and the traditional Arrhenius formula. The species concentration of nonuniform mixing of fuel and air in turbulent flow owning to the random and fluctuating eddies is found to be better resolved by an ensemble average. Especially, the chemical production terms have been solved by the transport equation for probability density function (PDF) involving velocity and mass fractions of chemical species.

Measurements of supersonic combustion for scramjet were obtained for the first time in a supersonic hydrogen diffusion flame [15]. The experimental data are collected by combining spontaneous Raman scattering with laser-induced pre-dissociative fluorescence and laser-induced fluorescence polarization detection techniques (LIPF). In the supersonic flame, the Kolmogorov length and time scales are determined to be typically at $10-20$ μm and $0.5-10$ μs, respectively. The temporal resolution (20 ns) by the Raman system is well within the requirement for this experiment, however, the spatial resolution (~0.4 mm) is beyond the theoretical scope.

To shorten the flow path within the combustor; an axial or streamwise vortex generator enhances the fuel-air mixture becomes paramount. The most common design for vortex generation has been focused on the step, protruding or receding ramp, and cavity along the flow path. Some design is implemented by a throttling effect by air injection to create a blockage to the flow path and to enhance mixing by a horse-shoe vortex [60]. The discontinuous configuration will always induce a strong vortical formation downstream, which increases turbulent mixing, in turn, sustains the supersonic nonequilibrium turbulent combustion. In most analyses, the finite-rate chemical reactions are modeled by chemical kinetic theory based on the law of mass action and gas kinetics theory for transport properties determination. Together with the turbulent model becomes the weakest link for scramjet design and analysis.

A jet stream injected into a crossflow is frequently utilized for supersonic combustion [85]. The mechanism of enhanced fuel-air mixing is built on the Kelvin-Helmholtz instability [21]. The vortex rolling up and stretching is the consequence of the instability. The stability is described by the Taylor-Goldstein equation by linearizing the Navier-Stokes equation; $u = U(y) + u' + w'$.

$$(U-c)^2 \partial\left[(\partial\varphi/\partial y) - k^2\varphi\right]/\partial y + \left[N^2/(U-c) - U\right]\varphi = 0 \qquad (12.10a)$$

where the notation c denotes the speed of sound. The Richardson number is derived from Equation (12.10a) predicts when the value is less than a quarter $(N^2/U) < 1/4$, leads to the onset of the instability transition of the mixing flow medium into the turbulent flow of different densities and velocity.

Under the supersonic combustion condition, the flow field created by the jet normal to the flow is further complicated by the interaction of the bow shock wave and the jet-induced shock wave system. The inducted shock wave in turn generates a complex multi-structure, three-dimensional, separated flow region that surrounds the jet aperture. The axial vortical formation increases the mixing and entraining process. The exit jet expands rapidly, followed by deceleration to form a Mach disk and deflecting to align with the external stream. The generated horseshoe vortical by mixing enhancing jet on the axial symmetrical plane leads to a formation consisting of four separated vortices upstream over the jet aperture which is displayed in Figure 12.12. The separate flow structure is dominated by the two pairs of counterclockwise and clockwise vortices. The outstanding twin counter-rotating vortical formation is similar to that of the horseshoe vortex ahead of a blunt fin or a strut. However, the counterclockwise vortex entrained and reinforced by the exiting jet is no longer a secondary structure and eventually intersects the solid surface ahead of the jet to form the primarily nodal attachment. The intrigue vortical formation and the limiting surface streamline on the surface satisfy the topological rule of a three-dimensional separated flow field and verify by experimental observations [61,80].

Another axial vortex enhancer in the flow path of the scramjet is developed by the mechanism of the Rayleigh or the Rayleigh-Taylor instability, which states that any shear layer velocity profile possessing an inflection point will lead to instability [57]. The instability in the laminar boundary layer has been derived for compressible flows as [43]

$$a^2 \partial\left\{\left[(U-c)(\partial\varphi/\partial y) - (\partial U/\partial y)\varphi\right]/\left[a^2 - (U-c)^2\right]\right\}/\partial y - \alpha^2(U-c)\varphi = 0 \qquad (12.10b)$$

In Equation (12.10b), the perturbation velocity component is a spherical harmonic function of space and time, $v = \varphi(y)e^{i\alpha(x-ct)}$ and c is designated as the speed of sound.

When the shear layer separated from the forward bulkhead of a shallow cavity, a recirculation flow will form within the cavity. The free shear layer impinges on the rear bulkhead of the open cavity will generate a pressure pulse that travel upstream

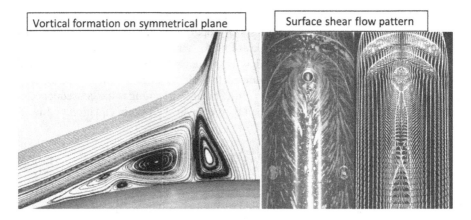

FIGURE 12.12   Vortical formation of the jet in the supersonic cross stream.

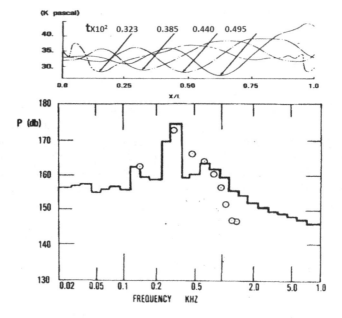

FIGURE 12.13   Pressure pulsation pattern and verification with theory.

through the nearly quiescent cavity upstream to close a feedback loop. As the consequence, the commensurable waves with distinctive frequencies at a low supersonic Mach number will ingest and repulse the air-fuel mixture by the cavity flow. The computational simulated wave pattern in time of the pressure pulsations over the entire length of the shallow open cavity and the comparison with theory is presented in Figure 12.13.

The most efficient mechanisms for generating the strongest axial vortices for supersonic combustion have been demonstrated by combining the Rayleigh-Taylor and Kelvin-Helmholtz instability. It has been demonstrated that the streamwise vorticity, the mixing rate, and the penetration depth of a transverse jet are the most effective by implementing a cavity upstream to the transverse jet. In this arrangement, the loss stagnation pressure can be minimized [34].

The propulsion requirement for a scramjet varies greatly from the acceleration phase to the final cruising condition, which must be efficiently supported by a fixed inlet configuration with tailored fuel injection. To sustain propulsion over the entire operation range, modification of the inlet may be the most cost-effective technique available. However, once a variable configuration inlet is implemented by an array of compression ramps and boundary-layer control such as suction slots, a complicated mechanical flow control system must be adopted that leads to a weight penalty. Meanwhile, when operating beyond the design condition, managing and eliminating parasitic effects increases the complicity of the propulsive system. For this reason, the aerodynamic design of a scramjet offers a formidable challenge to achieve an acceptable aerodynamic contraction ratio by a fixed geometry and variable fuel supply at varying flight speeds.

12.7 ELECTROMAGNETIC THRUSTER

The first application of plasma to aerospace engineering is attributed to Goddard, he actually conducted the first experimental of an ion engine in the time frame of 1916–1917 [36]. The ion thruster technology becomes one of the most sustained engineering developments using plasma for interplanetary exploration and satellite station keeping. Different types of ion engines become the single most practical plasma dynamics application to aerospace engineering. The research and development effort to improve the propulsive efficiency of ion thrusters is continuing even today.

The propulsion efficiency for a flight vehicle involves only the vehicle flight velocity u and the propellant exhaust velocity u_{ex} normal for the engine exhausting plane to appear as, $\eta_p = 2(u/u_{ex})/[1+(u/u_{ex})]$ [32]. The estimated propulsion efficiency is valid only as the flight vehicle has a slower speed than the exhaust propellant speed $u < u_{ex}$. Speeds of orbiting satellites and deep space exploitation vehicles are much greater than the convenient crafts powered by the air-breathing propulsive system. A higher propulsive efficiency of space travel must be derived from a higher jet exhaust speed. The velocity of electrically charged particles can easily meet the desired objective. Another measure of rocket engine efficiency is the specific impulse that relates to fuel consumption and the total thrust, thus it has also referred to as the propellant mass efficiency; $I_{sp} = T/\dot{m}g$ where g is the standard gravitational acceleration at sea level. This performance parameter has a physical dimension in seconds. A comparison of the ion thrusters with convenient chemical-propellant rockets is presented in Table 12.3.

It is important to realize that electromagnetic propulsion systems are limited by the thrust per unit area of the engine exhaust, and the avoidable degradation of performance due to space charge accumulation by Child's law [36]. In any event, the specific impulse of a typical ion thruster has a range from 1,000 to 10,000 seconds using propellants of xenon, bismuth, liquid cesium, and argon are able to attain

TABLE 12.3

Comparison of performance characteristics of propulsion system

Engine type	Exhaust speed (km/s)	Specific impulse (s)
Chemical rocket	2.5–4.4	250–400
electromagnetic thruster	20.0–50.0	1,700–4,200
Impulse magneto-thruster	42.0–210.0	3,000–12,000

a propulsive efficiency of up to 80%. From the third law of Newton's mechanics, the total thrust is proportional to the product of the exhaust velocity vector and the exhaust mass flow rate. The magnitude of the ion thrust in operations is however limited to hundreds of mille-Newton and the output power is around tens of kilowatts. In order to achieve the necessary change to flight velocity, a continuous thrust from an ion thruster often has a lifetime operation more than 20,000 hours.

There are two basic types of ion thrusters known as the gridded electrostatic ion thruster and the electromagnetic thruster or Hall-effect thruster. The major components of any ion engine consist of a plasma generator, accelerator grids, and neutralizers. The ionization is achieved by electron impact into cesium, argon, krypton, and most commonly xenon, all of which have relatively low atomic ionization potentials. The positively charged ions are extracted from the ionization chamber and are accelerated downstream of the engine by a system of grids. The last device of an ion thruster is the neutralizer which prevents charge accumulation on the vehicle.

The electromagnetic thruster or the Hall effect thruster relies on the interaction of conducting propellant-charged motion with an electromagnetic field to provide the acceleration by the Lorentz force $J \times B$. The applied electric field E and the magnetic flux density B are perpendicular to each other and to the ionized propellant velocity u, as shown in Figure 12.14. The ionized gas is accelerated by electric potential and a radial magnetic field between a cylindrical anode and a cathode. At the center of the thruster, a magnetic field is generated by an electric coil winding over a spike, at the end of the spike the electrons are trapped by the magnetic field and attracted to the anode. Some electrons spiral toward the anode to become the Hall current and close the electric circuit. While the major portion of the propellant is introduced near the anode after ionization, the electrons move toward the cathode, and ions finally leave the thruster at high speed. The streamwise body force accelerates the propellant along the thruster and transfers streamwise momentum to neutral particles by collisions and by a microscopic polarized field. As a consequence, the thruster's charged particle number densities are greater than that of the electrostatic ion thruster. The ionized propellant in an electromagnetic thruster is basically macroscopically neutral; therefore it is not overly constrained by the space charge limitations as in the electrostatic accelerators.

The operating environments for ion engines are in a rarified gas domain. The typical operating conditions in an ionization chamber of ion thruster usually have the maximum electron number density around $10^{13}/cm^3$ ($10^{19}/m^3$), and the electron temperature is on the order of 3 eV or greater. Thus, the Debye length in the high number density region of an ionization chamber is just a few microns and the plasma

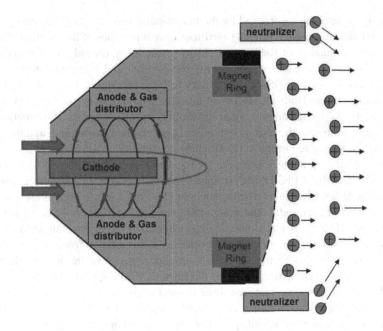

FIGURE 12.14   Schematic of the electromagnetic thruster.

frequency is estimated to be about 10^{11}/s. The mean free path λ a value of about 1.4 m within a typical ionization chamber, thus the Knudsen number $K_n = \lambda/L$ is greater than unity. Therefore, the gas mixture in the ion thruster is always in a rarefied condition. The computational simulation is beyond the validated range of the governing equations in the continuum regime and must be generated by a particle dynamics formulation; usually through a combination of Monte-Carlo techniques. Meanwhile, electron confinement is achieved by setting up a magnetic field along the walls of the ionization chamber. The applied magnetic field for axisymmetric configurations is by utilizing magnet rings that reflect electrons back into the discharge chamber giving them a much longer accelerating trajectory.

Most current methods of computational electromagnetic-aerodynamics in the rarefied gas domain within an ionization chamber adopt the particle-in-cell (PIC) approach in combination with a direct simulation Monte Carlo (DSMC) technique [7,81]. In a full PIC-DSMC simulation, all particles including the neutral and charged species are tracked by the macro particle assumption. The PIC-DSMC scheme simulates the dynamics of the electrical charge carried and neutral particles. The basic governing equations consist of the classic Newton-Lorentz equations for charged particle motion and the electric field distribution by the Poisson equation of charge conservation.

$$m_i \frac{d^2 x_i}{dt^2} = q\left(E + \frac{dx_i}{dt} \times B \right)$$

$$\nabla^2 \varphi = \frac{e}{\varepsilon}(n_- - n_+)$$

(12.11)

All charged particles are tracked by the macro-particle assumption. In theory, each macro particle represents billions to trillions of real particles of the rarefied plasma, thus the total number of the tracked particles can easily exceed 10^8. The required huge computational resource must satisfy the peculiar and stringent computational stability criteria of PIC-DSMC algorithms in time and space. The numerical procedure also suffers from the statistical noise of a large number of sampled macro particles. In short, the computational simulations by the Lagrangian formulation are essential to analyzing the plasma dynamics in the rarefied gas regime, and the computational resource requirement is extremely demanding.

The outstanding numerical results for the electrostatic ion thrust (NSTAR) have been generated by a PIC-DSMC scheme [46]. The considered particles consist of xenon, single- and double-charged xenon ions (Xe^+ and Xe^{++}), as well as electrons. A total of 1.267 million individual macro particles were carried out on 8,383 uneven-spaced cells. The numerical simulation is conducted at the steady state at the TH-15 operation condition. The charge number density spans a range from 10^{16} to 10^{19} per cubic meter (10^{10} to 10^{13}/cm^3), and the high concentration of the ions is clustered along the axis of symmetry. The lines of magnetic field intensity are also superimposed on the ion contour plot; the three magnet rings function well to confine the charged particles away from the solid chamber surface. The charged particle trajectories have the signature helical spiral pattern displayed in Figure 12.15.

The Electromagnetic thruster is one of the very few practicing technologies using electromagnetic forces to accelerate ionized gas for propulsion. The first working ion thruster is the SERT1 (SERT, Space Electric Rocket Test) was tested on a suborbital flight in 1964 [18]. In 1998 the Deep Space (DS1) was launched, it was the first

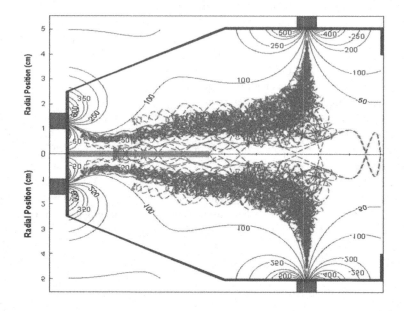

FIGURE 12.15 Charged particle trajectory in an electromagnetic thruster.

spacecraft using electromagnetic propulsion to reach another planetary body. Even so, Ion engines are routinely used for station-keeping to commercial and military communication satellites in geosynchronous orbits. More recently, a future launch has been planned for a 200 kW VASIMR (Variable Specific Impulse Magnetoplasma Rocket) electromagnetic thruster to be placed and tested on the International Space Station. The VASIMR is essentially an electro-thermal plasma thruster using either argon or xenon as the propellant. Plasma is generated by an electromagnet and heated by a helical RF antenna, then uses strong electromagnets as a convergent-divergent nozzle to accelerate the ions and electrons. Through these arrangements, a variable-specific impulse ion thruster is achievable and is capable to achieve an engine exhaust speed up to 50,000 m/s [45].

REFERENCES

1. Alfven, H., *Cosmical electromagnetics*, Clarendon Press, Oxford, England, 1950.
2. Anderson, D.A., Tannehill, J.C., and Pletcher, R.H., *Computational fluid dynamics and heat transfer*, Taylor & Francis, Bristol, PA, 1984.
3. Balsara, D. and Shu, C.W., Monotonicity preserving weighted essentially non-oscillatory schemes with increasingly high order of accuracy, *Journal of Computational Physics*, Vol. 160, 2000, pp. 405–452.
4. Bardeen, J., Copper, L., and Schriffer, J.R., Theory of superconductivity, *Physical Review*, Vol. 8, 1957, p. 1178.
5. Beam, R.M. and Warming, R.F., An implicit factored scheme for the compressible Navier-Stokes equations, *AIAA Journal,* Vol. 16, pp. 1978, 393–401.
6. Billig, F.S., Dugger, G.L., and Waltrup, P.J., Inlet-combustor interface problem in scramjet engines, *Proceeding of the first International Symposium on Air Breathing Engines*, Internal Society for Airbreathng Engine, June 1972.
7. Birdsall, C.K. and Langdon, A.B., *Plasma physics via computer simulation*, McGraw-Hill, Singapore, 1985.
8. Bityurin, V.A., Linberry, J.T., Litchford, R.J., and Cole J.W., Thermodynamic analysis of the AJAX propulsion concept, AIAA Preprint 2000-0445, Reno, NV, 2000.
9. Brandt, A., Multilevel adaptive solutions to boundary value problem, *Mathematics of Computation*, Vol. 31, 1977, pp. 333–390.
10. Brio, M. and Wu, C.C., An upwind differencing scheme for the equations of ideal magnetohydrodynamics, *Journal of Computational Physics*, Vol. 75, 1988, pp. 400–422.
11. Buning, P.G., Parks, S.J., Chan, W.M., and Renze, K.J., Application of the Chimera overlapped grid scheme to simulation of space shuttle accent flows, *Proceedings of the 4th International Symposium on CFD*, Davis, CA, 1991.
12. Cabannes, H., *Theoretical magnetofluiddynamics*, Academic Press, New York, 1970.
13. Cauchon, D.L., Radiative heating results from the fire II flight experiment at a reentry velocity of 11.4 km/s, 1967. NASA TM-X-1402.
14. Chen, Y.K. and Milos, F.S., Ablation and thermal response program for spacecraft heatshield analysis, *Journal of Spacecraft and Rockets*, Vol. 36, 1999, pp. 475–483.
15. Cheng, T., Wehrmeyer, J., Pitz, R., Jarrett, Jr., O., and Northam, G., Finite-rate chemistry effects in a Mach 2 reacting flow, *27th Joint Propulsion Conference*, AIAA Paper-91-2320, 1991.
16. Collatz, L., *The numerical treatment of differential equations*, Springer-Verlag, New York, 1966, p. 538.
17. Curran, E.T., Heiser, W.H., and Pratt, D.T., Fluid phenomena in scramjet combustion systems, *Annual Review of Fluid Mechanics*, Vol. 28, 1996, pp. 323–360.

18. Cybulski, R.J., Domino, E.J., Kotnik, J.T., Lovell, R.R., and Shellhammer, D.M., Results from CERT1 in rocket flight test, NASA TN-2718, Lewis Research Center, Cleveland, 1966.
19. Desai, P.N., Lyons, D.T., Tooley, J., and Kangas, J., Entry, descent, and landing operations analysis for the Stardust entry capsule, *Journal of Spacecraft and Rockets*, Vol. 45, 2008, pp. 1262–1268.
20. Ferri, A., Mixing-controlled supersonic combustion, *Annual Review of Fluid Mechanics*, Vol. 5, 1973, pp. 301–338.
21. Funda, T. and Joseph, D., Viscous potential flow analysis of Kelvin-Helmholtz instability in a channel, *Journal of Fluid Mechanics*, Vol. 45, 2001, pp. 263–283.
22. Fraishtadt, V.L., Kuranov, A.L., and Sheikin, E.G., Use of MHD systems in hypersonic aircraft, *Technical Physics*, Vol. 43., No. 11, 1998, p. 1309.
23. Gaitonde, D.V., High-order solution procedure for the three-dimensional nonideal magnetohydrodynamics, *AIAA Journal*, Vol. 39, No. 11, 2001, pp. 2111–2120.
24. Gaitonde, D.V., Magnetohydrodynamic energy-bypass procedure in a three-dimensional scramjet, *Journal of Propulsion and Power*, Vol. 22, No. 2, 2006, pp. 498–510.
25. Gaitonde, D. and Shang, J.S., Optimized compact-difference-based finite-volume schemes for linear wave phenomena, *Journal of Computational Physics*, Vol. 138, 1997, pp. 617–643.
26. Gaitonde, D.V. and Poggie, J., An implicit technique for 3-D turbulent magneto-aerodynamics, *AIAA Journal*, Vol. 41, No. 11, 2003, pp. 2179–2293.
27. Godunov, S.K., Finite-difference method for numerical computational of discontinuous solution of the equations of fluid dynamics, *Matematicheskii Sbornik*, Vol. 47, 1959, pp. 271–306.
28. Gulhan, A.B., Koch, U., Siebe, F., and Riehmer, J., Experimental verification of heat-flux mitigation by electromagnetic fields in partially-ionized-argon flows, *Journal of Spacecraft & Rockets*, Vol. 46, No. 2, 2009, pp. 274–282.
29. Gurijanov, E.P. and Harsha, P.T., Ajax: New directions in hypersonic technology, AIAA 1996–4609, November 1996.
30. Harten, A., High-resolution schemes for hyperbolic conservation laws, *Journal of Computational Physics*, Vol. 49, 1983, pp. 375–385.
31. Hartmann, J., Theory of the laminar flow of an electrically conductive liquid in a homogeneous magnetic field. *Mathematisk-fysiske Meddelelser*, Vol. 15, 1937, pp. 1–27.
32. Hill, P.G. and Peterson, C.R., *Mechanics and thermodynamics of propulsion*, Addison-Wesley Publishing Co., Reading, MA, 1965.
33. Huang, P.G., Shang, J.S., and Stanfield, S.A., Periodic electrodynamic field of dielectric barrier discharge, *AIAA Journal,* Vol. 49, No.1, 2011, pp. 119–127.
34. Huang, W., Du, Z.-B., Yan, L., and Xia, Z.-X., Supersonic mixing in air breathing propulsion systems for hypersonic flight, *Progress in Aerospace Science*, Vol. 109, 2019, pp. 100545.
35. Jackson, J.D., *Classical electrodynamics*, 2nd Ed., John Wiley & Sons, New York, 1975.
36. Jahn R.G., *Physics of electric propulsion*, McGraw-Hill, New York, 1968.
37. Johnston, C.O., Hollis, B.R., and Sutton, K., Nonequilibrium stagnation-line radiative heating for fire-II, *Journal of Spacecrafts and Rocket*, Vol. 45, 2008, pp. 1185–1195.
38. Khan, O.U. and Hoffmann, K.A., Numerical investigation of decoupled magnetofluid dynamics equations, *AIAA Journal*, Vol. 47, 2009, pp. 2666–2675.
39. Kimmel, R.L., Hayes, J.L., Menart, J.A., and Shang, J., Effect of magnetic fields on surface plasma discharges at Mach 5, *Journal of Spacecraft & Rockets*, Vol. 42, No. 6, 2006, pp. 1340–1346.
40. Kong, J.A., *Electromagnetic wave theory*, John Wiley & Sons, New York, 1986.
41. Krause, J.D., *Electromagnetics*, 1st Ed., McGraw-Hill, New York, 1953.

42. Lax, P.D., Weak solutions of nonlinear hyperbolic equations and their numerical computation, *Communications on Pure and Applied Mathematics*, Vol. 7, 1954, pp. 159–193.

43. Lees, L. and Lin, C.C., Investigation of stability of the laminar boundary layer in a compressible fluid, NACA TN-1115, 1946.

44. Lele, S.K., Finite difference schemes with spectral-like resolution, *Journal of Computational Physics*, Vol. 103, 1992, pp. 16–14.

45. Longmier, B., Squire, J., Carter, M., Cassady, L., Glover, T., Chancery, W., Olsen, C., Ilin, A., McCaskill, G., Chang Diaz, F.R., and Bering, E., Ambipolar ion acceleration in the expanding magnetic nozzle of the VASIMR VX-200i, AIAA Preprint 2009-5359, 2009.

46. Mahalingam, S. Particle-Based Plasma Simulation for an Ion Engine Discharge. Ph.D. Thesis, Wright State University, Dayton, OH, 2007.

47. Mitchner, M. and Kruger, C.H., *Partially ionized gases*, John Wiley & Sons, New York, 1973.

48. Matsuda, A., Otsu, H., Kawamura, M., Konigorski, D., Takizawa, Y., and Abe, T., Model surface conductivity effect for the electromagnetic heat shield in re-entry flight, *Physics of Fluids*, Vol. 20, 2008, pp. 1–10.

49. Olynick, D.R., Henline, W.D., Hartung L.C., and Candler, G.V., Comparison of Coupled radiative Navier-Stokes flow solutions with the project fire-II flight data, AIAA 94-1955, 1994.

50. Olynick, D., Chen Y.-K., and Tauber M.E. Aerothermodynamics of the stardust sample return capsule, *Journal Spacecraft and Rockets*, Vol. 36, No. 3, 1999, pp. 442–462.

51. Park, C., Calculation of stagnation-point heating rates associated with Stardust vehicle, *Journal Spacecraft & Rockets*, Vol. 44, No. 1, 2007, pp. 24–32.

52. Park, C., Bandanoff, D., and Mehta, V., Mganrtohydrodynamics energy bypass scramjet performance with real gas effects, *Journal Propulsion & Power*, Vol. 17, 2001, pp. 1049–1054.

53. Park, C., Chemical-kinetics parameters of hypersonic earth entry, *Journal Thermophysics and Heat Transfer*, 2001, Vol. 15, pp. 76–90.

54. Pitsch, H., Large-Eddy simulations of turbulent combustion, *Annual Review of Fluid Mechanics*, Vol. 38, 2006, pp. 454–486.

55. Poggie, J. and Gaitonde, D.V., Magnetic control of flow past a blunt body: Numerical validation and exploration, *Physics of Fluids*, Vol. 14, 2002, pp. 1730–1731.

56. Raizer, Y.P., *Gas discharge physics*, Springer-Verlag, Berlin, 1991.

57. Rayleigh, J.W.S., On the stability or instability of certain fluid motion, *Scientific Papers*, Vol. 1, 1880, pp. 474–484.

58. Resler, E.L. and Sears, W.R., The prospects for magneto-aerodynamics, *Journal of the Aerospace Sciences*, Vol. 25, No. 4, 1958, pp. 235–245 & 258.

59. Roe, P.L., Approximate Riemann solvers, parameter vectors and difference schemes, *Journal of Computational Physics*, Vol. 43, 1981, pp. 357–372.

60. Selezneev, R.K., Surahikov, S.T., and Shang, J.S., A review of the scramjet experimental data base, *Progress in Aerospace Sciences*, Vol. 106, 2019, pp. 43–70.

61. Shang, J.S., McMaster, D.L., Scaggs, N., and Buck, M., Interaction of jet in hypersonic cross stream, *AIAA Journal*, Vol. 27, 1989, pp. 323–329.

62. Shang, J.S., Recent research in Magneto-Aerodynamics, *Progress in Aerospace Science*, Vol. 37, No. 1, 2001, pp. 1–20.

63. Shang, J. S., Solving schemes for computational Magneto-Aerodynamics, *Journal of Scientific Computing*, Vol. 25, 2005, pp. 289–306.

64. Shang, J.S., Some flow-structure features of scramjet isolator, AIAA 2008-0772, 2008.

65. Shang, J.S., Shared knowledge in computational fluid dynamics, electromagnetics, and magneto-aerodynamics, *Progress in Aerospace Sciences*, Vol. 38, No. 6–7, 2002, pp. 449–467.

66. Shang, J.S., Computational fluid dynamics application to aerospace science, *The Aeronautical Journal*, Vol. 113, No. 1148, 2009, pp. 619–632.

67. Shang, J.S., *Computational electromagnetic-aerodynamics*, John Wiley & Sons, Hoboken, NJ, 2016.

68. Shang, J.S., Kimmel, R.L., Menart, J., and Surzhikov, S.T., Hypersonic flow control using surface plasma actuator, *Journal of Propulsion and Power*, No.5 September–October 2008, pp. 923–934.

69. Shang, J.S. and Surzhikov, S.T., Nonequilibrium radiative hypersonic flow simulation, *Progress in Aerospace Sciences*, Vol. 53, August 2012, pp. 46–65.

70. Shercliff, J.A., *A textbook of magnetohydrodynamics*, Pergamon Press, Oxford, England, 1965.

71. Shu, C.W. and Osher, S., Implementation of essentially nonoscillatory shock capture scheme II, *Journal of Computational Physics*, Vol. 83, 1989, pp. 32–78.

72. Sotnick, J.P., Kandula, M., and Buning, P., Navier-Stokes simulation of the space shuttle launch vehicle flight transonic flowfield using a large scale chimera grid system, AIAA 1994-1860, June 1994.

73. Steger, J.L. and Warming, R.F., Flux vector splitting of the inviscid gas dynamics equations with application to finite difference methods. *Journal of Computational Physics*, Vol. 20, No. 2, 1987, pp. 263–293.

74. Stone, H.L., Iterative solution of implicit approximations of multidimensional partial differential equations, *SIAM Journal on Numerical Analysis*, Vol. 5, No. 3, 1968, pp. 530–558.

75. Strikwerda, J.C., Initial boundary value problems for incompletely parabolic systems, PhD Thesis, Stanford University, Stanford, CA, 1976.

76. Surzhikov, S.T. and Shang, J.S., Two-component plasma model for two-dimensional glow discharge in magnetic field, *Journal of Computational Physics*, Vol. 199, No. 2, 2004, pp. 437–464.

77. Surzhikov, S.T. and Shang, J.S., Coupled radiation-gasdynamic model for Stardust earth entry simulation, *Journal of Spacecraft and Rockets*, Vol. 49, No. 5, 2012, pp. 875–888.

78. Surzhikov, S.T. and Shang, J.S., Fire-II flight data simulations with different physical-chemical kinetics and radiation models, *Frontier Aerospace Engineering*, Vol. 4, 2015, pp. 70–92.

79. Sutton, G.W. and Sherman, A., *Engineering magnetohydrodynamics*, McGraw-Hill, New York, 1965.

80. Tobak, M., and Peaks D.J., Topology of three-dimensional separated flows, *Annual Review of Fluid Mechanics*, 1982, Vol. 14, pp. 61–85.

81. Verboncoeur, J.P., Particle simulation of plasmas: Review and advances, *Plasma Physics and Controlled Fusion*, Vol. 47, 2005, pp. 231–260.

82. Vincenti, W. and Kruger, C., *Introduction to physical gas dynamics*, John Wiley & Sons, New York, 1965.

83. Yee, H.C., Linearized form of implicit TVD scheme for the multidimensional Euler and Navier-Stokes equations, *Computers & Mathematics with Applications*, Vol. 12A, 1986, pp. 413–432.

84. Ziemer, R.W., Experimental investigation in magneto-aerodynamics, *Journal of the American Rocket Society*, Vol. 29, 1959, pp. 642–647.

85. Zukoski, E.E. and Spaid, F.W., Secondary injection of gases into a supersonic flow, *AIAA Journal*, Vol. 2, 1964, pp. 1689–1696.

Index

Note: **Bold** page numbers refer to tables and *italic* page numbers refer to figures.

A

Abett, M. 99
ab initio (first principle) approach 143, 185–188, 190, 246
ablation 207–210, *208*
adjoint integro-differential equation 261
aerodynamic interactions 177–181, **180,** *181*
alternating direction implicit (ADI) 101–102, 113, 283
alternative direction explicit (ADE) 101–102, 112
ambipolar diffusion 13, 194–196
Ampere equation 268
Ampere's circuit law 267, 268, 276, 278
Arrhenius equations 174, 176–177, 294
Avogadro's law 34
Avogadro's number 153, 168

B

Beam-Warming algorithm 283
Bernoulli equation 48, 77
bifurcation 115–120
binary elastic collision 24–28, *26–27,* 37–38
Biot-Savart law 138, 222
black-body radiation 240–241
blast wave theory 59–62, *61*
Bohr magneton 133
Bohr's theory 125–127, 131, 133–134
Boltzmann: constant 6, 30, 168, 214, 242; equation 8, 21–24, 28–38, 153, 193, 240, 245–246; factor 174
Boltzmann-Maxwell equation 21–22, 34
Born-Oppenheimer approximation 15, 185
boundary-layer formulation 65–68, *68*
boundary-layer theory 69, 93, 96, 99, 103–104
bound-free transition 238
Bradshaw, P. 107

C

Candler, G. V. 149
Capitelli, M. 202, 205
captured shock wave 100–101, *101*
Carter, J. E. 107
Cartesian coordinate system 30, 252, 277, 281
CD404 (fourth-order approximation with low-pass filter) 102
Cebice-Smith model 107

CFD *see* computational fluid dynamics (CFD)
Chapman-Cowling theory 202, 204–206
Chapman-Enskog theory 193
Chapman-Rubesin parameter 69–70, 73–74
charged particles, dynamic motion of 221–223
charge exchange 219–220
chemical kinetics 172, 177–181, *179*
Cheng, H. K. 106
Chernyi-Losev-Macheret-Potapkin model 142
Cohen, C. B. 70, 75
coherent dispersion 246
collision cross section 199–202, *201*
collision integral 202–204, *203–204*
collision rate 6, 137, 153, 222
compressible turbulent boundary layer 84–88, *86–87*
computational fluid dynamics (CFD) 80, 99, 101, 107, 189–190
computational simulations 99–103, *101–102*
conservation equations 275–280, **276**
conservation law for energy 57–58
conservation law for momentum 57
Coulomb logarithm 141–142
Coulomb's law 138, 222, 277–278
counterflow jet 116–119, *117, 119*
coupled vibration and dissociation (CVD) 141, 183
critical point of transition 187–190
Crocco 72, 92, 94
cyclotron frequency 223, 230–231

D

Dalton's law 34, 165
Damkohler number distribution 10–11, *11*
de Broglie wavelength 16, 125, 159, 186, 193
Debye length 194, 221–222, 298
Deep Space (DS1) 300–301
Detra, R. W. 78
dielectric barrier discharge (DBD) 220, 224, 228–230, *229*
dielectric recombination mechanism 221
diminishing residual return (DRR) formulations 102
direct current discharge (DCD) 220, 224–228, *225–228,* 233
direct numerical simulation (DNS) 83, 88–90, 93, 96

305

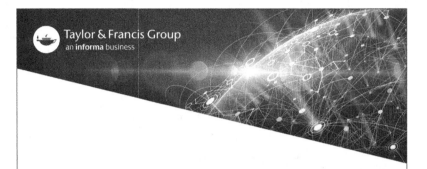

Printed in the United States
by Baker & Taylor Publisher Services